# FUNCTIONAL NANOMATERIAL FOR PHOTOENERGY CONVERSION

## Nanoelectronic Technology

# FUNCTIONAL NANOMATERIAL FOR PHOTOENERGY CONVERSION

## Nanoelectronic Technology

Editors

### Likun Pan
East China Normal University, China

### Xinjuan Liu
China Jiliang University, China

### Guang Zhu
Suzhou University, China

NEW JERSEY · LONDON · SINGAPORE · BEIJING · SHANGHAI · HONG KONG · TAIPEI · CHENNAI · TOKYO

*Published by*

World Scientific Publishing Co. Pte. Ltd.
5 Toh Tuck Link, Singapore 596224
*USA office:* 27 Warren Street, Suite 401-402, Hackensack, NJ 07601
*UK office:* 57 Shelton Street, Covent Garden, London WC2H 9HE

**British Library Cataloguing-in-Publication Data**
A catalogue record for this book is available from the British Library.

FUNCTIONAL NANOMATERIAL FOR PHOTOENERGY CONVERSION
Nanoelectronic Technology

ISBN 978-981-122-239-9 (hardcover)
ISBN 978-981-122-240-5 (ebook for institutions)
ISBN 978-981-122-241-2 (ebook for individuals)

For any available supplementary material, please visit
https://www.worldscientific.com/worldscibooks/10.1142/11887#t=suppl

Desk Editor: Nur Syarfeena Binte Mohd Fauzi

Typeset by Stallion Press
Email: enquiries@stallionpress.com

Printed in Singapore

# Contents

# Part II: Functional Nanomaterials for Solar Cells   153

# Preface

The increasing demand for energy and growing concerns about air pollution and global warming have called for intense research on energy conversion from alternative energy sources. Sustainable and renewable energy resources are being intensively pursued owing to the diminishing supply of fossil fuels and climate change. Solar energy photoelectric conversion systems, such as solar cells and photocatalysis, are considered promising methods to solve the energy crisis problems. In recent times, functional nanomaterials have attracted considerable attention due to their superior properties, such as large surface area, favorable transport properties, altered physical properties and confinement effects. Functional nanomaterials became the key elements in the development of advanced devices of photoelectric energy conversion in the search for a sustainable energy strategy. In this book, we highlight the recent progress in exploring various functional nanomaterials for applications in photoelectric energy conversion for solar cells and photocatalysis.

This book with eight chapters summarizes the current state of functional nanomaterials for photoelectric energy conversion. Most of the chapters have been designed to include (i) a review of the actual knowledge and (ii) cutting-edge research results. Thus, this book is an essential source of reference for scientists in the research fields of energy, physics, chemistry and materials. It is also suitable reading material for graduate students.

We would like to thank everyone who kindly contributed their high-quality manuscripts for this book and the editors at World Scientific for their kind help and cooperation. We are also indebted to the World Scientific editorial office, publishing and production.

# About the Authors

**Likun Pan** received his Ph.D. in 2005 from Nanyang Technological University, Singapore, and currently works as a professor at the School of Physics and Electronic Science, East China Normal University. His research interests include the synthesis and properties of nanomaterials and their applications in the energy and environmental fields. He has published over 300 SCI-cited papers with more than 13000 citations and his current H-index is 62. Dr. Pan has served as a member of the advisory/editorial board of several SCI-cited journals and as a technical chair or general secretary in several international conferences.

**Xinjuan Liu** received her Ph.D. in 2013 from East China Normal University, and currently works as an associate professor at the college of Optical and Electronic Technology, China Jiliang University. Her research interests include the synthesis and mechanical, thermal, vibrational and optical properties of nanomaterials and their applications in photocatalysis. She has published more than 120 journal articles with over 4000 citations and her current H-index is 34.

**Guang Zhu** received his Ph.D. in 2012 from East China Normal University. He was a postdoctoral fellow at Northwestern University, USA, and later joined Suzhou University, China, as an professor. His current research interests include materials synthesis, characterization and applications in photocatalysis, electrosorption and solar cells.

PART I

# Functional Nanomaterials for Photocatalysis

Chapter 1

# Functional Nanomaterials for Photocatalysis

Jiewei Chen, Peng Peng, Bi Luo, Yanjiao Guo, Rushui Cao,
Shangyi Dou, Yancong He and Meicheng Li*

*State Key Laboratory of Alternate Electrical Power System with
Renewable Energy Sources, North China Electric Power University,
Beijing 102206, China*
*mcli@ncepu.edu.cn*

## 1. Introduction

### 1.1. *Functional nanomaterials*

Solid materials can be divided into structural materials and functional materials based on their properties. The application target of structural materials is based on their strength and toughness, whereas functional materials have always had special functions, such as electrical, thermal, magnetic or light properties. The performance of functional materials differs from that of structural materials; functional materials depend on the electronic structure and electron motion (rotation, scattering, excitation and transitions), while the bonding of atoms (metal bonds, ionic bonds, covalent bonds and hydrogen bonds) and microstructures (crystal structures, grain size, morphology, dislocation substructure and second-phase properties) influence the properties of structural materials.

## 1.2. *Nanomaterial*

Nanomaterials are those in which at least one dimension is in nanometer size (0.1–100 nm) or as a basic unit in a three-dimensional space, which is approximately equal to 10–100 atoms.

Nanomaterials have some special properties in terms of melting point, vapor pressure, optical properties, chemical reactivity, magnetic superconductivity, plastic deformation and many other physical and chemical aspects. These properties are caused by the effects of nanomaterials, such as the volume effect, surface effect and quantum size effect. Therefore, functional nanomaterials have a wide range of uses.

## 1.3. *Photocatalysis*

A photocatalyst can convert photon energy into chemical energy to promote a catalytic reaction. Such a reaction is called a photocatalytic reaction. A photocatalytic reaction can excite the oxygen and water molecules into free oxidative ions, accelerating the reduction of organics and inorganics.

## 1.4. *The application of functional nanomaterials in photocatalysis*

In 1972, Fujishima and Honda [1] found that irradiated $TiO_2$ can decompose water to produce hydrogen, making $TiO_2$ an important photocatalytic material that captured the world's attention. With further exploration of photocatalytic materials, cadmium sulfide and transition metal sulfides were found to have excellent photocatalytic properties.

The nanomaterials are characterized by their small size, with more than half the total number of atoms exposed to the surface. In other words, they have a large specific surface area, due to which, nanomaterials also have greater surface tension, greater surface activity and better light absorption capacity. A large number of surface defects also provide nanomaterials with excellent dispersion performance, which is also a reason for their ultrahigh photocatalytic performance.

Taking titanium dioxide as an example, the photocatalytic reaction usually occurs on the surface of the catalyst. If the particle size is smaller, there will be larger specific surface, which leads to higher absorption efficiency. Further, there will be more photocatalytic activity sites, increasing the photocatalytic activity. With the smaller particle size of nanomaterials, their absorption spectrums will blueshift. The quantum effect will occur in the catalytic reaction, with a change in chemical properties and enhancement in catalytic activity. The reason is mainly the quantum confinement effect limiting the motion of electrons as analyzed by quantum mechanics in the material. Since the free movement of electrons in all three directions is limited, the energy distribution will become non-continuous, that is, quantized, and the quantized electron energy level will lead to higher electron excitation energy. The reduction ability becomes stronger, and the corresponding holes in the valence band also have higher energy and stronger oxidation capacity, which is due to the nature of the nanomaterials' stronger photocatalytic property.

## 2. Photocatalysis mechanism of functional nanomaterials

### 2.1. *General surface photocatalytic reactions of semiconductors*

Since titanium dioxide ($TiO_2$) has been recognized as the material with most potential for basic investigation and application because it exhibits higher photoactivity (about 10%) and is cheap, nontoxic, and chemically and biologically inactive [2], it has been chosen as the model material to demonstrate general surface photocatalytic reactions of semiconductors. It is generally accepted that the main reaction responsible for photocatalysis is the interfacial redox reaction of carriers generated when a certain amount of energy is absorbed by the semiconductor catalyst [3–5] (see Fig. 1).

Main steps of the photocatalysis mechanism:

(i) Carrier formation by absorbing photon;
(ii) Carrier recombination;

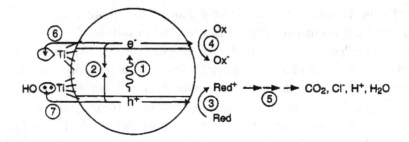

**Fig. 1.**  Primary steps in the mechanism of photocatalysis.

*Note*: The detailed process can be described as follows: (i) formation of charge carriers by photon absorption; (ii) charge carrier recombination; (iii) trapping of a conduction band electron at a Ti(IV) site to yield Ti(III); (iv) trapping of a valence band hole at a surficial titanyl group; (v) initiation of an oxidative pathway by a valence band hole; (vi) initiation of a reductive pathway by a conduction band electron; and (vii) further thermal (e.g., hydrolysis or reactions with active oxygen species) and photocatalytic reactions to yield mineralization products [6].

*Source*: Adapted with permission from Ref. [66]. Copyright 1995 American Chemical Society.

(iii) Capturing conduction band electrons at the Ti (IV) site to produce Ti (III);

(iv) Capturing the valence band on the surface of the titanyl ester group;

(v) The oxidation pathway through the valence band hole;

(vi) The reduction pathway through the conduction band electrons;

(vii) Degradation of organic matter.

## 2.2.  *Photoinduced reduction reactions at the* $TiO_2$ *surface*

Figure 2 shows that the photogenerated electrons cannot directly obtain the desired reaction on the surface, but the $TiO_2$ should be changed to activate the photogenerated electrons for the reduction reaction. This can be promoted by noble metal surface modification with nanocontact. In the commonly used photodeposition process, the metal is supported on the surface of $TiO_2$ by the reduction of electrons in the photocatalytic system by metal ions. Therefore,

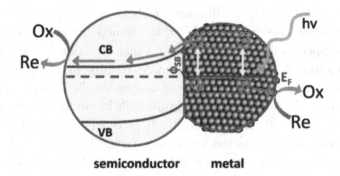

**Fig. 2.** A metal–semiconductor heterojunction for plasma-excited charge separation and related photochemical schemes. Here, charge separation is achieved in the metal by transfer of electrons to the semiconductor [7].

according to the oxidation state of metal ions, the reduction step is carried out by a single, double or tertiary electron transfer reaction according to the following reaction sequence [8–11]:

$$xe_{TiO_2^-} + M_{ads}^{x+} \rightarrow M^0 \text{ (reduction)}, \tag{1}$$

$$nM^0 \rightarrow M_n^0 \text{(nucleation)}, \tag{2}$$

$$M_n^0 + M_{ads}^+ \rightarrow M_{n+1}^0 \text{ (autocatalytic growth)}, \tag{3}$$

$$M_{n+1}^+ + e_{TiO_2}^- \rightarrow M_{n+1}^0, \tag{4}$$

$$M_{n+1}^0 \rightarrow M_x^0, \tag{5}$$

$$M_x^0 + M_y^0 \rightarrow M_{x+y}^0 \text{ (coalescence)}, \tag{6}$$

$$M_{x+y}^0 + Ze_{TiO_2}^- \rightarrow (M_{x+y}^0)^{Z-} \text{ (further transfer of electrons).} \tag{7}$$

By noble metal modification, $TiO_2$ nanoparticles with higher charge separation typically exhibit enhanced photocatalytic activity.

## 2.3. *Photoinduced oxidation reactions at the TiO₂ surface*

Under photoexcitation, metal nanoparticles can act not only as electron acceptors but also electron donors. However, more efficient electron donors, such as different alcohols, are commonly used in photocatalytic systems. Tamaki *et al.* studied the reaction kinetics of the trapped wells by TAS and found that the transient absorption of the captured cavities decayed more rapidly in the presence of alcohols, clearly indicating that the trapped wells could react with these alcohols [13]. The lives of capture wells in methanol and ethanol were 300 and 1000 ps, respectively. It is reported that the charge transfer rate is also affected by the adsorption behavior of the $TiO_2$ surface adsorbate. For example, it has been found that methanol and ethanol are mainly adsorbed by the dissociation pathway on the rutile surface, forming surface alkoxide and hydroxyl groups, increasing the adsorption intensity from the charge transfer from the alkoxide groups to the surface [14, 15]. In the case of ultraviolet irradiation of $TiO_2$ particles, the alkenyl groups of these surfaces are used as effective hole collection centers. In general, it is assumed that the hole or $\bullet$ OH-induced alcohol oxidation is carried out by two reaction steps according to Eqs. (8) and (9): The first step involves cracking the CH bond, forming individual $\alpha$-hydroxyalkyl groups, when electrons are injected into $TiO_2$. In the second step, in the conduction band, the corresponding aldehyde is formed, called "current doubling" [16–18].

$$CH_3OH \xrightarrow{\;\; E_{1/2}=0.72V \;\;} \bullet CH_3OH$$

$$+ e^- \xrightarrow{\;\; E_{1/2}=-0.98V \;\;} CH_2O + e^- E_{1/2} = 1.7\,eV, \qquad (8)$$

$$CH_3CH_2OH \xrightarrow{\;\; E_{1/2}=0.72V \;\;} \bullet CH_3 \bullet CHOH$$

$$+ e^- \xrightarrow{\;\; E_{1/2}=-1.18V \;\;} CH_3CHO + e^- E_{\frac{1}{2}} = 1.9\,eV, \qquad (9)$$

$$(CH_3)_2CHOH \xrightarrow{\;\; E_{1/2}=0.9V \;\;} (CH_3)_2 \bullet COH$$

$$+ e^- \xrightarrow{\;\; E_{1/2}=-1.3V \;\;} (CH_3)_2CO + e^- E_{1/2} = 2.2\,eV. \qquad (10)$$

The thermodynamic driving forces used to oxidize different alcohols are very good in the kinetic relationship as observed by Tamaki *et al.* [19] who proposed the overall photooxidation mechanism of

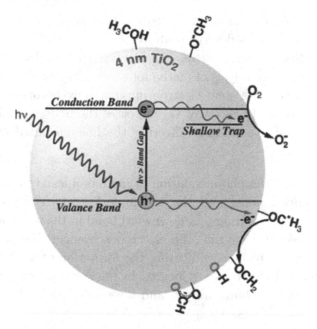

**Fig. 3.** Photooxidation mechanism of methanol on rutile $TiO_2$ nanoparticles in the presence of molecular oxygen [12].

*Source*: Reprinted with permission from Ref. 196. Copyright 2012 American Chemical Society.

methanol on the surface of $TiO_2$ nanoparticles, as shown in Fig. 3. Their experimental data show the following:

(i) Methanol, through the dissociation pathway on the rutile nanoparticles surface, is mainly absorbed to generate the methoxy and hydroxyl groups.

(ii) Surface methoxy groups are used as effective hole collection sites under UV light.

(iii) The holes captured by the methoxy group lead to efficient charge separation, which is supported by the IR absorption of the electrons with a long lifespan of trapped light.

(iv) The generation of a free radical anion by the direct hole oxidation of the methanol molecule is then initiated and then rapidly injected from the radical anion into the double-electron transfer process in the $TiO_2$ conduction band (i.e., the current doubling process described above).

(v) In addition, Morris *et al.* [20] showed that the role of $O_2$ in promoting photolysis of methanol was to remove free electrons, which opened a new electron-implanted receptor site during ethoxylation. In this way, $O_2$ increases the photon efficiency by a factor of five without affecting the hole-mediated oxidation mechanism.

In summary, the photooxidation of methanol leads to the migration of trapped pores and the accumulation of electrons on the surface of $TiO_2$, resulting in a downward bend in the semiconductor. Without molecular oxygen, the electron scavenger, $TiO_2$, changes color from white to blue, indicating the formation of a $Ti^{3+}$ center. Photogenerated electrons are located in the corresponding traps on the surface of the photocatalyst, and are now available for different reduction reactions [8, 9].

## 3. Photocatalysis under different light spectra

### 3.1. *UV-light-active photocatalysis*

#### 3.1.1. *The status of UV-light-active photocatalysis*

Ultraviolet (UV) light catalysis is a new type of purification technology developed in recent years. It can activate the molecular oxygen method to produce active material in the environment, destroy and remove pollutants that are in the environment through the effective use of ultraviolet light and it has broad application prospects. UV light catalysis can overcome the general high-cost investment of traditional processing methods that are likely to cause shortcomings such as secondary pollution; at the same time, due to the light energy utilization rate being higher in the gas phase system suitable for the purification of organic waste gases, there are an increasing number of researchers both at home and abroad dedicating effort to such research. But the nature of UV light catalysis as a catalyst, dealing with the issue of gas, still needs further research and development to be realized for industrial-scale applications [21, 22].

### 3.1.2. *The commonly used UV semiconductor photocatalyst*

For UV light catalysis, some commonly used UV semiconductor catalysts are ZnO, $ZrO_2$, $SrTiO_3$ and $TiO_2$, which is one of the most important kinds of UV light catalysts [23–25]. Titanium dioxide has attracted extensive attention due to the following few factors: (i) $TiO_2$ is a kind of commonly used chemical material, has a rich source and has been widely applied as a white pigment; (ii) the chemical stability and photochemical stability are very high, and it has corrosion resistance and light corrosion resistance; (iii) under the condition of UV light, it has high redox ability; and (iv) $TiO_2$ is non-toxic to the human body and has good environmental compatibility [26].

With sufficient oxygen and ultraviolet light, it can theoretically degrade almost all organic pollutants in the air and convert them into $CO_2$, $H_2O$ and other inorganic substances since the photo-redox effect of electrons and holes, thereby making the fruit complete Mineralization. Theoretically, especially for the difficult degradation of some VOCs, it can take advantage of the photocatalytic technology to greatly improve its inherent nature, making it suitable for subsequent processing of the material, thereby reducing running cost and not producing secondary pollution effects. At present, $TiO_2$ photocatalytic technology is mainly used for pollution control of drinking water, pollution control of VOCs and environmental remediation of persistent pollutants. Now, the photocatalytic technology for air pollution control is still in its infancy. Some researchers of the $TiO_2$-modified instruments for preparing chlorine volatile organic compound photocatalysts, aromatic volatile organic compounds, aldehydes and ketones of volatile organic compounds, such as instrument decontamination, achieved good results, which lays a theoretical foundation for technology promotion and application.

### 3.1.3. *Factors influencing photocatalytic degradation effectiveness*

The pH of the sample and $TiO_2$ concentration (surface area) have different effects on degradation rate and extent. The oxidation process occurs on the catalyst surface. The affinity/rejection effect of compound $TiO_2$ surface contaminants, pH value of solution and the point of zero charge of photocatalyst affect the order trend degradation rate constant. The effect of $TiO_2$ with different concentrations on the photocatalytic degradation of the five different contaminants was also observed. In the water samples, the increase of the concentration of $TiO_2$ slurry resulted in the enlargement of the surface area and oxidation in the active zone. However, to some extent, due to the attenuation of UV light and/or by the dispersion of solid particles, increasing the slurry concentration will reduce the ability of UV radiation on the surface of $TiO_2$. Other parameters, such as the initial concentration of pollutants, the UV radiation flux, the UV wavelength and the type of photocatalyst, remain unchanged [27].

## 3.2. *Visible-light-active photocatalysis*

### 3.2.1. *Visible-light-active photocatalysts*

General semiconductor photocatalysts can only respond to ultraviolet sunlight, and plasma metal nanostructures, such as gold, silver and copper nanorods, nanostars, balls, nanowires/nanoblocks and nanoparticles, can be visible through surface plasma resonance absorption such as infrared wavelengths of sunlight; they have potential application value in the fields of photocatalytics, optoelectronics, light detection and Raman enhancement.

### 3.2.2. *Visible-light-responsive junctions*

There are several kinds of visible-light-responsive junctions: $BiVO_4$-based junctions, visible-light-driven photocatalyst/OEC junctions, $Fe_2O_3$-based junctions, $Ag_3PO_4$-based junctions, $WO_3$-based junctions, CdS-based junctions, $C_3N_4$-based junctions and $Cu_2O$-based junctions [28].

$BiVO_4$, which has a bandgap of $2.4\,eV$, is a photocatalyst with high visible light response, attracting much attention. There are several kinds of $BiVO_4$-based junctions: (a) $BiVO_4$–$WO_3$,

(b) $BiVO_4$–ZnO, (c) $BiVO_4$–$TiO_2$ and (d) $BiVO_4$–carbon-based materials.

Unlike the evolutionary photocatalyst $BiVO_4$, CdS is attractive for it is evolutionarily efficient in the water, because it has a bandgap of 2.4 eV and its conduction band is sufficiently negative. However, the problems of particle aggregation, the poor stability of liquor and the high recombination rate of carriers seriously limit its practical applications.

Visible-light-driven photocatalyst/OEC junctions include $BiVO_4$–OEC, $Fe_2O_3$–OEC and $WO_3$–OEC. As it possesses a lot of good properties such as good stability, non-toxicity and is cheaply available, hematite ($\alpha$-$Fe_2O_3$) has great potential in photocatalysis under visible light irradiation.

Graphitic (polymeric) carbon nitride (g-$C_3N_4$) has the best stability in allotropes of carbon nitride. Its bandgap is 2.7 eV (visible region) and it has a sufficient negative conduction band position. Liu *et al.* found that the three-phase g-$C_3N_4$/g-$C_3N_4$/m-$LaVO_4$ heterojunction composed of m-$LaVO_4$ and mixed g-$C_3N_4$ showed the greatest photocatalytic activity [29].

### 3.3. *Full-solar-spectrum (UV–Vis–NIR) photocatalysis*

#### 3.3.1. *NIR-light-active photocatalysis*

As is well known, in the solar spectrum, UV content is less than 5%, visible content is less than 45% and the infrared (NIR) energy accounts for about 50% of the solar energy to achieve efficient use, due to which there is an urgent need for research and development on UV, visible light to NIR light response, even full-spectrum light catalysts. In the last twenty years or so, certain progress has been made in expanding the scope of the utilization efficiency of sunlight and near-infrared light semiconductor catalysts:

(i) NIR-light-active catalysis is based on infrared photon energy transforming the response;

(ii) Near-infrared light-active semiconductor photocatalyst;
(iii) Based on the energy band control design of the new NIR light
      catalyst, the synthesis of the near-infrared light response of the
      narrow bandgap semiconductor photocatalyst is conducted;
(iv) NIR-light-active heterojunction photocatalyst.

### 3.3.2. *Infrared-light-responsive heterojunction*

There are few reports on infrared-light-responsive heterostructures. PaK *et al.* were the first to build gold nanoparticles
and PbSe, CMSe–C multilayer core–shell-structured nanoheterojunctions, where infrared light response was obtained by the
heterojunction stepwise level distribution, implemented in infrared
methylene blue. Then, Zheng *et al.* built Pt nanoparticles modified
by differential growth of Au nanorods, using a preparation of
Pt–Au heterojunction double-metal nanorods. Cui *et al.* prepared
for the first time a conductive metal oxide heterostructure material
($WO_2$–$Na_xWO_3$) [30–32].

### 3.3.3. *Materials used in full-solar-spectrum photocatalysis*

It is of great importance to find a photocatalyst utilizing the
full solar spectrum, and many people have conducted the related
research. Hu *et al.* built a heterostructure ($Ag_2S$ quantum dots–
$TiO_2$ nanobelt) and enhanced the photocatalytic performance [33].
Wang *et al.* carried out plasma generation and separation by
using plasmons, and photocatalytic hydrolysis was enhanced by
synergism using a Schottky diode to achieve the $H_2$ generation
rate of $637 \, mol \, g^{-1} \, h^{-1}$ in the whole spectrum [34]. Wu *et al.*
found that $In_{1-x}Ga_xN$, of which the bandgap energy is tunable
between 0.7 and 3.4 eV, could became a remarkable material in
photovoltaic and optoelectronic devices [35]. Er : $Bi_2WO_6$ and
$Bi_2WO_6$ nanosheets were reported to possess NIR photocatalytic
activity [36]. Besides, Sang *et al.* reported for the first time that
$WS_2$ nanosheets are active photocatalysts under either the visible or
NIR light irradiation [37].

## 4. Nanostructured functional materials for photocatalysis

### 4.1. *The development of functional nanomaterials for photocatalysis*

Functional materials refer to those materials with special properties, such as electricity, magnetism, sound, light, heat and so on, or those materials that have a special function [38]. One kind of functional material with a special structure is nanostructured materials used for bulk materials, comparable with traditional materials, with many special properties, such as surface effects, small size effects and macroscopic quantum tunneling effects, which have gained much attention. Nanomaterials have unique light absorption, light emission and nonlinear optical properties, which are expected in optical storage. According to the dimensions of space, nanostructures can be divided into four types: (i) nanoatomic groups of zero dimension; (ii) 1D nanofibrous structures, with length significantly greater than the width, such as carbon nanotubes; (iii) 2D layered nanostructures, with length and width at least much greater than thickness, and grain size in one direction for the nanometer level; (iv) 3D nanosolids, which are used in hydrogen storage materials because of their structure and performance. Hydrogen storage alloys are mainly magnesium alloys, rare earth alloys, titanium and zirconium alloys, and are used in hydrogen storage, purification and recovery of hydrogen fuel engines, thermal pressure sensors and thermal fluid actuators, hydrogen isotope separation and nuclear reactors, air conditioning, heat pumps and heat storage, hydrogenation and dehydrogenation reaction catalysts, and hydride nickel batteries.

Photocatalytic technology has many green, non-polluting characteristics. Due to the mild reaction conditions, low consumption of energy, high-energy electrons and holes, use of sunlight as the light source and other advantages, photocatalysis has an incomparable edge over other traditional techniques in environmental governance, such as the splitting of water, the reduction of carbon dioxide and heavy metal ions, the degradation of organic pollutants and organic synthesis [39]. Photocatalytic technology has great strategic significance in the use of solar energy and in environmental protection. The most commonly used photocatalysts are metal oxides or sulfides and other semiconductor materials, such as $TiO_2$, $ZnO$, $SnO_2$ and $Fe_2O_3$. Compared with traditional catalytic technology, photocatalytic technology has many unique advantages: (i) it is energy saving [40–42]; (ii) it has mild reaction conditions [43–46]; and (iii) the reaction is easy to control. Photocatalytic technology is caused by light. Only when light is available can it be used for catalysis; when there is no light, the catalyst has no photocatalytic activity, so we can control the reaction by precisely controlling the light. The traditional catalyst is controlled by temperature, or the reaction between the reactant substrate and the catalyst. Compared with the control of light photography, this is more difficult. At present, the photocatalytic conversion of solar energy is divided into three main research directions: (i) to transform solar energy into electrical energy in dye-sensitized solar cells; (ii) to transform solar energy into chemical energy decomposition of water to hydrogen, as hydrogen combustion produces only water, does not pollute the environment and is convenient for storage and transportation of [47]; and (iii) to transform solar energy into chemical energy in organic synthesis, $CO_2$ photocatalytic conversion and preparation of organic compounds [48].

## 4.2. *Titanium dioxide materials*

Photocatalytic research into titanium dioxide originated in 1972, when Japanese scientists Fujishima and Honda [1] used photon energy to decompose water with titanium dioxide as the electrode. The papers they published can be seen as a sign of the new era of polyphase photocatalysis. Because $TiO_2$ has resistance to chemical corrosion and light corrosion, and has properties of stability, non-toxicity, high catalytic activity, low cost and good resistance to refractory organic matter, organic and inorganic pollutants, it is widely used.

$TiO_2$ has three crystal structures: a plate titanium phase, an anatase phase and a rutile phase. Anatase $TiO_2$ is widely used in photoelectrochemical cells, photocatalytic decontamination, photo catalytic sterilization, water solubility and other fields, because photocatalytic activity of anatase $TiO_2$ is better than other phases.

### 4.2.1. *Synthetic method*

Two substantially different methods are commonly used to synthesize nanoparticles. The so-called top-down approach begins with a large amount of material and uses various techniques such as mechanical grinding, laser ablation and sputtering to finally achieve the desired particle size, while the so-called bottom-top method of chemical synthesis starts from a suitable molecular precursor. Here, the emphasis will be on the latter method, as this has been shown to produce materials with high photocatalytic activity.

As a method for developing visible light in response to the titanium dioxide photocatalyst, different kinds of adsorbed/supported photosensitizing dyes on $TiO_2$ were studied.

As an alternative preparation method for doped $TiO_2$, the electronic properties of the $TiO_2$ photocatalyst were modified by high-temperature metal ion bombardment by implanting metal ions. In particular, the Anbao Group found that the implantation of various transition metal ions such as V, Cr, Fe, Co and Ni into $TiO_2$ through high-voltage acceleration at 50–200 KeV can cause the absorption of these photocatalysts, moving the photocatalyst to the visible region bandwidth, with different levels of effectiveness [19, 49–54].

### 4.2.2. *Preparation of visible-light-responsive $TiO_2$ photocatalysts by chemical doping*

Due to the large band gap, intrinsic $TiO_2$ can only react with UV light. Hence, in order to make $TiO_2$ active to visible light, various modifications such as impurity doping (chemical and physical), semiconductor coupling, dye sensitization and the like have been used in recent years.

In the above modification, although dye-sensitized photocatalysts are generally considered to be unstable under ultraviolet light irradiation, they are widely used in the preparation of solar cells due to the strong visible light absorption capacity of dye-sensitized materials.

Except for dye sensitization, it was found that other modifications of the $TiO_2$ matrix, such as modifying $TiO_2$ by impurity doping, showed visible-light-responsive photocatalytic reactivity and showed more stability under UV irradiation. Therefore, this method has also attracted much attention, and has been applied to organic pollutants in photodegradation and water decomposition reactions.

In addition to different synthesis methods, chemical doping modification of $TiO_2$ can be divided into metal doping and non-metal doping.

In the metal doping method, a certain amount of metal ions, such as $Fe^{3+}$ [55–60], $Cr^{3+}$ [11, 60], $Ru^{2+}$ [61, 62], $Ce^{4+}$ [63–65, 65–67] and $V^{5+}$ [68–70], are introduced into the $TiO_2$ matrix, where they are likely to form active "small oxide islands", thereby increasing the life of the carriers. The rate of the transfer process leads to an increase in the observed photocatalytic activity of the doped $TiO_2$. In addition, impurities induced by metal doping to $TiO_2$ can effectively narrow the bandgap and extend the visible light absorption range. However, metal doping also shows several disadvantages: undesirable thermal stability, the electron capture of the metal center and the introduction of the electron/hole recombination center [71].

It is worth mentioning that, in addition to "traditional" impurity metal ion doping, many investigations have been conducted on $Ti^{3+}$ self-doped $TiO_2$. For example, it has been observed that an excess of $Ti^{3+}$ self-doping will not lead to a recombination center in $TiO_2$ [72].

On the contrary, non-metallic doping is another technique for the modification of $TiO_2$, which can replace lattice oxygen by non-metallic elements [71, 73–75].

In 2001, Asahi et al. [76] found that nitrogen doping can enhance the photocatalytic activity of $TiO_2$ due to photodegradation of methylene blue and gaseous acetaldehyde under VIR, but the photocatalytic activity of the UV region decreases. Since then, a large number of research studies have focused on various non-metallic doped $TiO_2$ photocatalysts such as N [71, 77], B [78–80], C [81–84],

F [78, 85–87], S [78, 88] and P [22]. Although non-metallic doping may change the band structure of $TiO_2$, which affects the transfer of electrons and holes, the origin of visible light activity is still debated, especially in relation to nitrogen-doped photocatalytic mechanisms.

### 4.2.3. *Titanium dioxide self-structural modifications*

$TiO_2$ performance in the abovementioned application depends to a large extent on its optical, electronic, structural, morphological and surface properties as well as its size, crystallinity and surface [89–91], and great efforts have been made to adjust these properties to improve their performance. In their 0D, 1D, 2D and 3D nanomaterials and surface facets, the nanostructures are synthesized and understood in terms of their structural, mechanical, thermodynamic, optical, electronic, surface and interfacial properties [89–91]. However, the white and large bandgap limits its practical applications in photocatalytic hydrogen production, carbon dioxide reduction and environmental pollution removal [51, 71, 76, 92, 93].

Visible light activity can also be triggered by intrinsic defects, such as oxygen vacancies (VO) [94, 95]. After hydrogen plasma treatment, electrons trapped in oxygen vacancies are detected under VIR.

Hydrogen heat treatment of $TiO_2$ nanocrystals has been shown to be a new concept in enhancing solar absorption in visible and near-infrared regions [96–99].

This treatment leads to the formation of black $TiO_2$ nanoparticles, which contain structural disorders near the surface.

These black $TiO_2$ nanoparticles exhibit excellent photoactivity and stability in photocatalytic hydrogen production.

In addition, the modification of the structural factor of $TiO_2$ has been shown, including phase [100], morphology [101–104], structure [105–107] and porosity [108, 109], affecting its photochemical activity. For example, the mixed anatase–rutile phase of $TiO_2$ is well known and has a higher photoactivity than the pure phase [96, 110, 111].

Degussa P25, at present a widely used mixed phase material (including anatase–rutile–amorphous) [112], shows enhanced photocatalytic and photovoltaic properties in theory [113] and from experiments [114], which are believed to benefit from the interphase synergy effect.

Electron characteristics of nanomaterials depend on the high structural, morphological characteristics. The surface morphology of the 124-nm nanocrystalline semiconductors usually leads to an increase in chemical activity. Compared with the batch, the melting point decreases, the phase transition pressure increases and the solubility is higher [115].

Here, our aim is to summarize the above efforts in the structural modification of $TiO_2$ in addition to the introduction of any foreign element or compound or any substance other than the normal valence state of $Ti^{4+}$ and $O^{2-}$ in $TiO_2$.

From the chemical stoichiometric volume that is truncated, the atoms on the surface are usually unsaturated and have many hatch bonds. One of the effects of these unsaturated bonds is that they cause high surface activity to react with adjacent molecules or particles to form new compounds [116–119]. Another significant effect is that they introduce an electronic state in the bandgap, resulting in long tail light absorption and red light emission.

### 4.3. *Transition metal dichalcogenide materials*

Transition metal dichalcogenides (TMDs) include pyrite phase-structured TMDs ($CoS_2$, $FeS_2$, $NiS_2$, $NiSe_2$ and so on) and 2D layered materials, such as $MoS_2$ and $WS_2$ [120, 121]. Due to their excellent stability and bandgap, a large number of transition metal two-sulfur elements with hierarchical nanostructures have been investigated. In particular, 2D-layered $MX_2$ nanosheets of atomic thickness and WD morphology exhibit some unique physical properties, particularly their excellent chemical or electronic properties compared to their bulk counterparts [122]. Each layer consists of three atomic layers with covalently bonded X–M–X and adjacent $MX_2$ layers, which are coupled by weak Van der Waals forces to develop bulk crystals [123].

The synthesis of TMD with large-area uniformity and layer control is an integral requirement in practical applications of water splitting. In the meantime, the site of the catalytic activity of exposure to TMD should be considered through the engineering preparation process. Nowadays, solvothermal and CVD processes (both of which are bottom-up approaches) and chemical exfoliation (a top-down approach) are the frequently reported methods. These methods can be used individually or cohesively to form a best performing TMD.

### 4.3.1. $MoS_2$

Among all TMDs, $MoS_2$ has been paid most attention recently. Enlightened by the catalytic capabilities of hydrogenase and nitridation for hydrogen evolution reactions, researchers have made biomimetic speculations in Mo-based materials more active in HER [124]. Figure 4 shows a comparison of the free energy ($\Delta$GH) of atomic hydrogen bonding to the catalyst on different materials; it shows that those materials that do not combine or develop strong bonds of atomic H (Ni and Mo) with atomic hydrogen (Au) materials should be eliminated from the expected good HER catalysts. Because it resists a resemblance to the enzymes, $MoS_2$ is one of the most promising candidates to replace Pt, for which there are active sites on the edges.

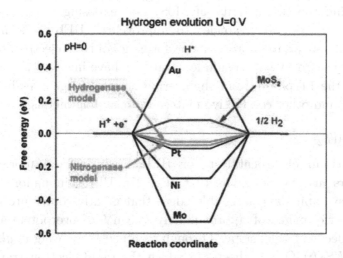

**Fig. 4.** Calculated free energy diagram for hydrogen evolution [125].

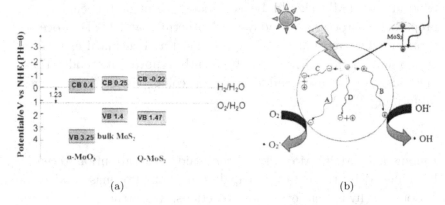

**Fig. 5.** (a) The redox potential diagram of the conduction band and valence band of different samples. (b) Photocatalytic mechanism diagram [126].

$MoS_2$, as an indirect bandgap semiconductor, by adjusting the width of the bandgap configuration, can extend its scope to 1.8 eV. In theory, 1.2 and 1.8 eV spectra correspond to photon energy cutoff wavelengths of 1050 and 688 nm, so they can effectively absorb visible light. On the contrary, because of the special properties, $MoS_2$ has been applied to photocatalytic hydrogen production generally, increasing the conductivity of $MoS_2$ and exposing more edges to the two most used strategies to improve the HER performance. Generally, so far, to improve the performance of HER electrocatalysts of $MoS_2$-based hydrogen sulfide, strategies have involved the formation of the 1T phase, from engineering or active-edge sites between hybrids and other conductive materials, or adding metals.

### 4.3.2. $WS_2$

$WS_2$, as an electrocatalyst for HER, was reported more than 25 years ago by Sobczynski et al. [127]. Ultrathin nanoflakes of $WS_2$ have abundant accessible edges that greatly contribute to the HER performance of approximately 100 mV of overpotential and 48 mV dec$^{-1}$ of Tafel slope. Through modified $WS_2$, Yang et al. [128] made $WS_2$/rGO nanosheets, in which the rapid electron transport contributed to a Tafel slope of 58 mV dec$^{-1}$ and an overpotential of

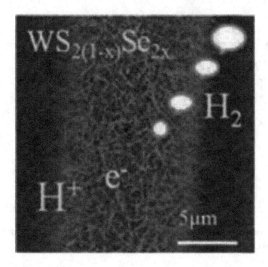

**Fig. 6.** HRTEM image with many defects and dislocations indicated by red arrows in $\text{WS}_{2(1-x)}\text{Se}_{2x}\text{NT}$ [125].

approximately 180 mV. Further, researchers [125] synthesized ternary $\text{WS}_2(1-x)\text{Se}_{2x}$ NTs [129] as active sites, on the nanotubes' walls.

### 4.3.3. *CdS*

Cadmium sulfide is a direct semiconductor with a bandgap of 2.42 eV, which has a good response to visible light. Cadmium sulfide semiconductor materials have been applied in a variety of technical fields, such as optical devices, diodes, light-emitting diodes and solar cells. The synthesis of CdS semiconductor nanomaterials refers to the basic methods of synthesizing nanomaterials: direct precipitation method, solvothermal method and ion exchange method. There are many semiconductor materials that can be used for photocatalytic cracking of hydrogen, but their bandgap width is greater than 3 eV, which means that they can only be excited by UV light. The bandgap of CdS is about 2.4 eV, which can be excited by visible light. Therefore, cadmium sulfide is an ideal material for photocatalytic hydrogen production.

CdS has attracted much attention due to its narrow bandgap and suitable band position in photocatalytic degradation and hydrogen

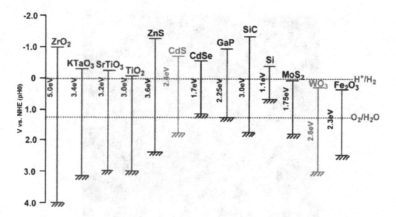

**Fig. 7.** Relationship between band structure of semiconductor and redox potentials of water splitting [130].

production. However, many studies have shown that CdS is prone to light corrosion, and its presence makes light-induced electrons and holes easy to compound and the particles easy to agglomerate. Therefore, to improve the photocatalytic efficiency of CdS, the separation and transfer efficiency of photogenerated electron–hole pairs should be improved. The former is related to the crystal structure and band structure of the photocatalyst, while the latter is related to the surface properties of the photocatalyst and the interfacial environment. In order to optimize the crystal structure, band structure, surface properties and interface environment of cadmium sulfide, several important modification methods have been summarized from the literature, such as crystal structure and morphology optimization, band structure optimization, supported noble metal catalysis, new material modifications and so on. In this chapter, we focus on the last one.

Transition metals and their compounds:

With the development of the modification process of CdS, MoS$_2$ was used as a catalyst modifier. CdS has become one of the prominent topics of research in recent years. Thomas *et al.* reported that the edge of layered MoS$_2$ has high photocatalytic activity, and its layered structure is favorable for electron transport [132]. After the excitation of CdS under the visible light, the conduction band electrons can be transferred to the conduction band of $MoS_2$, so as to effectively separate the photogenerated hole electron and reduce the corrosion of CdS. A $MoS_2$–CdS composite was prepared by the hydrothermal method and its photocatalytic activity was studied by Xu et al. [88]. $MoS_2$–CdS composites were prepared by ball milling and the calcination method by Chen et al. [133]. A $MoS_2$–CdS composite material was prepared using industrial CdS with layered $MoS_2$ by Chang et al. [134], and the Mo layers were made of 1–112 layers. In the photocatalytic hydrolysis hydrogen production experiments, the hydrogen production rate of pure CdS was quite low, only $(0.22 \, \text{mmol} \, g^{-1} \, h^{-1})$. When the mass ratio of $MoS_2$ increased from 0.2% to 2%, the photocatalytic hydrogen production rate increased rapidly, and the 2 wt.% $MoS_2$–CdS composite hydrogen production rate $(2.11 \, \text{mmol} \, g^{-1} \, h^{-1})$ was 10 times that of pure CdS. When the content of $MoS_2$ exceeds 2%, the rate of $MoS_2$–CdS composites for hydrogen reduces, as shown in Fig. 8. The results show that suitable

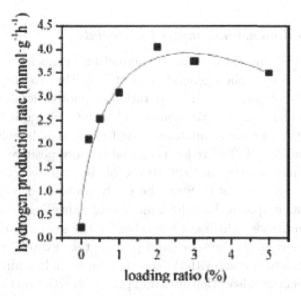

**Fig. 8.** Hydrogen production rate of $MoS_2$–CdS composite catalysts with different $MoS_2$ loading amounts [126].

*J. Chen et al.*

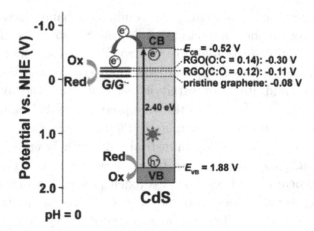

**Fig. 9.** Relative potential positions of CdS band edge and graphene–graphene level, and schematic illustration for the charge transfer in the CdS–graphene system [131].

content of $MoS_2$ can enhance the photocatalytic activity of $MoS_2$–CdS composites [135].

### 4.4. *Two-dimensional layered material*

Graphene [87] can be used as a conducting channel, receiver and storage carrier of photogenerated electrons. Due to the excellent conductivity of graphene, the separation of photogenerated carriers is greatly enhanced, and the transfer of surface charge is improved, such that the lifetime of photogenerated electron–hole pairs (Xiang *et al.*) is prolonged. The redox potential is more positive than that of the conduction band of CdS, so the photogenerated electrons on the conduction band of CdS can be easily transferred to graphene. In addition, graphene has the function of capturing, storing and shuttling electrons, which are favorable for the separation of photogenerated carriers in the space and provide the electrons with more opportunities to contact with the reaction material. In addition, there are a lot of active adsorption sites and photocatalytic reaction centers on the graphene surface, when the semiconductor catalyst and graphene composite catalyst can make the photocatalytic reaction occur on the surface of a semiconductor material and graphene surface.

## 5. Summary and outlook

Functional nanomaterials have attracted an increasing amount of attention for photocatalysis due to their excellent properties. In the past few decades, development of efficient, available and stable photocatalyst–photoelectrode functional nanomaterials has been the key to the commercialization of photocatalysts. Light absorption, charge separation and transport, and surface reaction are very important factors for photocatalysis. Based on the development and understanding of the photocatalytic mechanism and the structure design, functional nanomaterials have been considered useful. Functional nanomaterials could be widely used for pollution control, green synthesis, water splitting to $H_2$ or $O_2$, $CO_2$ reduction and photothermal catalysis in the near future.

To design efficient and available functional nanomaterials for photocatalysis, there are promising ways: full-spectrum solar-light-activated photocatalysts and noble metal-free photocatalysts. The realization of efficient solar energy conversion is an ideal goal in renewable energy studies, and it still needs much effort to synthesize photocatalysts activated by the full solar spectrum. Moreover, noble metal-free photocatalysts can really reduce the cost of the photocatalytic process and they are attracting a lot of attention for sustainable production of energy from water.

Photocatalysis has contributed much to the development of society and it will continue to play an increasingly important role in solving the challenges in energy cost and sustainability.

## References

[1] A. Fujishima, K. Honda, Electrochemical photolysis of water at a semiconductor electrode, *Nature*, 238 (1972) 37–38.
[2] A.R. Khataee, M.B. Kasiri, Photocatalytic degradation of organic dyes in the presence of nanostructured titanium dioxide: Influence of the chemical structure of dyes, *J. Mol. Catal. A: Chem.*, 328 (2010) 8–26.

[3]  K. Maeda, K. Domen, Photocatalytic water splitting: recent progress and future challenges, *J. Phys. Chem. Lett.*, 1 (2010) 2655–2661.

[4]  R. Abe, Recent progress on photocatalytic and photoelectrochemical water splitting under visible light irradiation, *J. Photochem. Photobiol. C*, 11 (2010) 179–209.

[5]  V. Knox, Review of Aurivillius photocatalysts, *Am. Ceram. Soc. Bull.*, 89 (2010) 31–31.

[6]  M.R. Hoffmann, S.T. Martin, W. Choi, D.W. Bahnemann, Environmental applications of semiconductor photocatalysis, *Chem. Rev.*, 95 (1995) 69–96.

[7]  Z. Zhang, J.T. Yates, Band bending in semiconductors: chemical and physical consequences at surfaces and interfaces, *Chem. Rev.*, 112 (2012) 5520–5551.

[8]  H.H. Mohamed, C.B. Mendive, R. Dillert, D.W. Bahnemann, Kinetic and mechanistic investigations of multielectron transfer reactions induced by stored electrons in $TiO_2$ nanoparticles: a stopped flow study, *J. Phys. Chem. A*, 115 (2011) 2139–2147.

[9]  H.H. Mohamed, R. Dillert, D.W. Bahnemann, Kinetic and mechanistic investigations of the light induced formation of gold nanoparticles on the surface of $TiO_2$, *Chem. Eur. J.*, 18 (2012) 4314–4321.

[10] L. Huang, C. Qizhong, B. Zhang, W. Xiaojing, G. Peng, J. Zibin, L. Yingliang, Preparation of sodium tantalate with different tructures and its photocatalytic activity for $H_2$ Evolution from water splitting, *Chin. J. Catal.*, 32 (2011) 1822–1830.

[11] M. Anpo, M. Takeuchi, The design and development of highly reactive titanium oxide photocatalysts operating under visible light irradiation, *J. Catal.*, 216 (2003) 505–516.

[12] D.A. Panayotov, S.P. Burrows, J.R. Morris, Photooxidation mechanism of methanol on rutile $TiO_2$ nanoparticles, *J. Phys. Chem. C*, 116 (2012) 6623–6635.

[13] Y. Tamaki, A. Furube, M. Murai, K. Hara, R. Katoh, M. Tachiya, Direct observation of reactive trapped holes in $TiO_2$ undergoing photocatalytic oxidation of adsorbed alcohols: Evaluation of the reaction rates and yields, *J. Am. Chem. Soc.*, 128 (2006) 416–417.

[14] J.M.R. Muir, Y. Choi, H. Idriss, Computational study of ethanol adsorption and reaction over rutile $TiO_2(110)$ surfaces, *Phys. Chem. Chem. Phys.*, 14 (2012) 16788–16788.

[15] J. Zhao, J. Yang, H. Petek, Theoretical study of the molecular and electronic structure of methanol on a $TiO_2(110)$ surface, *Phys. Rev. B*, 80 (2009) 235416.

[16] R. Memming, *Photoinduced charge transfer processes at semiconductor electrodes and particles* (Springer, Berlin, 1994).

[17] N. Hykaway, W. Sears, H. Morisaki, S.R. Morrison, Current-doubling reactions on titanium dioxide photoanodes, *J. Phys. Chem.*, 90 (1986) 6663–6667.

[18] S. Nakabayashi, S. Komuro, Y. Aoyagi, A. Kira, Transient grating method applied to electron-transfer dynamics at semiconductor/liquid interface, *J. Phys. Chem.*, 91 (1987) 1696–1698.

[19] M. Anpo, Y. Ichihashi, M. Takeuchi, H. Yamashita, Design of unique titanium oxide photocatalysts by an advanced metal ion-implantation method and photocatalytic reactions under visible light irradiation, *Res. Chem. Intermed.*, 24 (1998) 143–149.

[20] D.A. Panayotov, S.P. Burrows, J.R. Morris, Photooxidation mechanism of methanol on rutile TiO2 nanoparticles, *J. Phys. Chem. C*, 116 (2012) 6623–6635.

[21] J. Biomorgi, E. Oliveros, Y. Coppel, F. Benoit-Marquie, M.T. Maurette, Effect of V-UV-radiation on VOCs-saturated zeolites, *J. Photochem. Photobiol. A*, 214 (2010) 194–202.

[22] S. Qianhong, C. Zhenqian, G. Jiwei, Y. Hui, A preparation method to obtain porous anatase $TiO_2$ films at room temperature, *Chin. J. Catal.*, 28 (2007) 153–157.

[23] J.H. Luo, P.A. Maggard, Hydrothermal synthesis and photocatalytic activities of $SrTiO_3$-coated $Fe_2O_3$ and $BiFeO_3$, *Adv. Mater.*, 18 (2006) 514–517.

[24] S. Polisetti, P.A. Deshpande, G. Madras, Photocatalytic activity of combustion synthesized $ZrO_2$ and $ZrO_2$-$TiO_2$ mixed oxides, *Ind. Eng. Chem. Res.*, 50 (2011) 12915–12924.

[25] H.B. Zeng, W.P. Cai, P.S. Liu, X.X. Xu, H.J. Zhou, C. Klingshirn, H. Kalt, ZnO-based hollow nanoparticles by selective etching: Elimination and reconstruction of metal-semiconductor interface, improvement of blue emission and photocatalysis, *ACS Nano*, 2 (2008) 1661–1670.

[26] M. Ni, M.K.H. Leung, D.Y.C. Leung, K. Sumathy, A review and recent developments in photocatalytic water-splitting using $TiO_2$ for hydrogen production, *Renew. Sustain. Energy Rev.*, 11 (2007) 401–425.

[27] J.R. Alvarez-Corena, J.A. Bergendahl, F.L. Hart, Advanced oxidation of five contaminants in water by $UV/TiO_2$: Reaction kinetics and byproducts identification, *J. Environ. Manage.*, 181 (2016) 544–551.

[28] S.J.A. Moniz, S.A. Shevlin, D.J. Martin, Z.X. Guo, J.W. Tang, Visible-light driven heterojunction photocatalysts for water splitting-a critical review, *Energy Environ. Sci.*, 8 (2015) 731–759.

[29] X.P. Liu, H. Qin, W.L. Fan, Enhanced visible-light photocatalytic activity of a g-$C_3N_4$/m-$LaVO_4$ heterojunction: band offset determination, *Sci. Bull.*, 61 (2016) 645–655.

[30] G.W. Cui, W. Wang, M.Y. Ma, J.F. Xie, X.F. Shi, N. Deng, J.P. Xin, B. Tang, IR-driven photocatalytic water splitting with $WO_2$-$NaxWO_3$ hybrid conductor material, *Nano Lett.*, 15 (2015) 7199–7203.

[31] Z.K. Zheng, T. Tachikawa, T. Majima, Single-particle study of Pt-modified Au Nanorods for plasmon-enhanced hydrogen generation in visible to near-infrared region, *J. Am. Chem. Soc.*, 136 (2014) 6870–6873.

[32] C. Pak, J.Y. Woo, K. Lee, W.D. Kim, Y. Yoo, D.C. Lee, Extending the limit of low-energy photocatalysis: dye reduction with PbSe/CdSe/CdS Core/Shell/Shell Nanocrystals of varying morphologies under infrared irradiation, *J. Phys. Chem. C*, 116 (2012) 25407–25414.

[33] X.L. Hu, Y.Y. Li, J. Tian, H.R. Yang, H.Z. Cui, Highly efficient full solar spectrum (UV-vis-NIR) photocatalytic performance of $Ag_2S$ quantum dot/$TiO_2$ nanobelt heterostructures, *J. Ind. Eng. Chem.*, 45 (2017) 189–196.

[34] X.N. Wang, R. Long, D. Liu, D. Yang, C.M. Wang, Y.J. Xiong, Enhanced full-spectrum water splitting by confining plasmonic Au nanoparticles in N-doped $TiO_2$ bowl nanoarrays, *Nano Energy*, 24 (2016) 87–93.

[35] J. Wu, W. Walukiewicz, K. Yu, W. Shan, J. Ager Iii, E. Haller, H. Lu, W.J. Schaff, W. Metzger, S. Kurtz, Superior radiation resistance of $In_{1-x}Ga_xN$ alloys: full-solar-spectrum photovoltaic material system, *J. Appl. Phys.*, 94 (2003) 6477–6482.

[36] J. Tian, Y.H. Sang, G.W. Yu, H.D. Jiang, X.N. Mu, H. Liu, A $Bi_2WO_6$-based hybrid photocatalyst with broad spectrum photocatalytic properties under UV, visible, and near-infrared irradiation, *Adv. Mater.*, 25 (2013) 5075–5080.

[37] Y.H. Sang, Z.H. Zhao, M.W. Zhao, P. Hao, Y.H. Leng, H. Liu, From UV to near-infrared, $WS_2$ nanosheet: a novel photocatalyst for full solar light spectrum photodegradation, *Adv. Mater.*, 27 (2015) 363–369.

[38] S. Changxu, *Dictionary of Materials* (Chemical Industry Press, 1994).

[39] X.W.W. Xun, Application and research development of photocatalytic oxidation technique in water treatment, *Shanxi Architecture*, 35 (2009) 182–183.

[40] F.S. Yu, F. Li, B.B. Zhang, H. Li, L.C. Sun, Efficient electrocatalytic water oxidation by a copper oxide thin film in borate buffer, *ACS Catal.*, 5 (2015) 627–630.

[41] M.A. Esteruelas, A.M. Lopez, M. Mora, E. Onate, Ammonia-Borane dehydrogenation promoted by an osmium dihydride complex: kinetics and mechanism, *ACS Catal.*, 5 (2015) 187–191.

[42] A. Savini, A. Bucci, M. Nocchetti, R. Vivani, H. Idriss, A. Macchioni, Activity and recyclability of an iridium-EDTA water oxidation catalyst immobilized onto rutile $TiO_2$, *ACS Catal.*, 5 (2015) 264–271.

[43] J.G. Wang, Z.F. Bian, J. Zhu, H.X. Li, Ordered mesoporous $TiO_2$ with exposed (001) facets and enhanced activity in photocatalytic selective oxidation of alcohols, *J. Mater. Chem. A*, 1 (2013) 1296–1302.

[44] R.F. Chong, J. Li, X. Zhou, Y. Ma, J.X. Yang, L. Huang, H.X. Han, F.X. Zhang, C. Li, Selective photocatalytic conversion of glycerol to hydroxyacetaldehyde in aqueous solution on facet tuned $TiO_2$-based catalysts, *Chem. Commun.*, 50 (2014) 165–167.

[45] D. Tsukamoto, Y. Shiraishi, T. Hirai, Selective side-chain oxidation of alkyl-substituted aromatics on $TiO_2$ partially coated with $WO_3$ as a photocatalyst, *Catal. Sci. Technol.*, 3 (2013) 2270–2277.

[46] Z.Z. Zhang, Z.S. Luo, Z.P. Yang, S.Y. Zhang, Y. Zhang, Y.G. Zhou, X.X. Wang, X.Z. Fu, Band-gap tuning of N-doped $TiO_2$ photocatalysts for visible-light-driven selective oxidation of alcohols to aldehydes in water, *RSC Adv.*, 3 (2013) 7215–7218.

[47] P.D. Tran, L.H. Wong, J. Barber, J.S.C. Loo, Recent advances in hybrid photocatalysts for solar fuel production, *Energy Environ. Sci.*, 5 (2012) 5902–5918.

[48] C. Bruckmeier, M.W. Lehenmeier, R. Reithmeier, B. Rieger, J. Herranz, C. Kavakli, Binuclear rhenium(I) complexes for the photocatalytic reduction of $CO_2$, *Dalton Trans.*, 41 (2012) 5026–5037.

[49] M. Anpo, Y. Ichihashi, H. Yamashita, $TiO_2$ photocatalysts able to operate under visible light irradiation-new development by ion-implantation method, *Optronics*, 186 (1997) 161.

[50] M. Anpo, Utilization of $TiO_2$ photocatalysts in green chemistry, *Pure Appl. Chem.*, 72 (2000) 1265–1270.

[51] M. Anpo, S. Kishiguchi, Y. Ichihashi, M. Takeuchi, H. Yamashita, K. Ikeue, B. Morin, A. Davidson, M. Che, The design and development of second-generation titanium oxide photocatalysts able to operate under visible light irradiation by applying a metal ion-implantation method, *Res. Chem. Intermed.*, 27 (2001) 459–467.

[52] M. Anpo, Preparation, characterization, and reactivities of highly functional titanium oxide-based photocatalysts able to operate under UV-visible light irradiation: Approaches in realizing high efficiency in the use of visible light, *Bull. Chem. Soc. Jpn.*, 77 (2004) 1427–1442.

[53] V. Ramamurthy, K.S. Schanze, *Semiconductor Photochemistry and Photophysics*, Vol. 10 (CRC Press, 2003).

[54] A. Stepanov, Applications of ion implantation for modification of $TiO_2$: a review, (2012).

[55] J.F. Zhu, F. Chen, J.L. Zhang, H.J. Chen, M. Anpo, $Fe^{3+}$-$TiO_2$ photocatalysts prepared by combining sol-gel method with hydrothermal treatment and their characterization, *J. Photochem. Photobiol. A*, 180 (2006) 196–204.

[56] Y. Yang, C. Tian, Effects of calcining temperature on photocatalytic activity of Fe-doped sulfated titania, *Photochem. Photobiol.*, 88 (2012) 816–823.

[57] J.-W. Shi, J.-T. Zheng, Y. Hu, Y.-C. Zhao, Influence of $Fe^{3+}$ and $Ho^{3+}$ co-doping on the photocatalytic activity of $TiO_2$, *Mater. Chem. Phys.*, 106 (2007) 247–249.

[58] J.F. Zhu, W. Zheng, H.E. Bin, J.L. Zhang, M. Anpo, Characterization of Fe-$TiO_2$ photocatalysts synthesized by hydrothermal method and their photocatalytic reactivity for photodegradation of XRG dye diluted in water, *J. Mol. Catal. A: Chem.*, 216 (2004) 35–43.

[59] T. Tong, J. Zhang, B. Tian, F. Chen, D. He, Preparation of $Fe^{3+}$-doped $TiO_2$ catalysts by controlled hydrolysis of titanium alkoxide and study on their photocatalytic activity for methyl orange degradation, *J. Hazard. Mater.*, 155 (2008) 572–579.

[60] J.F. Zhu, Z.G. Deng, F. Chen, J.L. Zhang, H.J. Chen, M. Anpo, J.Z. Huang, L.Z. Zhang, Hydrothermal doping method for preparation of $Cr^{3+}$-$TiO_2$ photocatalysts with concentration gradient distribution of $Cr^{3+}$, *Appl. Catal. B*, 62 (2006) 329–335.

[61] N. Hamzah, N.M. Nordin, A.H.A. Nadzri, Y.A. Nik, M.B. Kassim, M.A. Yarmo, Enhanced activity of $Ru/TiO_2$ catalyst using bisupport, bentonite-$TiO_2$ for hydrogenolysis of glycerol in aqueous media, *Appl. Catal. A*, 419 (2012) 133–141.

[62] P. Panagiotopoulou, D.I. Kondarides, X.E. Verykios, Mechanistic study of the selective methanation of CO over $Ru/TiO_2$ catalyst: identification of active surface species and reaction pathways, *J. Phys. Chem. C*, 115 (2011) 1220–1230.

[63] S. Yuan, Y. Chen, L. Shi, J. Fang, J. Zhang, J. Zhang, H. Yamashita, Synthesis and characterization of Ce-doped mesoporous anatase with long-range ordered mesostructure, *Mater. Lett.*, 61 (2007) 4283–4286.

[64] T. Tong, J. Zhang, B. Tian, F. Chen, D. He, M. Anpo, Preparation of Ce-$TiO_2$ catalysts by controlled hydrolysis of titanium alkoxide based on esterification reaction and study on its photocatalytic activity, *J. Colloid Interface Sci.*, 315 (2007) 382–388.

[65] H.T. Gao, W.C. Liu, B. Lu, F.F. Liu, Photocatalytic activity of La, Y Co-doped $TiO_2$ nanoparticles synthesized by ultrasonic assisted sol-gel method, *J. Nanosci. Nanotechnol.*, 12 (2012) 3959–3965.

[66] M.Y. Xing, D.Y. Qi, J.L. Zhang, F. Chen, One-step hydrothermal method to prepare carbon and lanthanum co-doped $TiO_2$ nanocrystals with exposed {001} facets and their high UV and visible-light photocatalytic activity, *Chem. Eur. J.*, 17 (2011) 11432–11436.

[67] S. Yuan, Q. Sheng, J. Zhang, F. Chen, M. Anpo, Q. Zhang, Synthesis of $La^{3+}$ doped mesoporous titania with highly crystallized walls, *Microporous Mesoporous Mater.*, 79 (2005) 93–99.

[68] H.B. Liu, Y.M. Wu, J.L. Zhang, A new approach toward carbon-modified vanadium-doped titanium dioxide photocatalysts, *ACS Appl. Mater. Interfaces*, 3 (2011) 1757–1764.

[69] B. Tian, C. Li, F. Gu, H. Jiang, Y. Hu, J. Zhang, Flame sprayed V-doped $TiO_2$ nanoparticles with enhanced photocatalytic activity under visible light irradiation, *Chem. Eng. J.*, 151 (2009) 220–227.

[70] W.-C. Lin, Y.-J. Lin, Effect of vanadium(IV)-doping on the visible light-induced catalytic activity of titanium dioxide catalysts for methylene blue degradation, *Environ. Eng. Sci.*, 29 (2012) 447–452.

[71] J. Zhang, Y. Wu, M. Xing, S.A.K. Leghari, S. Sajjad, Development of modified N doped $TiO_2$ photocatalyst with metals, nonmetals and metal oxides, *Energy Environ. Sci.*, 3 (2010) 715–726.

[72] M. Xing, W. Fang, M. Nasir, Y. Ma, J. Zhang, M. Anpo, Self-doped $Ti^{3+}$-enhanced $TiO_2$ nanoparticles with a high-performance photocatalysis, *J. Catal.*, 297 (2013) 236–243.

[73] G. Liu, C. Han, M. Pelaez, D. Zhu, S. Liao, V. Likodimos, A.G. Kontos, P. Falaras, D.D. Dionysiou, Enhanced visible light photocatalytic activity of C N-codoped $TiO_2$ films for the degradation of microcystin-LR, *J. Mol. Catal. A: Chem.*, 372 (2013) 58–65.

[74] K. Selvam, M. Swaminathan, Nano $N$-$TiO_2$ mediated selective photocatalytic synthesis of quinaldines from nitrobenzenes, *RSC Adv.*, 2 (2012) 2848–2855.

[75] W. Zhang, B. Yang, J. Chen, Effects of calcination temperature on preparation of boron-doped $TiO_2$ by sol-gel method, *Int. J. Photoenergy*, (2012) 528637.

[76] R. Asahi, T. Morikawa, T. Ohwaki, K. Aoki, Y. Taga, Visible-light photocatalysis in nitrogen-doped titanium oxides, *Science*, 293 (2001) 269–271.

[77] X.-L. Yuan, J.-L. Zhang, M. Anpo, D.-N. He, Synthesis of $Fe^{3+}$ doped ordered mesoporous $TiO_2$ with enhanced visible light photocatalytic activity and highly crystallized anatase wall, *Res. Chem. Intermed.*, 36 (2010) 83–93.

[78] X. Wang, M. Blackford, K. Prince, R.A. Caruso, Preparation of boron-doped porous titania networks containing gold nanoparticles with enhanced visible-light photocatalytic activity, *ACS Appl. Mater. Interfaces*, 4 (2012) 476–482.

[79] X.N. Lu, B.Z. Tian, F. Chen, J.L. Zhang, Preparation of boron-doped $TiO_2$ films by autoclaved-sol method at low temperature and study on their photocatalytic activity, *Thin Solid Films*, 519 (2010) 111–116.

[80] Y.B. Tang, L.C. Yin, Y. Yang, X.H. Bo, Y.L. Cao, H.E. Wang, W.J. Zhang, I. Bello, S.T. Lee, H.M. Cheng, C.S. Lee, Tunable band gaps and p-type transport properties of boron-doped graphenes by controllable ion doping using reactive microwave plasma, *ACS Nano*, 6 (2012) 1970–1978.

[81] M. Xing, J. Zhang, F. Chen, B. Tian, An economic method to prepare vacuum activated photocatalysts with high photo-activities and photosensitivities, *Chem. Commun.*, 47 (2011) 4947–4949.

[82] Y. Wu, J. Zhang, L. Xiao, F. Chen, Properties of carbon and iron modified $TiO_2$ photocatalyst synthesized at low temperature and photodegradation of acid orange 7 under visible light, *Appl. Surf. Sci.*, 256 (2010) 4260–4268.

[83] S.K. Parayil, H.S. Kibombo, C.-M. Wu, R. Peng, J. Baltrusaitis, R.T. Koodali, Enhanced photocatalytic water splitting activity of carbon-modified $TiO_2$ composite materials synthesized by a green synthetic approach, *Int. J. Hydrogen Energy*, 37 (2012) 8257–8267.

[84] J. Zhong, F. Chen, J. Zhang, Carbon-Deposited $TiO_2$: Synthesis, Characterization, and visible photocatalytic performance, *J. Phys. Chem. C*, 114 (2010) 933–939.

[85] Y. Kuwahara, K. Maki, Y. Matsumura, T. Kamegawa, K. Mori, H. Yamashita, Hydrophobic modification of a mesoporous silica surface using a fluorine-containing silylation agent and its application as an advantageous

host material for the $TiO_2$ photocatalyst, *J. Phys. Chem. C*, 113 (2009) 1552–1559.

[86] S. Tosoni, D. Fernandez Hevia, O. Gonzalez Diaz, F. Illas, Origin of optical excitations in fluorine-doped titania from response function theory: relevance to photocatalysis, *J. Phys. Chem. Lett.*, 3 (2012) 2269–2274.

[87] Q.J. Xiang, J.G. Yu, M. Jaroniec, Synergetic effect of $MoS_2$ and graphene as cocatalysts for enhanced photocatalytic $H_2$ production activity of $TiO_2$ nanoparticles, *J. Am. Chem. Soc.*, 134 (2012) 6575–6578.

[88] J. Xu, X.J. Cao, Characterization and mechanism of $MoS_2/CdS$ composite photocatalyst used for hydrogen production from water splitting under visible light, *Chem. Eng. J.*, 260 (2015) 642–648.

[89] X. Chen, S.S. Mao, Titanium dioxide nanomaterials: Synthesis, properties, modifications, and applications, *Chem. Rev.*, 107 (2007) 2891–2959.

[90] X. Chen, S. Shen, L. Guo, S.S. Mao, Semiconductor-based photocatalytic hydrogen generation, *Chem. Rev.*, 110 (2010) 6503–6570.

[91] X. Chen, C. Li, M. Grätzel, R. Kostecki, S.S. Mao, Nanomaterials for renewable energy production and storage, *Chem. Soc. Rev.*, 41 (2012) 7909–7937.

[92] W. Choi, A. Termin, M.R. Hoffmann, The role of metal ion dopants in quantum-sized $TiO_2$: correlation between photoreactivity and charge carrier recombination dynamics, *J. Phys. Chem.*, 98 (1994) 13669–13679.

[93] X.B. Chen, C. Burda, The electronic origin of the visible-light absorption properties of C-, N- and S-doped $TiO_2$ nanomaterials, *J. Am. Chem. Soc.*, 130 (2008) 5018–5019.

[94] I. Nakamura, N. Negishi, S. Kutsuna, T. Ihara, S. Sugihara, K. Takeuchi, Role of oxygen vacancy in the plasma-treated $TiO_2$ photocatalyst with visible light activity for NO removal, *J. Mol. Catal. A: Chem.*, 161 (2000) 205–212.

[95] X. Pan, M.-Q. Yang, X. Fu, N. Zhang, Y.-J. Xu, Defective $TiO_2$ with oxygen vacancies: synthesis, properties and photocatalytic applications, *Nanoscale*, 5 (2013) 3601–3614.

[96] Y.K. Kho, A. Iwase, W.Y. Teoh, L. Madler, A. Kudo, R. Amal, Photocatalytic $H_2$ Evolution over $TiO_2$ nanoparticles. the synergistic effect of anatase and rutile, *J. Phys. Chem. C*, 114 (2010) 2821–2829.

[97] X. Chen, L. Liu, P.Y. Yu, S.S. Mao, Increasing solar absorption for photocatalysis with black hydrogenated titanium dioxide nanocrystals, *Science*, 331 (2011) 746–750.

[98] Y. Tachibana, L. Vayssieres, J.R. Durrant, Artificial photosynthesis for solar water-splitting, *Nat. Photon.*, 6 (2012) 511–518.

[99] U. Diebold, Photocatalysts Closing the gap, *Nat. Chem.*, 3 (2011) 271–272.

[100] T.-S. Kang, A.P. Smith, B.E. Taylor, M.F. Durstock, Fabrication of highly-ordered $TiO_2$ nanotube arrays and their use in dye-sensitized solar cells, *Nano Lett.*, 9 (2009) 601–606.

[101] F. Sauvage, F. Di Fonzo, A.L. Bassi, C.S. Casari, V. Russo, G. Divitini, C. Ducati, C.E. Bottani, P. Comte, M. Graetzel, Hierarchical $TiO_2$

photoanode for dye-sensitized solar cells, *Nano Lett.*, 10 (2010) 2562–2567.

[102] D. Kuang, J. Brillet, P. Chen, M. Takata, S. Uchida, H. Miura, K. Sumioka, S.M. Zakeeruddin, M. Graetzel, Application of highly ordered TiO$_2$ nanotube arrays in flexible dye-sensitized solar cells, *ACS Nano*, 2 (2008) 1113–1116.

[103] W. Jiao, L.Z. Wang, G. Liu, G.Q. Lu, H.M. Cheng, Hollow anatase TiO$_2$ single crystals and mesocrystals with dominant {101} facets for improved photocatalysis activity and tuned reaction preference, *ACS Catal.*, 2 (2012) 1854–1859.

[104] X.W. Zhao, W.Z. Jin, J.G. Cai, J.F. Ye, Z.H. Li, Y.R. Ma, J.L. Xie, L.M. Qi, Shape- and size-controlled synthesis of uniform anatase TiO$_2$ nanocuboids enclosed by active {100} and {001} facets, *Adv. Funct. Mater.*, 21 (2011) 3554–3563.

[105] W.G. Yang, F.R. Wan, Q.W. Chen, J.J. Li, D.S. Xu, Controlling synthesis of well-crystallized mesoporous TiO$_2$ microspheres with ultrahigh surface area for high-performance dye-sensitized solar cells, *J. Mater. Chem.*, 20 (2010) 2870–2876.

[106] D. Chen, F. Huang, Y.-B. Cheng, R.A. Caruso, Mesoporous anatase TiO$_2$ beads with high surface areas and controllable pore sizes: a superior candidate for high-performance dye-sensitized solar cells, *Adv. Mater.*, 21 (2009) 2206–2210.

[107] L. Robben, A.A. Ismail, S.J. Lohmeier, A. Feldhoff, D.W. Bahnemann, J.C. Buhl, Facile Synthesis of highly ordered mesoporous and well crystalline TiO$_2$: Impact of different gas atmosphere and calcination temperatures on structural properties, *Chem. Mater.*, 24 (2012) 1268–1275.

[108] N. Tetreault, E. Horvath, T. Moehl, J. Brillet, R. Smajda, S. Bungener, N. Cai, P. Wang, S.M. Zakeeruddin, L. Forro, A. Magrez, M. Graetzel, High-efficiency solid-state dye-sensitized solar cells: fast charge extraction through self-assembled 3D fibrous network of crystalline TiO$_2$ nanowires, *ACS Nano*, 4 (2010) 7644–7650.

[109] F. Sauvage, D. Chen, P. Comte, F. Huang, L.-P. Heiniger, Y.-B. Cheng, R.A. Caruso, M. Graetzel, Dye-sensitized solar cells employing a single film of mesoporous TiO$_2$ beads achieve power conversion efficiencies over 10%, *Acs Nano*, 4 (2010) 4420–4425.

[110] Q. Zhu, J.S. Qian, H. Pan, L. Tu, X.F. Zhou, Synergistic manipulation of micro-nanostructures and composition: anatase/rutile mixed-phase TiO$_2$ hollow micro-nanospheres with hierarchical mesopores for photovoltaic and photocatalytic applications, *Nanotechnology*, 22 (2011).

[111] W.J. Zheng, X.D. Liu, Z.Y. Yan, L.J. Zhu, Ionic Liquid-assisted synthesis of large-scale TiO$_2$ nanoparticles with controllable phase by hydrolysis of TiCl$_4$, *ACS Nano*, 3 (2009) 115–122.

[112] B. Ohtani, O. Prieto-Mahaney, D. Li, R. Abe, What is Degussa (Evonik) P25? Crystalline composition analysis, reconstruction from isolated pure particles and photocatalytic activity test, *J. Photochem. Photobiol. A*, 216 (2010) 179–182.

[113] P. Deak, B. Aradi, T. Frauenheim, Band lineup and charge carrier separation in mixed rutile-anatase systems, *J. Phys. Chem. C*, 115 (2011) 3443–3446.

[114] T.A. Kandiel, A. Feldhoff, L. Robben, R. Dillert, D.W. Bahnemann, Tailored titanium dioxide nanomaterials: anatase nanoparticles and brookite nanorods as highly active photocatalysts, *Chem. Mater.*, 22 (2010) 2050–2060.

[115] D.P. Macwan, P.N. Dave, S. Chaturvedi, A review on nano-$TiO_2$ sol-gel type syntheses and its applications, *J. Mater. Sci.*, 46 (2011) 3669–3686.

[116] C. Burda, X.B. Chen, R. Narayanan, M.A. El-Sayed, Chemistry and properties of nanocrystals of different shapes, *Chem. Rev.*, 105 (2005) 1025–1102.

[117] U. Diebold, The surface science of titanium dioxide, *Surf. Sci. Rep.*, 48 (2003) 53–229.

[118] E. Farfan-Arribas, R.J. Madix, Characterization of the acid-base properties of the $TiO_2(110)$ surface by adsorption of amines, *J. Phys. Chem. B*, 107 (2003) 3225–3233.

[119] G. Martra, Lewis acid and base sites at the surface of microcrystalline $TiO_2$ anatase: relationships between surface morphology and chemical behaviour, *Appl. Catal. A*, 200 (2000) 275–285.

[120] V. Nicolosi, M. Chhowalla, M.G. Kanatzidis, M.S. Strano, J.N. Coleman, Liquid exfoliation of layered materials, *Science*, 340 (2013) 1420.

[121] L. Yang, H. Hong, Q. Fu, Y.F. Huang, J.Y. Zhang, X.D. Cui, Z.Y. Fan, K.H. Liu, B. Xiang, Single-crystal atomic-layered molybdenum disulfide nanobelts with high surface activity, *ACS Nano*, 9 (2015) 6478–6483.

[122] F. Wang, Z.X. Wang, Q.S. Wang, F.M. Wang, L. Yin, K. Xu, Y. Huang, J. He, Synthesis, properties and applications of 2D non-graphene materials, *Nanotechnology*, 26 (2015).

[123] B. Peng, P.K. Ang, K.P. Loh, Two-dimensional dichalcogenides for light-harvesting applications, *Nano Today*, 10 (2015) 128–137.

[124] B. Hinnemann, P.G. Moses, J. Bonde, K.P. Jorgensen, J.H. Nielsen, S. Horch, I. Chorkendorff, J.K. Norskov, Biornimetic hydrogen evolution: $MoS_2$ nanoparticles as catalyst for hydrogen evolution, *J. Am. Chem. Soc.*, 127 (2005) 5308–5309.

[125] F.M. Wang, T.A. Shifa, X.Y. Zhan, Y. Huang, K.L. Liu, Z.Z. Cheng, C. Jiang, J. He, Recent advances in transition-metal dichalcogenide based nanomaterials for water splitting, *Nanoscale*, 7 (2015) 19764–19788.

[126] M.Z. Zhong, Thesis, Zhejiang Normal University, 2015.

[127] A. Sobczynski, A. Yildiz, A.J. Bard, A. Campion, M.A. Fox, T. Mallouk, S.E. Webber, J.M. White, Tungsten disulfide: a novel hydrogen evolution catalyst for water decomposition, *J. Phys. Chem.*, 92 (1988) 2311–2315.

[128] J. Yang, D. Voiry, S.J. Ahn, D. Kang, A.Y. Kim, M. Chhowalla, H.S. Shin, Two-dimensional hybrid nanosheets of tungsten disulfide and reduced graphene oxide as catalysts for enhanced hydrogen evolution, *Angew. Chem. Int. Ed.*, 52 (2013) 13751–13754.

[129] K. Xu, F.M. Wang, Z.X. Wang, X.Y. Zhan, Q.S. Wang, Z.Z. Cheng, M. Safdar, J. He, Component-controllable $WS_{2(1-x)}Se_{2x}$ nanotubes for efficient hydrogen evolution reaction, *ACS Nano*, 8 (2014) 8468–8476.

[130] Kudo, Akihiko, and Yugo Miseki, Heterogeneous photocatalyst materials for water splitting, *Chemical Society Reviews*, 38.1 (2009): 253–278.

[131] D. Lang, Thesis, Huazhong Agricultural University, 2016.

[132] T.F. Jaramillo, K.P. Jorgensen, J. Bonde, J.H. Nielsen, S. Horch, I. Chorkendorff, Identification of active edge sites for electrochemical $H_2$ evolution from $MoS_2$ nanocatalysts, *Science*, 317 (2007) 100–102.

[133] G.P. Chen, D.M. Li, F. Li, Y.Z. Fan, H.F. Zhao, Y.H. Luo, R.C. Yu, Q.B. Meng, Ball-milling combined calcination synthesis of $MoS_2$/CdS photocatalysts for high photocatalytic $H_2$ evolution activity under visible light irradiation, *Appl. Catal. A*, 443 (2012) 138–144.

[134] K. Chang, M. Li, T. Wang, S.X. Ouyang, P. Li, L.Q. Liu, J.H. Ye, Drastic Layer-number-dependent activity enhancement in photocatalytic $H_2$ evolution over $nMoS_2$/CdS($n \geq 1$) under visible light, *Adv. Energy Mater.*, 5 (2015) 1402279.

[135] X. Zong, H.J. Yan, G.P. Wu, G.J. Ma, F.Y. Wen, L. Wang, C. Li, Enhancement of photocatalytic $H_2$ evolution on CdS by loading $MOS_2$ as cocatalyst under visible light irradiation, *J. Am. Chem. Soc.*, 130 (2008) 7176–7177.

# Chapter 2

# Functional Metal Sulfide Nanomaterials for Photocatalytic Hydrogen Evolution

Da Chen[*,‡] and Xiang Li[†,§]

*College of Materials Science and Engineering,
China Jiliang University,
Hangzhou, Zhejiang 310018, China
†Parabon Nanolabs, Huntington, WV 25701, USA
‡dchen_80@hotmail.com
§xiangli.purdue@gmail.com

## 1. Introduction

The rapid development of the world economy has caused enormous energy consumption as well as severe environmental pollution, which has become an urgent problem and a growing threat to the sustainable development of mankind. A vast array of problems caused by energy shortage and environmental pollution has forced countries to make great effort in searching for clean, green and renewable energy. Solar energy is a clean and renewable energy source with unlimited and inexhaustible supply from the sun, but its direct utility efficiency is very low. Therefore, the conversion of solar energy into storable energy is essential for its use. Semiconductor photocatalysis has become an increasingly promising technology for solar energy conversion and environmental pollution mediation. On the basis of solar energy conversion and storage, semiconductor photocatalysis can not only convert low-density solar energy into high-density stored chemical or electrical energy but can also directly decompose the organic pollutants into small inorganic molecules

($H_2O$, $CO_2$, etc.) [1, 2]. Thus, semiconductor photocatalysis holds huge advantages and great promise in addressing the worldwide energy crisis and environmental issues [3, 4]. With respect to photocatalysis, the photocatalyst is the key factor to determine the photocatalytic properties. Until now, several photocatalysts have been explored to meet specific requirements for highly efficient photocatalytic reactions. Among the various photocatalysts developed, $TiO_2$ is undoubtedly the most investigated and widely used photocatalyst because of its high photocatalytic activity, chemical and photochemical stability, innocuity as well as low cost [5]. However, the practical applications of $TiO_2$ in association with solar photocatalysis are seriously restricted by its poor utilization efficiency of solar energy due to its wide bandgap (3.2 eV), which can only respond to ultraviolet light occupying about 4% of the total solar energy [6]. Therefore, it is a highly crucial task to improve the solar energy utilization efficiency and photocatalytic activities of the photocatalyst. To achieve this goal, two common strategies have been proposed. One is to modify the wide bandgap photocatalysts (such as $TiO_2$ and ZnO) by doping or by coupling them with other semiconductors to form heterostructures [7]. The other is more important and involves the exploration of novel visible-light-driven photocatalysts, such as $BiVO_4$ [8], $Bi_2WO_6$ [9], $g$-$C_3N_4$ [10] and $Ag_3PO_4$ [11].

In recent years, metal sulphide photocatalysts have attracted much attention due to their excellent visible light responses and the outstanding photocatalytic properties derived from their suitable bandgap structure. Numerous sulphides have been reported as efficient photocatalysts for hydrogen evolution through photocatalytic water splitting and photocatalytic degradation of organic pollutants [12–14], and the number of the related published papers is increasing year by year. In spite of this, there are few review articles involving metal sulphide photocatalysts for photocatalytic applications. In this regard, the present chapter will briefly overview the recent research advances in the field of metal sulphide photocatalysts and their photocatalytic applications with a particular emphasis on photocatalytic hydrogen evolution from water splitting.

## 2. Photocatalytic hydrogen evolution

### 2.1. *Mechanism*

Since the discovery of $TiO_2$ photoelectrodes for $H_2$ and $O_2$ evolution from water splitting upon ultraviolet light irradiation in 1972 by Fujishima and Honda [15], the focus has been on how to improve the solar energy utilization efficiency and photocatalytic water-splitting efficiency, which are still considered great challenges. The intrinsic mechanism for photocatalytic hydrogen production from water splitting is the photoelectric effect that happens on semiconductor photocatalytic materials, and the corresponding schematic illustration of the photocatalytic mechanism [16] and reaction process [17] for hydrogen evolution from water splitting is shown in Fig. 1. Upon irradiation with photon energy equal to or larger than the bandgap energy, the valence electrons in the photocatalyst will be excited and leap from the valence band (VB) to the conduction band (CB), thus leading to the formation of photogenerated electrons and holes at CB and VB, respectively. These photogenerated electrons and holes have strong reducing and oxidation abilities, respectively. For a given photocatalyst, to realize photocatalytic $H_2$ and $O_2$ production from water splitting upon sunlight irradiation, the reducing ability of the photogenerated electrons must be strong enough to react with

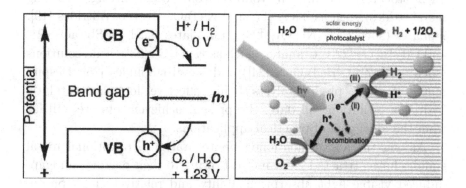

**Fig. 1.** Schematic illustration of photocatalytic mechanism.

*Source*: Reprinted with permission from Ref. [16] and reaction process reprinted with permission from Ref. [17] for hydrogen evolution from water splitting.

water for $H_2$ production, while the photogenerated holes must have sufficient capability to oxidize water for $O_2$ production. To meet this demand, the CB potential position of the photocatalyst must be less than the $H_2O/H_2$ electrode potential ($0\,V$ vs. NHE) (*note*: a more negative CB potential means a stronger reducing ability), and the VB potential position must be higher than the $O_2/H_2O$ electrode potential ($1.23\,V$ vs. NHE) (*note*: a more positive VB potential means a stronger oxidation ability). That is, the bandgap structure of a photocatalyst determines the range of solar spectrum response: the narrower the bandgap, the wider the range of the solar spectrum response that the photocatalyst possesses. Meanwhile, the positions of the CB and VB potentials should also meet the electrode potential for the redox reaction with $H_2O$. Thus, the energy structure of a photocatalyst has a decisive role in its optical absorption behavior as well as photocatalytic activities. So, the development of novel, highly efficient semiconductor photocatalysts is pivotal to realize the photocatalytic $H_2$ evolution from water splitting under visible light irradiation, and it also becomes an important research area and a new research hotspot in the field of photocatalysis.

## 2.2. *Current status and challenge*

Until now, a number of photocatalytic materials have been reported for photocatalytic hydrogen evolution from water splitting, including metal oxides (such as $TiO_2$ [17] and $NaTaO_3$ [18]), oxynitrides [19], niobates (such as $K_4Nb_6O_{17}$ [20]), tantalates (such as $SbTaO_4$ [21] and $Cd_2Ta_2O_7$ [22]), titanates (such as $K_2Ti_4O_9$ [23]), solid solutions (such as $BaZrO_3$-$BaTaO_2N$ [24]) and metal sulphides [25]. Despite many of these photocatalysts being effective for photocatalytic hydrogen evolution, till date, the present achievements are still far from the ideal goal of practical applications.

Among the above mentioned photocatalysts, transitional metal sulfides have attracted great interest owing to their narrow bandgap-induced visible light absorption ability and relatively high photo-catalytic activity. For instance, CdS has a bandgap of only $2.4\,eV$, which allows effective absorption and utilization of visible light, and has been regarded as a promising visible-light-driven photocatalyst for $H_2$ evolution owing to its narrow bandgap (2.4 eV) and good band positions for reducing $H_2O$ to $H_2$ [26]. However, CdS is prone to photocorrosion upon irradiation [27], thus leading to poor photocatalytic stability. Meanwhile, cadmium is also hazardous to the environment, which also hinders the practical applications of CdS. To address these drawbacks of CdS, multi-component metal sulphides (such as $ZnIn_2S_4$, $CaIn_2S_4$ and $CuInS_2$) [28–30] have been developed in recent years owing to their excellent optical properties and catalytic stability, and have been considered as one of the most promising photocatalysts. Though much concern has been expressed about these photocatalysts, their photocatalytic efficiencies are still low because of their easy recombination of photogenerated electron–hole pairs. Thus, how to further improve the photocatalytic properties of multi-component metal sulphides has become one of the research hot spots in the field of photocatalysis. At the present time, research on the creation of metal sulphide-based photocatalysts for photocatalytic hydrogen production is continuing actively around the world, and in the present review an attempt is made to briefly summarize the significant recent progress in the development of sulphide photocatalysts for photocatalytic hydrogen production in the following sections.

## 3. Binary metal sulfides for photocatalytic hydrogen evolution

### 3.1. *ZnS nanomaterials*

#### 3.1.1. *Structure and features*

ZnS is a classical IIB-VIA sulfide with two polymorphs, namely, the zinc blende structure and the wurtzite structure. As shown in Fig. 2, the zinc blende structure (cubic) has an FCC lattice and the wurtzite structure (hexagonal) has a CPH lattice [31]. The two structures are pretty close to each other and are different on the stacking sequences of packed planes: ABCABCABC packing pattern for the zinc blende structure and ABABAB packing pattern for the wurtzite structure.

ZnS is a widely investigated photocatalyst with a set of advantages, such as low toxicity, low cost and fast generation of highly

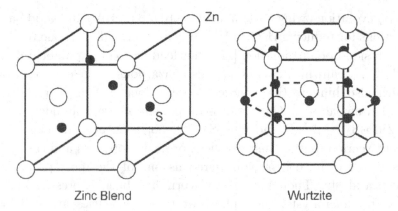

**Fig. 2.** Schematic illustration of zinc blende and wurtzite ZnS crystal structures. *Source*: Reprinted with permission from Ref. [31].

active photoexcited electrons and holes [32]. Moreover, it possesses high photoactivity even in the absence of noble metal co-catalysts [33]. However, one major issue for ZnS-based photocatalysts, which is also a common problem for many metal sulfides, is the strong photocorrosion during photocatalytic reaction, which can decompose the catalyst. Thus, sacrificial reagents are usually needed in the system to scavenge the photoexcited holes. Another problem of ZnS is its relatively wide bandgap (3.66 eV), and hence ZnS alone can only absorb light in the UV range (< 340 nm). This limit remarkably reduces the efficiency of the utilization of the solar spectrum and restricts ZnS from practical applications. Thus, various efforts have been carried out to improve the visible light response of ZnS.

### 3.1.2. *Structural modification*

#### 3.1.2.1. Hetero-element doping

Hetero-element doping is an efficient method to engineer the bandgap of ZnS with a redshift in adsorption while maintaining its high CB position. Different foreign metal cations, such as $Ni^{2+}$ [34], $Cu^{2+}$ [35] and $Pb^{2+}$ [36], have been doped with ZnS, resulting in significantly improved visible light response compared with pristine ZnS, and the edge of absorption could move to over 500 nm upon doping.

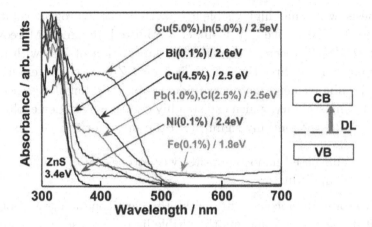

**Fig. 3.** Diffuse reflection spectra of ZnS photocatalysts upon doping with different cations.

*Source*: Reprinted with permission from Ref. [15].

The diffuse reflection spectra of some metal ion-doped ZnS photocatalysts are presented in Fig. 3 [15]. More importantly, the high CB of ZnS is still maintained in the doped catalysts, and noble metal cocatalysts are usually not necessary for efficient hydrogen evolution. Recently, Sun's group [37] prepared Cu-doped ZnS nanospheres with a hierarchical surface structure, which realizes a high hydrogen evolution rate of 1.03 mmol h$^{-1}$ with a quantum efficiency of 26.2% at 425 nm. Chang *et al.* [38] have reported the Ni doping and decorated ZnS on graphene with the highest photocatalytic activity of 8683 μmol h$^{-1}$ g$^{-1}$. Shamsuddin's group [39] have fabricated a ZnS photocatalysts with In and Cu co-doping, and have achieved a hydrogen production rate of 131.32 μmol h$^{-1}$ under visible light irradiation, which is about 8 times higher than ZnS doped with only In.

Besides metal elements, non-metal doping of ZnS as a photocatalyst has also been studied, but with fewer cases. Muruganandham *et al.* [40] first reported the synthesis of C, N co-doped hierarchical porous ZnS nanomaterial as a photocatalyst for efficient degradation of AO7. Mani and co-workers [41] have developed another C, N-doped ZnS photocatalyst with rice grain morphology through a single-step synthesis, which had high visible light activity for hydrogen evolution (up to $10,000\,\mu\text{mol}\,\text{h}^{-1}\,\text{g}^{-1}$ under optimized doping). Moreover, Zhou et al. [42] recently reported the fabrication of ZnS with only N doping for the first time. Such N-doped ZnS not only exhibited obvious enhancement in activity under visible light but also gained an outstanding stability, which can steadily catalyze hydrogen evolution for over 12 h without any significant drop in performance.

### 3.1.2.2. ZnS-based nanocomposites with other semiconductors

Fabrication of nanocomposites with other semiconductors is another useful strategy for promoting the visible light response and photoactivity of ZnS. The formation of heterostructures in such composites usually possesses at least three major advantages: (1) modulating the bandgap to improve the light absorption of the solar spectrum[43]; (2) reducing the recombination of photoexcited electrons and holes by efficient charge transfer at the interfaces of the heterostructures, thus leading to improved photocatalytic yield [44]; and (3) producing more active sites for photocatalytic reaction. Among all candidates, ZnO/ZnS are the most widely studied heterostructures [45–49] due to the advantages of physical and chemical properties of each individual material as well as the synergic improvement of photocatalytic properties upon combination. For example, Wang's group [45] prepared a series of ZnO/ZnS core/shell nanorods with different molar ratios via a water bath route, and the products resulted in a maximum hydrogen production of $388.4\,\mu\,\text{mol}\,\text{h}^{-1}\,\text{g}^{-1}$ under solar-simulated light irradiation. The study further confirmed that the n-p heterojunction between ZnO and ZnS could contribute to the increased surface areas, the enhanced photoabsorption and the efficient separation as well as extended lifetime of photoexcited charge carriers, as illustrated in Fig. 4. Other nanostructures of ZnO/ZnS, such as heterostructured nanowires [46], nanoplates [47], nanorod arrays [48] and cluster microspheres [49], have also been prepared and all exhibited extended absorption under visible light and significant improved activities for hydrogen evolution. Besides ZnO, a few other materials have also been reported to couple

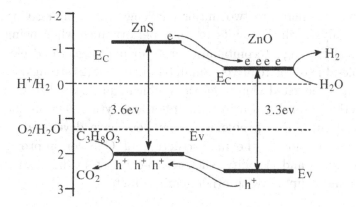

**Fig. 4.** An illustration of the charge transfer process in ZnO/ZnS hetero-structures.

*Source*: Reprinted with permission from Ref. [45].

with ZnS for improving its performance on hydrogen evolution, including g-$C_3N_4$ [50], polyaniline [51], carbon [52], $Fe_3O_4$ [53] and $NiCo_2O_4$ [53]. Compared with ZnS alone, these nanocomposites not only remarkably enhance the photocatalytical performance to different extents but also bring some new advantages upon modification, such as improved dispersing stability in water [51] and magnetic recyclability [53].

## 3.2. *CdS nanomaterials*

### 3.2.1. *Structure and features*

CdS is another IIB-VIA sulfide and similar to ZnS, and there are usually two types of CdS crystal structures: the cubic zinc blende structure and hexagonal wurtzite structure. For each structure, the atom arrangement of CdS is identical to the corresponding ZnS structure as discussed in earlier but with the replacement of Zn by Cd. CdS is one of the most popular candidates among all semiconductors for photocatalytic hydrogen evolution due to its suitable band positions [54] and narrow bandgap (2.4 eV) [3]. That is, CdS has a high CB to reduce protons to hydrogen and can efficiently utilize solar light (< 510 nm) [3], which is a broad portion of the solar spectrum.

However, there are two major challenges in applying CdS as a photocatalyst, which results in poor performance when using CdS alone: (1) the high recombination rate of photogenerated electrons and holes due to the relatively small bandgap and (2) strong corrosion by the photoexcited charge carriers which can destabilize CdS and lose activity quickly during the photocatalytic reaction [55]. To overcome the above problems, researchers have developed various strategies to modify the nanostructure of CdS for improving its photoactivity and stability, such as co-catalyst loading, structural control and coupling with other semiconductors.

### 3.2.2. *Co-catalyst loading*

When applying CdS as the photocalalyst, co-catalysts are widely used to lower the overpotential for water reduction, thus resulting in a significantly improved performance. Noble metals, such as Pt, Pd and Au, are the most commonly used co-catalysts for CdS [56–62]. For example, Nosaka *et al.* [56] loaded Pt on CdS nanoparticles as the co-catalyst, and observed increased photoactivity of CdS by a factor of ∼20. Besides different metal species, many other factors of the co-catalyst, such as the size, shape and hybrid structure, could affect the overall hydrogen evolution efficiency. Wu *et al.* [58] explored the size effect of Pt on CdS nanoparticle-based photocatalysts, and revealed both experimentally and theoretically that a sub-nm Pt cluster has better performance for hydrogen evolution compared with 5-nm Pt particles. Ma *et al.* [59] have developed a novel Au–Pt–CdS hybrid system with direct contact among each constituent, which results in excellent photocatalytic activity by the multiple electron transfer pathways.

Owing to the high cost of noble metals, many non-noble metal co-catalysts have been actively developed recently as alternatives for noble metals to reduce the cost for more practical use [63–66]. Yue *et al.* [63] have applied MoP as a highly active co-catalyst for CdS, which dramatically improves the hydrogen evolution, enabling it to reach a rate of $163.2 \, \mu \, \mathrm{mol \, h^{-1} \, mg^{-1}}$ under visible light illumination. Ma *et al.* [64] have reported $Mo_2N$ as a promising co-catalyst for

CdS with excellent visible light response, high activity and stability. The prepared 2.0% $Mo_2N$/CdS could reach an activity that was 6.1 times that of CdS alone and 1.4 times that of 1.0% Pt/CdS, and the longevity of the catalyst could be over 42 h. Many sulfides have also been reported to act as efficient co-catalysts for CdS and will be discussed in later sections with more details.

### 3.2.3. *Structural control*

The structure factors of CdS, such as the size, morphology, crystal structure and defects, can also have significant influence on the photoactivity. Understanding the effects of these structure factors provides another aspect for improving hydrogen evolution of CdS, and much effort has been reported along this route. Tsubomura's group [67] discovered that Pt-loaded CdS powder with a hexagonal crystal structure presented as a much more efficient photocatalyst for hydrogen evolution than the one with a cubic crystal structure, indicating that the crystal structure can have considerable influence on the activity. Li's group [68] successfully achieved the phase control of CdS crystal to maximize the wurtzite phase, which has been proved to be a facile and low-cost method to improve the photocatalytic performance. Zhang's group [69] reported the fabrication of ultrathin CdS nanosheets stabilized by L-cysteine, as shown in Fig. 5. The unique nanosheet morphology not only resulted in dramatic improvement of CdS for hydrogen evolution to a rate of 41.1 mmol $g^{-1}$ $h^{-1}$ under visible light irradiation but could also achieve high stability for over 3 weeks under ambient conditions. Fierro's group [70] systematically studied the effect of CdS crystallinity on photocatalytic hydrogen production. Moreover, Li's group [71] prepared CdS nanocrystals with rod and granular shapes via a hydrothermal method. The morphologies of CdS depended on the hydrothermal temperature and the amount of ethylenediamine. With increased aspect ratio, CdS nanorods exhibited a higher activity than the nanograins due to better charge separation, and achieved the highest apparent quantum yield of 13.9% for hydrogen evolution under visible light irradiation.

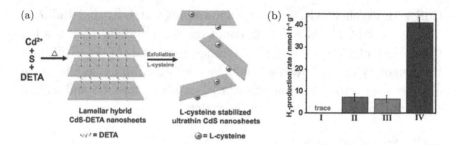

**Fig. 5.** Schematic illustration of the preparation of ultrathin CdS nanosheets (a) and their enhanced hydrogen evolution activity (b-IV) compared with different nanostructured CdS (CdS nanoparticles, b-I; CdS–DETA nanosheets, b-II; CdS nanosheet-based aggregates, b-III).

*Source*: Reprinted with permission from Ref. [69].

### 3.2.4. *Coupling with other semiconductors*

Coupling CdS with other semiconductors to form composites has been proved to be an efficient method to improve photocatalytic performance. Among all the candidates, metal oxides are promising semiconductors because of their low cost, high photoactivity and high stability. The metal oxides usually have low visible light response due to the wide bandgaps, which can be compensated by CdS. Thus, a number of metal oxides/CdS composites have been prepared and studied for photocatalytic hydrogen evolution [72–81]. Zou *et al.* [72] synthesized 1D porous CdS/ZnO with tunable aspect ratios via a facile one-pot cation exchange reaction. The activity of the nanocomposite was 15 times higher than that of pure CdS under visible light irradiation, which was attributed to the overall more conducive structural features for hydrogen evolution arising from nanocomposite formation, as schematically illustrated in Fig. 6. Shen and co-workers [73] prepared $MoO_3$/CdS core/shell nanospheres by a template-free sonochemistry method under ambient conditions. The core–shell structure of $MoO_3$ and CdS could efficiently harvest visible light, significantly enhance the charge separation and prolong electron–hole lifetime. In the absence of noble metal as co-catalysts, the $MoO_3$/CdS nanocomposite could reach a high hydrogen evolution yield of $5.25 \, \text{mmol} \, \text{h}^{-1} \, \text{g}^{-1}$ with a high apparent

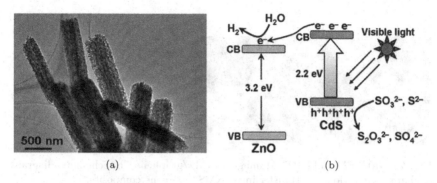

**Fig. 6.** TEM image of 1D porous CdS/ZnO (a) and the schematic illustration of charge transfer mechanism in the composite (b).

*Source*: Reprinted with permission from Ref. [72].

yield of 28.86% at 420 nm. Huang *et al.* [74] developed a rapid, one-pot method to synthesize $TiO_2$/CdS porous hollow microspheres via salt-assisted aerosol decomposition. The prepared photocatalyst exhibited high activity under visible light and could maintain an activity of 162 $\mu$mol h$^{-1}$ per 50 mg photocatalysts even without any noble metal co-catalyst. Such improved photocatalytic performance could be attributed to a combination of different factors, including the formation of heterostructures, unique surface morphology and high crystallinity.

The nanocomposites based on CdS and carbon materials have become popular in recent years and attracted great attention [82–87]. Due to the unique morphology and high conductivity, carbon nanomaterials, such as carbon nanotube, graphene and carbon nanofibers, can significantly increase the surface area, efficiently suppress the charge carrier recombination and increase the stability of CdS. For example, Li and co-workers [82] reported the fabrication of CdS/graphene with CdS clusters dispersed on graphene nanosheets (Fig. 7). Due to the high surface area, excellent supporting ability and high efficiency of electron–hole transportation of graphene, the composite could dramatically improve the photocatalytic activity and the hydrogen production rate could reach 1.12 mmol h$^{-1}$ upon visible light irradiation under optimized conditions. Ye *et al.* [83] prepared the nanocomposites of CdS/graphene and CdS/carbon

(a)                                                    (b)

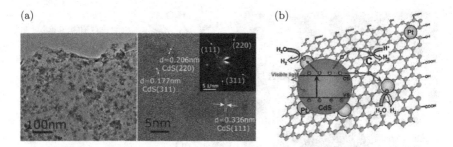

**Fig. 7.** (a) TEM and HRTEM images of CdS/graphene. (b) Schematic diagram of charge separation and transfer in the CdS/graphene composite.

*Source*: Reprinted with permission from Ref. [82].

nanotubes, and compared their photocatalytic performance for hydrogen evolution. While both nanocomposites showed significant enhancement in both activity and stability as compared to pure CdS, the CdS/graphene resulted in a higher rate ($70\,\mu\,\mathrm{mol\,h^{-1}}$ per $0.1\,\mathrm{g}$ photocatalyst) of hydrogen evolution than CdS/carbon nanotubes ($52\,\mu\mathrm{mol\,h^{-1}}$ per $0.1\,\mathrm{g}$ photocatalyst). The reason for the better activity of CdS/graphene could be the larger contact interfaces and the stronger interactions between CdS and graphene, which improved the efficiency of photoexcited electron–hole separation.

Also, other semiconductors have been studied recently to couple with CdS, and have resulted in promising performance for hydrogen evolution, such as $C_3N_4$/CdS [88–90], BNNS/CdS [91] and SiC/CdS [92].

### 3.3. *MoS₂ nanomaterials*

#### 3.3.1. *Structure and features*

$MoS_2$ is a 2D material with layered structures with numerous advantages, such as earth abundance, very low toxicity, good stability against photocorrosion and environmental friendliness. As shown in Fig. 8, in each distinct layer of $MoS_2$ crystal, both Mo and S atoms form hexagonal shaped layers, and the location of the Mo atom is in the center of a trigonal prism with one S atom sitting at the six vertexes [93]. $MoS_2$ has three polymorphs, trigonal, hexagonal

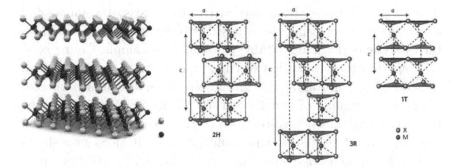

**Fig. 8.** Schematic illustration of MoS$_2$ crystal structures and the three poly-types based on layer stacking: 2H, 3R and 1T.

*Source*: Reprinted with permission from Ref. [93].

and rhombohedral, which are based on the arrangement of adjacent MoS$_2$ layers and usually named 1T, 2H and 3R, respectively. The (1010) Mo-edge structure of MoS$_2$ has a close resemblance to the active site of nitrogenase and its free energy of atomic hydrogen bonding is close to Pt [94]. Hence, MoS$_2$ has been considered a catalyst for hydrogen evolution reaction (HER) with good potential. One important feature of MoS$_2$ is that its electronic and optical properties are strongly affected by the size and number of layers due to the quantum confinement effect. For example, MoS$_2$ in bulk form has poor HER activity and nanosized MoS$_2$ has been proved to be a promising catalyst to replace Pt. The bulk MoS$_2$ has a bandgap of $\sim$1.2 eV and increases to $\sim$1.9 eV when down to a monolayer [95]. Due to such structure-dependent property, 2D ultrathin MoS$_2$ usually presents as a much more promising material for photocatalytic hydrogen evolution and has been widely studied recently.

### 3.3.2. *Structural influence of MoS$_2$ on hydrogen evolution reactions*

In 2007, Jaramillo *et al.* [96] firstly identified that the edge sites of MoS$_2$ were the active sites for hydrogen evolution and the catalytic ability of MoS$_2$ was directly related to the edge state length. Since

then, much effort has been carried out to increase the number of
active sites by increasing the exposed edge length, so as to improve
the catalytic performance of $MoS_2$ [97–99]. For example, Chung et al.
[97] reported the fabrication of edge-exposed $MoS_2$ nanoassembled
structures by controlling the formation of nanoassembled spheres
with the assembly of small-sized fragments of $MoS_2$, and the
nanostructured $MoS_2$ possessed high HER activity and long-term
stability. Wu et al. [98] designed and synthesized $MoS_2$ nanosheets
with rich exposed edge sites via a microdomain reaction method.
Excellent HER performance was observed due to high, active site
densities in the tailored $MoS_2$ nanostructures.

Another useful strategy to increase the number of active sites is by
introducing defective structures on $MoS_2$. Wang et al. [100] prepared
distorted $MoS_2$ nanosheets from bulk $MoS_2$ via a mechanical activation method. Compared with the commercial microcrystalline $MoS_2$,
significant enhancement in HER activity was realized, which could
be attributed to the crystalline defects and distortion generated as
reaction sites during the milling. Xie et al. [101] developed an efficient
method to engineer the defects on $MoS_2$ ultrathin nanosheets with
additional active edge sites by partial cracking the basal planes. With
the introduction of the extra active sites, the defect-rich $MoS_2$ sheets
outcompeted their defect-free counterpart for HER.

Besides the morphology, the crystal structure also plays an
important role in catalytic activity for $MoS_2$. The 1T phase of $MoS_2$
is metastable, and has more metallic features and better electrical
conductivity than the 2H phase [102], and it obtains more active sites
as well [103]. Both reasons result in an enhanced catalytic activity for
the 1T phase compared with the 2H phase. Thus, the stabilization
and the application of 1T $MoS_2$ for HER have attracted considerable
attention recently [104, 105]. For example, Song's group [104] has
developed a novel strategy to mass produce colloidal 1T-$MoS_2$ layers,
which were stabilized by the intercalated ammonium ions. Due
to outstanding electron–hole separation efficiency and more active
sites, the as-prepared 1T-$MoS_2$ exhibited obvious enhancement in
photocatalytic activity than 2H-$MoS_2$ by a factor of 3, as seen in
Fig. 9.

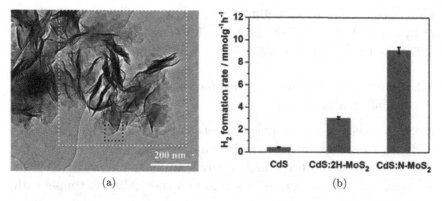

**Fig. 9.** TEM image of ammonium ion-intercalated and stabilized $MoS_2$ with 1T phase (a) and its photocatalytic hydrogen production rate as a co-catalyst compared with that of 2H phase.

*Source*: Reprinted with permission from Ref. [104].

### 3.3.3. *$MoS_2$ as host catalysts for photocatalytic hydrogen evolution*

The monolayer $MoS_2$ with a bandgap of $\sim$1.9 eV ensures its good response under visible light. However, while $MoS_2$ has been widely studied as an efficient host catalyst for electrocatalytic hydrogen production, there are few reports on using it as a host catalyst for photocatalytic HER. This might be because the CB position of $MoS_2$ is very close to the reduction potential of $H^+/H_2$. Thus, the ability of photoexcited electrons from $MoS_2$ to reduce $H^+$ is weak, and usually the results are inefficient. Recently, there has been some progress on applying $MoS_2$ as the host photocatalyst for hydrogen evolution [106–109]. Shi *et al.* [106] reported the fabrication of $NH_3$-treated $MoS_2$ nanosheets for HER with the highest rate of 190 $\mu$mol h$^{-1}$ g$^{-1}$. The $NH_3$ treatment resulted in a more electronegative and basic surface of $MoS_2$ nanosheets, which favored the adsorption of the reactants and improved the reaction efficiency. Zhang *et al.* [107] developed a novel multi-layered mesh-like $MoS_2$ hierarchical nanostructure composed of $MoS_2$ nanosheets grown on Ti foil with vertical alignment. The unique $MoS_2$ structure could enable more surface area and exposed active sites, and the internal meshes could improve the diffusion of sacrificial reagents and release of hydrogen bubbles. Such mesh-like $MoS_2$ reached an HER rate of $\sim$240 $\mu$mol g$^{-1}$ h$^{-1}$ under visible light irradiation.

### 3.3.4. *MoS$_2$ as co-catalysts for photocatalytic hydrogen evolution*

$MoS_2$ is considered a promising alternative of Pt for the photocatalytic hydrogen evolution. Its high catalytic efficiency, low cost and unique structure features makes $MoS_2$ suitable for practical HER applications, wherein it usually serves as a co-catalyst to couple with a series of other photocatalysts.

Multiple works have been reported on a $MoS_2/TiO_2$-based system [110–114]. Zhang's group [110] applied a $TiO_2$ nanobelt as the template to coat with few layers of $MoS_2$ by a simple hydrothermal method, which resulted in a 3D hierarchical core–shell heterostructure. The synthesized $TiO_2$ @$MoS_2$ heterostructure showed outstanding activity in hydrogen production and reached the highest rate of 1.6 mmol h$^{-1}$ g$^{-1}$ in the presence of 50 wt.% $MoS_2$. The excellent performance was attributed to the matched energy band of the $TiO_2$ @$MoS_2$ heterostructure, which could improve the transfer of the charge carriers and reduce the recombination. Recently, Xu's group [111] fabricated another novel $MoS_2/TiO_2$ heterostructure with few-layer $MoS_2$ nanosheets vertically grown on $TiO_2$ nanofibers (Fig. 10). Due to increased exposure of $MoS_2$ as well as the synergetic effect between the two components, this $MoS_2/TiO_2$ photocatalyst possessed excellent longevity and recycling performance, and achieved a hydrogen production rate of 1.68 and 0.49 mmol h$^{-1}$ g$^{-1}$ under UV–Vis and visible light illumination, respectively, which were among the best reported HER performance values with $MoS_2$.

Graphic carbon nitride (g-$C_3N_4$) is another semiconductor to couple with $MoS_2$ that has been widely studied for photocatalytic hydrogen generation [115–118]. Both g-$C_3N_4$ and $MoS_2$ have a 2D structure feature. The formation of such a layered heterostructure can efficiently increase the contact area to improve the charge carrier separation and shorten the charge transport distance, thus

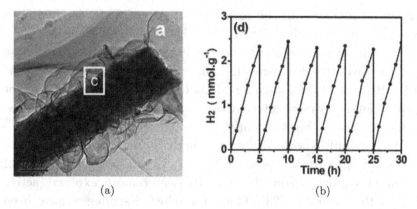

(a)                                    (b)

**Fig. 10.** (a) TEM image of TiO$_2$@MoS$_2$ heterostructures of few-layered MoS$_2$ grown vertically on porous TiO$_2$ nanofibers. (b) Photocatalytic activity of prepared TiO$_2$@MoS$_2$ ($\lambda > 420$ nm) with cyclic hydrogen evolution curves for over 30 h.

*Source*: Reprinted with permission from Ref. [111].

resulting in dramatic HER activity. For example, Chai's group [115] reported a MoS$_2$/g-C$_3$N$_4$ heterojunction photocatalyst via a simple impregnation and heating method, with an enhancement of 11.3 times the pure C$_3$N$_4$ for hydrogen evolution. Besides, MoS$_2$ has been successfully applied as an efficient co-catalyst with a number of other materials as well, such as CdS [119, 120], ZnO [121, 122], CdSe/NiO [123] and 6,13-Pentacenequinone [124].

### 3.4. WS$_2$ nanomaterials

#### 3.4.1. Structural features

As both belong to the family of transition metal dichalcogenides (TMDs), WS$_2$ has a structure analogous to MoS$_2$. WS$_2$ also consists of layered structures held together by weak Van der Waals force, and there are three different WS$_2$ crystal phases based on layer stacking: 1T (trigonal), 2H (hexagonal) and 3R (rhombohedral). Due to the structure similarity, WS$_2$ and MoS$_2$ share a number of important chemical and physical properties in common. One of the most important properties for WS$_2$ is its potential use as a catalyst for HER. Both computations [94] and experiments [96, 125] have been reported on the catalytic activity of TMD for hydrogen evolution at the active site on the edge of the 2D layer, which is due to the unique structure having a resemblance to nitrogenase and the favorable hydrogen binding energy of HER close to that of Pt. Because of the low cost and high earth abundance, $WS_2$ might serve as an inexpensive alternative catalyst of Pt for HER, which has attracted great interest recently.

$WS_2$ in bulk shows little HER activity but can turn into an efficient HER catalyst with few-layer thickness, as a result of increased bandgap from 1.4 to 2.0 eV and more exposed active sites at the nanoscale [93]. Thus, a number of strategies have been proposed to prepare different nanostructured $WS_2$ with improved catalytic activity, such as nanosheets [126], nanoplates [127] and nanoflakes [128]. Besides the morphology, some recent studies reveal that the crystal phase of $WS_2$ also plays an important role inits HER performance [129, 130]. The less stable 1T phase of $WS_2$ is predicted to be metallic with better conductivity, while the more stable 2H phase is semiconducting, and the phase transition from 2H to 1T can usually be induced by an exfoliation method. By increasing the content of the 1T phase in the $WS_2$ catalyst, the charge transfer kinetics is significantly improved, which further leads to enhanced HER activity.

### 3.4.2.  $WS_2$ as a co-catalyst for photocatalytic hydrogen evolution

$WS_2$ has been widely known to bean efficient electrocatalyst or photoelectrocatalyst for HER [130–133] and is also applied as a useful photocatalyst for degradation of a number of pollutants [134, 134, 136]. However, $WS_2$ has rarely been reported as a host catalyst for photocatalytic hydrogen evolution. This might be because of its relatively narrow bandgap and its CB position being pretty close to the hydrogen reduction potential even at the monolayer. Thus, the photogenerated electrons have very weak power to reduce $H^+$, and $WS_2$ alone is not efficient to serve as the photocatalyst for

HER. On the contrary, $WS_2$ has some unique advantages, such as good visible light response, unique 2D crystal shapes and relative good conductivity compared to many other catalysts. These features have enabled $WS_2$ to become a promising component to combine with other photocatalysts for improving the overall performance of photocatalytic hydrogen evolution. Jing *et al.* [137] have prepared $WS_2$-sensitized mesoporous $TiO_2$, and the incorporation of nanosized $WS_2$ led to a higher photocatalytic activity, which was attributed to the dramatic improvement of visible light absorption as well as the better stability of the composites. Hou *et al.* [138] loaded $WS_2$ on mesoporous graphitic carbon nitride (mpg-CN) for photocatalytic hydrogen evolution through an impregnation sulfidation process. A significant enhancement in catalytic activity was achieved as a result of both the good HER activity of $WS_2$ and the better electron transportation between the mpg-CN layers in the presence of $WS_2$. However, the performance of the $WS_2$/mpg-CN catalyst decreased every reaction cycle because of the oxidation/corrosion of $WS_2$. More recently, Zhang's group [139] reported another g-$C_3N_4$/$WS_2$ composite for photocatalytic hydrogen production prepared via a gas–solid reaction. As a co-catalyst, the $WS_2$ not only extended the light absorption of g-$C_3N_4$ from 450 to 800 nm but also effectively collected the CB electrons of g-$C_3N_4$ and reduced the recombination rates of photogenerated charge carriers, as schematically illustrated in Fig. 11(a). Therefore, a significantly enhanced visible light photo-catalytic hydrogen production rate ($101 \, \mu \text{mol g}^{-1} \, \text{h}^{-1}$) was observed for the g-$C_3N_4$/$WS_2$ composite, which was even better than the g-$C_3N_4$/Pt composite with the same loading amount of Pt, as shown in Fig. 11(b). Moreover, the g-$C_3N_4$/$WS_2$ had good stability under the working environment and could maintain its performance after multiple cycles. Besides, a number of studies have been reported on the $WS_2$/CdS composite for HER with excellent properties, which will be discussed in a later section. In general, compared with $MoS_2$, the studies on $WS_2$ for photocatalytic HER are still few, and more work might need to be carried out to explore on the application of $WS_2$ as an efficient co-catalyst.

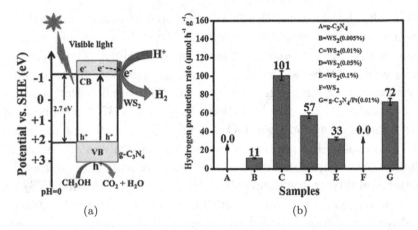

(a)          (b)

**Fig. 11.** (a) Proposed scheme of electron transfer of g-$C_3N_4$/$WS_2$ heterostructures for hydrogen production. (b) Photocatalytic hydrogen production performance of g-$C_3N_4$/$WS_2$ with different $WS_2$ loading amounts, with pristine g-$C_3N_4$, $WS_2$ and g-$C_3N_4$/Pt as references.

*Source*: Reprinted with permission from Ref. [139].

## 3.5. *Other binary metal sulfides*

### 3.5.1. *$Bi_2S_3$*

$Bi_2S_3$ is a direct semiconductor with a bandgap of $\sim$1.3 eV [140] and has been broadly used for various applications, such as the lithium ion battery fabrication [141], electrochemical hydrogen storage [142] and hydrogen sensing [143]. Owing to its good visible light response and high absorption coefficient, $Bi_2S_3$ also exhibits great potential to serve as a photocatalyst. Nevertheless, due to the relatively narrow bandgap, the recombination rate of photogenerated charge carriers in $B_2S_3$ is fast, which leads to significantly reduced activity [144]. Thus, efforts have been made to solve the problem to achieve efficient photocatalytic hydrogen production with $Bi_2S_3$. Abdi *et al.* [145] prepared a mesoporous $Bi_2S_3$/Y-09zeolite composite, and the hydrogen evolution rate was twice that of pristine $Bi_2S_3$, owing to the higher surface area and better charge separation in the mesoporous composite. Kadam *et al.* [146] synthesized a $Bi_2S_3$ quantum dot–glass nanosystem via a melt and quench method. The size of the $Bi_2S_3$ quantum dot could be tuned by controlling the heating temperature, which showed direct influence on the bandgap of the composite. Under the optimized conditions, the $Bi_2S_3$ particles with the size of 3–4 nm produced a hydrogen production rate of $6418.8 \, \mu\text{mol h}^{-1} \text{g}^{-1}$, which was dramatically improved compared with the bulk $Bi_2S_3$ under identical conditions. Moreover, much work has focused on $Bi_2S_3/TiO_2$ composites for two main reasons: (1) a significant enhancement in visible light region in the presence of $Bi_2S_3$ and (2) a good match in CB edges of the two materials, which would facilitate the separation of charge carriers [147–149]. For example, recently Garcia-Mendoza *et al.* [147] developed a facile method to prepare $Bi_2S_3/TiO_2$ nanorods by a one-step solvothermal reaction. The prepared composite presented good stability and achieved a hydrogen production rate of as high as $2460 \, \mu\text{mol h}^{-1} \text{g}^{-1}$, which was 3 times higher than the activity of $TiO_2$ alone.

### 3.5.2. *CuS*

CuS is a semiconductor with attractive optical and electronic properties, and has been applied in many fields for a number of purposes [150–153]. One unique feature of CuS is its high transmittance values [150]. Coupled with the features of low cost, low toxicity, efficient visible light absorption, and good physical and chemical stabilities, CuS has been considered a promising candidate for photocatalysis. For the photocatalytic hydrogen evolution, CuS alone is not sufficient to serve as the host catalyst, due to the unfit conductive band edge and the high recombination rate of photogenerated electrons and holes. On the contrary, CuS has been reported as an effective co-catalyst to combine with a set of classical host catalysts, and can significantly enhance the overall performance by improving the charge separation as well as accelerating the HER [154–157]. For example, the CuS/ZnO composite with CuS cluster loaded on the ZnO surface was prepared by the Kaneco group [154], and the experiment conditions, such as the concentration of the sacrificial anions, were optimized for maximizing the hydrogen evolution rate. Pure ZnO achieved a hydrogen production rate of $255 \, \mu\text{mol g}^{-1}$, and the rate was boosted 8.5 times in the presence of CuS. Wang *et al.* [155] synthesized a series of CuS/$TiO_2$ composites

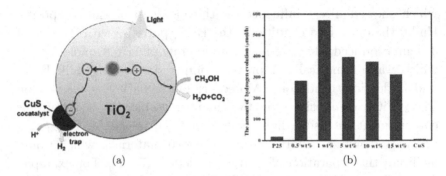

(a)                                          (b)

**Fig. 12.** (a) Schematic illustration of the charge separation process of CuS/TiO$_2$ composites for hydrogen generation. (b) Photoactivities of P25, CuS and CuS/TiO$_2$ composites with different loading amounts of CuS.

*Source*: Reprinted with permission from Ref. [155].

with varied compositions by a hydrothermal method (Fig. 12). The optimized 1 wt.% CuS/TiO$_2$ exhibited a maximum hydrogen production rate of 570 $\mu$mol h$^{-1}$, which dramatically enhanced the efficiency by 32 times compared with pure TiO$_2$.

### 3.5.3. *Ag$_2$S*

Ag$_2$S is a chalcogenide semiconductor with a narrow bandgap of $\sim$0.9–1.05 eV [158]. Thus, Ag$_2$S has a good response to a very wide region of light. Moreover, Ag$_2$S possesses a few other advantages such as a large absorption coefficient, good stability and excellent optical limiting properties [159, 160]. These features have enabled Ag$_2$S to become a useful material for a myriad of photovoltaic and photocatalytic applications [160–163]. As a photocatalyst for hydrogen evolution, pure Ag$_2$S is not very efficient, and it is necessary to form a composite with other materials in order to achieve a satisfactory performance. The role of Ag$_2$S in the composite photocatalyst is essentially important, as it can not only significantly enhance the efficiency of the solar absorption in the visible and near-visible light region but can also facilitate the charge separation of photoexcited electron–hole pairs. Jiang *et al.* [164] reported the fabrication of 5 wt.% Ag$_2$S/g-C$_3$N$_4$ composite photocatalysts with a hydrogen production rate of 10 $\mu$mol h$^{-1}$, which was $\sim$100-fold increase in photocatalytic activity compared with pure g-$C_3N_4$. Liu *et al.* [165] synthesized $ZnO/ZnS$-$Ag_2S$ core–shell nanorods with different molar ratios via a facile two-step method and studied their photocatalytic performance for hydrogen evolution. The incorporation of $Ag_2S$ into $ZnO/ZnS$ nanorods efficiently improved the hydrogen production rate, which obtained a maximum of 4942.9 and $650.4\,\mu\mathrm{mol}\,h^{-1}\,g^{-1}$ under UV and solar-simulated light irradiation, respectively. In addition, a few groups have studied $Ag_2S/TiO_2$ composites because of the good band edge match between $Ag_2S$ and $TiO_2$ as well as the good sensitization from $Ag_2S$, which overcomes the disadvantage of using $TiO_2$ as the photocatalyst [166–168]. For example, Liu *et al.* [166] prepared the $Ag_2S$-sensitized $TiO_2$ nanotube array with remarkably improved photocatalytic efficiency of hydrogen generation. A maximum photoconversion efficiency of 1.21% and the highest hydrogen production rate of $1.13\,\mathrm{mL/cm^2}$ were achieved, where the activity of pure $TiO_2$ nanotubes array could be negligible.

### 3.5.4. $In_2S_3$

$In_2S_3$ is a semiconductor with three polymorphs: $\alpha$-$In_2S_3$ with a defect cubic phase, $\beta$-$In_2S_3$ with a tetragonal defect spinel-like phase and $\gamma$-$In_2S_3$ with a layered structure [169]. Among them, $\beta$-$In_2S_3$ is the most stable form at room temperature and has been extensively investigated as an efficient photocatalyst because of its high photosensitivity, high photoconductivity and suitable band for solar light response [170–172]. Moreover, $In_2S_3$ has good stability and low toxicity, which enables it to become a promising alternative for some classical photocatalysts with toxicities, such as CdS. Several groups have explored the preparation of $In_2S_3$ with different morphologies [173, 174], and various nanostructured $In_2S_3$ have been reported for efficient degradation of dyes and pollutants [174–179]. As a photocatalyst for hydrogen evolution, one challenge for $In_2S_3$ is the fast recombination of photoexcited electron–hole pairs. To overcome the obstacle, a myriad of works have been reported to enhance the charge separation and photocatalytic hydrogen production by combining with suitable semiconductors [180–185], where $TiO_2$ is a major

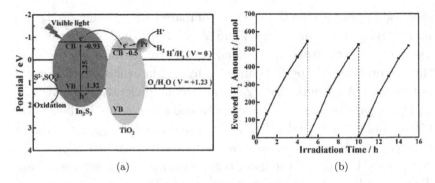

(a)                                                                    (b)

**Fig. 13.** The proposed electron transfer mechanism during photocatalytic hydrogen evolution (a) and time course of stable photocatalytic hydrogen evolution (b) for $In_2S_3/(Pt–TiO_2)$ nanocomposites.

*Source*: Reprinted with permission from Ref.[180].

candidate. For example, Chai *et al.* [180] synthesized the floriated $In_2S_3$-decorated $TiO_2$ nanoparticles loaded with Pt, which achieved a hydrogen production rate of $135\,\mu mol\,h^{-1}$. As schematically shown in Fig. 13, the electron pathway was from $In_2S_3$ as the sensitizer to $TiO_2$ then to Pt as the co-catalyst. Such a pathway was essentially crucial for improved catalytic performance, as reversing the role of $In_2S_3$ and $TiO_2$ resulted in a sharp decrease in hydrogen production rate. $In_2S_3$ has also been applied to form $In_2O_3–In_2S_3$ core–shell type I nanostructures via a simple, low-temperature hydrothermal process [181]. The $In_2O_3–In_2S_3$ nanocomposites exhibited a good hydrogen evolution rate of $61.4\,\mu mol\,h^{-1}\,g^{-1}$, which was significantly better than the plain $In_2O_3$ nanorods and the hollow $In_2S_3$. Besides forming composites, Fu *et al.* [186] reported the improved photocatalytic performance of $In_2S_3$ for hydrogen production by controlling the crystal structures. While the tetragonal $In_2S_3$ showed negligible activity under visible light irradiation, the cubic $In_2S_3$ could possess stable photocatalytic activity for hydrogen evolution with a maximum rate of $960.2\,\mu mol\,g^{-1}\,h^{-1}$ when Pd was applied as the co-catalyst. The distinct performances of $In_2S_3$ were attributed to the different crystal structures, as tetragonal $In_2S_3$ had ordered indium vacancies, while cubic $In_2S_3$ had disordered indium vacancies. The ordered indium vacancies were more likely to depress the charge separation than the disordered vacancies, thus resulting in much poorer activity.

### 3.5.5. *NiS*

NiS is a p-type semiconductor with a narrow bandgap [187]. Due to its good photoactivity, high earth abundance and low cost, NiS is a promising co-catalyst for photocatalytic hydrogen production, which can serve as an alternative to the noble metals. $NiS/C_3N_4$ composites have gained increasing attentions in recent years [188–192]. Hong *et al.* [188] first reported the preparation of noblemetal-free $NiS/C_3N_4$ for hydrogen production via a simple hydrothermal method. The optimized $NiS/C_3N_4$ photocatalyst (1.1 wt.% NiS) was 253 times more active than the pristine $C_3N_4$ and over 80% activity could be maintained after 96 hours. Zhong *et al.* [189] fabricated mpg-$C_3N_4/CNT/NiS$ composite photocatalysts with the highest hydrogen evolution rate of $\sim$521 $\mu$mol g$^{-1}$ h$^{-1}$ under visible light ($\lambda > 420$ nm), which was $\sim$148 times higher than that of mpg-$C_3N_4/CNT$. The dramatic enhancement in performance could be attributed to synergistic effects of both better separation of photoexcited charge carriers by CNT and the enhanced hydrogen evolution kinetics by NiS. Yuan *et al.* [190] developed g-$C_3N_4/CdS/NiS$ composite photocatalysts for hydrogen evolution and achieved a 1,582-fold increase in hydrogen evolution rate when compared with pure g-$C_3N_4$. From Fig. 14, it was revealed that synergistic effects of the formation of CdS/g-$C_3N_4$ 1D/2D nanoheterojunctions as well as the loading of NiS accounted for the remarkably enhanced hydrogen production activity, which not only improved the charge separation and visible light absorption but also favored the water reduction kinetics. NiS has also been combined with a couple of $TiO_2$ materials with different morphologies, such as $TiO_2$ nanoparticles [156], nanosheets [193], nanotubes [194] and nanotubearrays [195], to enhance their photocatalytic hydrogen production activities. As a co-catalyst, NiS can efficiently lower the overpotential of the HER and provide adsorption and redox reaction sites to improve the reaction kinetics. NiS can also expand the light adsorption, and the formation of $TiO_2$–NiS heterojunction structures can further suppress the recombination of photogenerated

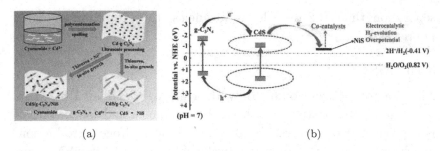

(a)                                                    (b)

**Fig. 14.**    Schematic illustrations of (a) the synthesis of CdS-C$_3$N$_4$/NiS nanocomposites and (b) the proposed photocatalytic hydrogen evolution mechanism.

*Source*: Reprinted with permission from Ref. [190].

charge carriers. For example, NiS–TiO$_2$ nanosheet films achieved an activity 21 times higher than bare TiO$_2$ films [193]. NiS-loaded TiO$_2$ nanotubes exhibited a maximum hydrogen production rate of $7486\,\mu\mathrm{mol\,h^{-1}\,g^{-1}}$, which was $\sim$79 times higher than that of pure TiO$_2$ nanotubes [194]. Recently, Peters *et al.* [196] reported the synthesis of a porous photocatalyst with NiS loaded on a Zr(IV)-based metal− organic framework via the atomic layer deposition method. The prepared catalyst achieved a hydrogen production rate of $3.1\,\mathrm{mmol\,g^{-1}\,h^{-1}}$ upon UV irradiation and $4.8\,\mathrm{mmol\,g^{-1}\,h^{-1}}$ upon visible light irradiation with the addition of rose bengal dye, whereas negligible photocatalytic activities were observed in the absence of NiS. Besides, a series of studies have also been reported combining NiS with other metal sulfides for efficient hydrogen production, such as CdS and ZnIn$_2$S$_4$, which will be discussed in a later section.

## 4.  Ternary metal sulfides for photocatalytic hydrogen evolution

### 4.1.  $ZnX_2S_4\,(X = In, Ga)$

#### 4.1.1.  $ZnIn_2S_4$

ZnIn$_2$S$_4$ has a layered structure with vacancies existing between two adjacent sulfur layers due to the weak interactions [197, 198]. Further, there are three distinct polymorphs of ZnIn$_2$S$_4$ based on the

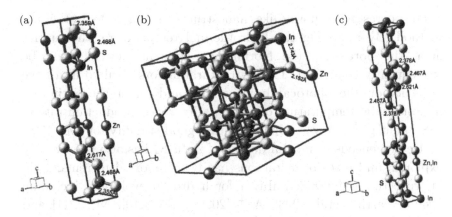

**Fig. 15.** Three polymorphs of $ZnIn_2S_4$ with different lattices: (a) hexagonal, (b) cubic and (c) rhombohedral.

*Source*: Reprinted with permission from Ref. [199].

layer stacking, which results in hexagonal, cubic and rhombohedral lattices (Fig. 15) [199]. As an n-type semiconductor, $ZnIn_2S_4$ has attracted great interest because of its outstanding electrical and optical properties as well as high chemical stability [200]. $ZnIn_2S_4$ is one of the most widely studied photocatalysts in the family of ternary metal sulfides. With a bandgap of 2.0–2.5 eV, $ZnIn_2S_4$ can efficiently respond to visible light [201]. Coupled with its suitable band positions and good photostability, $ZnIn_2S_4$ is an excellent candidate for photocatalytic hydrogen evolution.

Nanostructured $ZnIn_2S_4$ with different crystal structures and morphologies have been prepared and characterized previously [199, 202–207]. Shen *et al.* [199] have fabricated $ZnIn_2S_4$ with different crystal structures and studied their performance for photocatalytic hydrogen production under visible irradiation. With increased sulfidation temperatures, the prepared $ZnIn_2S_4$ gradually transferred from a cubic phase to a rhombohedral phase, and the corresponding photocatalytic activity enhanced from negligible to efficient hydrogen production. The photoluminescence study further indicated that such distinct performances can be attributed to the different recombination rates of photogenerated charge carriers within two different crystal structures. Tian *et al.* [202] synthesized a series of $ZnIn_2S_4$

microspheres with flower-like nanostructures via a hydrothermal method. By adjusting the hydrothermal temperature, the surface area, the pore size, the morphology and the defects could be subsequently controlled. These structural factors resulted in a huge influence on the photocatalytic activity and under the optimized hydrothermal temperature of 160°C, a hydrogen production rate of $248.9\,\mu\mathrm{mol\,g^{-1}\,h^{-1}}$ was achieved with a good stability for over 15 h.

Heterogeneous elemental doping of $ZnIn_2S_4$ has also been widely explored, and a set of metal ions has been reported to remarkably improve the activity of $ZnIn_2S_4$ for hydrogen evolution, including alkaline earth metals [208], $Ag^+$ [209], $Cu^{2+}$ [210], $Ni^+$ [211] and rare earth metals [212]. Recently, Yang et al. [213] synthesized a non-metal-doped $ZnIn_2S_4$ by substituting sulfur atoms with oxygen atoms. Upon oxygen doping, $ZnIn_2S_4$ exhibited a dramatic enhancement in hydrogen evolution with a rate up to $2120\,\mu\mathrm{mol\,h^{-1}\,g^{-1}}$, which was 4.5 times greater than the pure $ZnIn_2S_4$ nanosheets. An in-depth characterization by ultrafast transient absorption spectroscopy indicated that the oxygen doping increased the average recovery lifetime of photoexcited electrons by 1.53 times, and the separation of photogenerated electrons and holes was remarkably improved, which accounted for the better performance.

Moreover, $ZnIn_2S_4$ has been combined with a number of other materials to form heterostructured composites for more efficient hydrogen evolution, such as $ZnIn_2S_4$–metal oxides [214, 215], $ZnIn_2S_4$–carbon nitride [216, 217] and $ZnIn_2S_4$–carbon composites [218–225]. The as-formed heterojunctions can contribute to the charge carrier separation at the interface, thus suppressing their recombinations and extending their lifetime. Due to the unique layered structure of $ZnIn_2S_4$, coupling $ZnIn_2S_4$ with other 2D materials could raise particular interest as many 2D materials possess good conductivities, and the matching of morphology can further give rise to more intimate contacts at the heterojunctions for more efficient charge transfer. For example, Liu et al. [216] fabricated $ZnIn_2S_4$–g-$C_3N_4$ nanocomposites with sheet-on-sheet morphology by a facile hydrothermal method. The well-matched band structures between $ZnIn_2S_4$ and g-$C_3N_4$ enabled the photoexcited electrons to efficiently

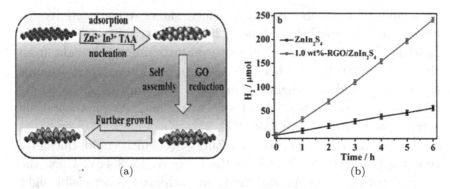

**Fig. 16.** (a) Schematic illustration of the synthesis of sheet-on-sheet RGO/ZnIn$_2$S$_4$ nanocomposites. (b) Photocatalytic activities of ZnIn$_2$S$_4$ before and after loading RGO.

*Source*: Reprinted with permission from Ref. [218].

transfer from g-C$_3$N$_4$ to ZnIn$_2$S$_4$ and vice versa for the photoexcited holes, hence resulting in an efficient charge separation. Moreover, the intimate contact of two layered composites elevated the charge transfer at the interface. Both reasons contributed to an enhanced hydrogen evolution rate under visible light irradiation, which was ~1.91 times higher than pristine ZnIn$_2$S$_4$. Ye *et al.* [218] reported another sheet-on-sheet composite of ZnIn$_2$S$_4$ combined with reduced graphene oxide (RGO), as shown in Fig. 16 Due to the close contact at the heterojunction interface and the good conductivity of RGO, photogenerated electrons can effectively transport from ZnIn$_2$S$_4$ to RGO to reduce the charge recombination. Meanwhile, the RGO could also serve as a co-catalyst with hydrogen evolution reactions being carried out at the defective carbon sites, which would further boost the photocatalytic activity. With the optimized ratio (1.0 wt.% - RGO), the ZnIn$_2$S$_4$/RGO composite produced 245.1 $\mu$mol hydrogen in 6 h, which was ~4.3 times that of pure ZnIn$_2$S$_4$.

### 4.1.2. *ZnGa$_2$S$_4$*

ZnGa$_2$S$_4$ is a semiconductor with a defective chalcopyrite structure and a bandgap of ~3.2–3.4 eV [226], and its optical and electronic properties have been previously studied [227, 228]. ZnGa$_2$S$_4$ has been applied as a photocatalyst, which shows a high activity for degradation of methylene blue [229]. However, due to the wide bandgap, $ZnGa_2S_4$ cannot efficiently absorb the visible light from the solar spectrum, which hinders its applications. A few attempt shave been made to engineer the band structure of $ZnGa_2S_4$ by the substitution of guest metal ions to enhance its photocatalytic activity for hydrogen evolution [230, 231]. Kaga *et al.* [230] prepared Cu(I) and Ga(III) co-substituted $ZnGa_2S_4$, which could efficiently tune the bandgap of $ZnGa_2S_4$ in the range of 2.5–3.4 eV. Upon the co-substitution, a high photoactivity was achieved under visible light with an apparent quantum yield of 15% at 420 nm and a hydrogen production rate of $1.4\,L\,h^{-1}$ was obtained in the presence of 0.5 wt.% Pd. Yang *et al.* [231] engineered the band structure of $ZnGa_2S_4$ via a series of $In^{3+}$-to-$Ga^{3+}$ and $(Cu^+/Ga^{3+})$-to-$Zn^{2+}$ substitutions, resulting in adjustable bandgaps from 3.36 to 3.04 eV and $\sim$2.5 eV, respectively. With optimized composition, the prepared substituted $ZnGa_2S_4$ presented as an efficient photocatalyst for hydrogen evolution under visible light irradiation with a rate of $629\,\mu mol\,h^{-1}\,g^{-1}$ when 0.5 wt.% Ru was loaded and a rate of $386\,\mu mol\,h^{-1}\,g^{-1}$ under the noble metal-free condition.

## 4.2. $XInS_2$ $(X = Cu, Ag)$

### 4.2.1. $CuInS_2$

$CuInS_2$ is a popular $I$-$III$-$VI_2$ ternary compound with three different crystals polymorphs, and the chalcopyrite structure is its most stable phase [232]. $CuInS_2$ is reported to have a bandgap of $\sim$1.5 eV, which is a good match with the solar spectrum [233]. Considering the additional advantages of being free from toxic elements, having good stability, a very high absorption coefficient [234] and tunable optical adsorption by size [235], $CuInS_2$ has been applied for solar energy conversion under multiple fields, such as solar cells, photoelectrochemical devices, photodegradation and photocatalytic hydrogen evolution [236–241].

Garskaite *et al.* [236] fabricated Zn-, Sb- and Ni-doped $CuInS_2$ films on ITO substrates by chemical bath deposition, and the Zn-doped $CuInS_2$ films exhibited the best hydrogen evolution rate, which was $33.26\,mLcm^{-2}$. Guo *et al.* [237] synthesized $TiO_2$–$CuInS_2$ core–shell nanoarrays, which achieved a high photocurrent density of $19.07\,mA\,cm^{-2}$ and a hydrogen generation efficiency of 11.48%. Li and co-workers [238] reported on the p–n-type $CuInS_2/TiO_2$ particles prepared for photocatalytic hydrogen evolution by a one-step solvothermal method. The loaded $CuInS_2$ acted as an efficient sensitizer to significantly enhance the light absorption of both UV and visible light. Meanwhile, the p–n-type heterojunction structures efficiently promoted the charge separation of photoexcited electrons and holes. Both factors contributed to the enhancement of the photocatalytic hydrogen generation and achieved the highest rate of $273.25\,\mu mol$ in 5 h, which was ~7-fold higher than that of pure $TiO_2$.

The structural factors of $CuInS_2$, such as the size, dispersity, morphology and crystallinity, can considerably influence its properties in many aspects, and a number of studies have investigated the preparation of $CuInS_2$ with different structures [239–243]. For achieving the goal of good photoactivity with $CuInS_2$, Zheng *et al.* [244] reported the fabrication of nearly monodispersed $CuInS_2$ with hierarchical microarchitectures via a one-pot solvothermal route, which applied the intermediate CuS, serving as a self-sacrificed template. The unique structure of $CuInS_2$ resulted in a change of bandgap from 1.53 eV of bulk $CuInS_2$ to 1.80 eV and a shift of absorption peak to ~620 nm. More importantly, with the wider bandgap, enlarged surface area and better surface permeability, the photocatalytic activity of prepared $CuInS_2$ for hydrogen evolution was dramatically improved to the highest rate of $84.0\,\mu mol\,h^{-1}\,g^{-1}$, whereas the rate was only $0.53\,\mu mol\,h^{-1}\,g^{-1}$ in previous reported work [245].

### 4.2.2. *AgInS₂*

$AgInS_2$ is another widely studied I–III–VI$_2$ ternary compound with two types of crystal structures, a tetragonal chalcopyrite-type phase and an orthorhombic wurtzite-like phase [246]. $AgInS_2$ with the tetragonal phase is the most extensively reported because of its good stability at room temperature, which can further convert to an orthorhombic phase at 620°C. The tetragonal $AgInS_2$ has shown great potential as a photocatalyst because of its suitable bandgap of 1.87 eV, which can efficiently utilize the visible solar light [247]. Moreover, the band edges of $AgInS_2$ are capable of $H_2O$ reduction, which enables it to become a catalyst for photocatalytic hydrogen evolution. $AgInS_2$ has been reported for use in photodegradation [248, 249] and photoelectrochemical hydrogen evolution [250]. For photocatalytic hydrogen evolution applications, so far the reported studies have majorly focused on the ZnS–$AgInS_2$ and ZnS–$AgInS_2$–$CuInS_2$ solid solutions [251–254]. Tsuji *et al.* [251] first reported the application of ZnS–$AgInS_2$–$CuInS_2$ solid solution for efficient hydrogen evolution with a rate of 2.3 mmol h$^{-1}$ (0.3 g catalyst and Ru as co-catalyst) and a comparatively stable activity for over 20 h. Tsuji *et al.* [252] further studied the effect of the composition of the solid solution on the photocatalytic performance. The results revealed that the composition of ZnS–$AgInS_2$–$CuInS_2$ could determine both the crystal structure and the energy band structure of the solid solution. Under the optimized condition, a high hydrogen production rate of 2320 $\mu$mol h$^{-1}$ (0.3 g catalyst and 0.5 wt.% Ru as co-catalyst) was achieved under simulated solar irradiation. More recently, Kameyama *et al.* [253] reported the preparation of ZnS/$AgInS_2$ solid solution in a highly controllable manner (as shown in Fig. 17), and successfully applied the composites for hydrogen evolution. Both the composition and the size of ZnS–$AgInS_2$ could be independently tuned by adjusting the amount of input chemicals, which further resulted in designable energy structures. Further, the ZnS/$AgInS_2$ particles with a size of 4.2–5.5-nm diameter and a bandgap of 2.3–2.4 eV exhibited the best photocatalytic activity. Jagadeeswararao *et al.* [254] also reported on another $(ZnS)_{0.4}(AgInS_2)_{0.6}$ composite, which exhibited high photoactivity with a hydrogen production rate of 5.0 mmol g$^{-1}$ h$^{-1}$ even in the absence of noble metal co-catalysts.

### 4.3. *Other ternary metal sulfides*

### 4.3.1. *AgGaS$_2$*

Similar to $AgInS_2$, $AgGaS_2$ is a p-type semiconductor with two types of crystal structures: a thermodynamically stable tetragonal

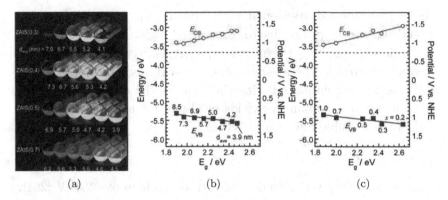

**Fig. 17.** Photoluminescence of $ZnS$–$AgInS_2$ solid solution nanocrystals with different averaged diameters and compositions (a), and the conduction band edge ($E_{CB}$) and valence band edge ($E_{VB}$) as a function of $E_g$: (b) $(AgIn)_{0.5}ZnS_2$ with different averaged diameters and (c) $(AgIn)_xZn_{2(1-x)}S_2$ nanocrystals with a diameter of $\sim 5\,nm$ and with different $x$ values.

*Source*: Reprinted with permission from Ref. [253].

phase [255] and a metastable orthorhombic phase [256]. So far, $AgGaS_2$ has mostly focused on the use of the tetragonal phase, which is also known as the chalcopyrite structure. With a direct bandgap of $\sim 2.6\,eV$, $AgGaS_2$ can efficiently respond to solar light and has been explored as a promising material for solar cells and photocatalysis. $AgGaS_2$ has also been reported to have considerable activity for photocatalytic hydrogen production [257, 258]. Jang *et al.* [257] prepared bulky $AgGaS_2$ as a photocatalyst for hydrogen production by a conventional solid-state reaction. By means of post treatment under $H_2S$ flow at $1123\,K$, the photocatalytic activity of $AgGaS_2$ under visible light was significantly elevated due to the removal of impurity phases and high crystallinity. Jang *et al.* [259] further reported the preparation of a $TiO_2$–$AgGaS_2$ composite for hydrogen production with highly crystalline $AgGaS_2$ decorated on $TiO_2$ nanoparticles. The $AgGaS_2$–$TiO_2$ heterostructures resulted in an improved charge separation at the interface due to the facile dissociation of photoexcited electrons from $AgGaS_2$ to $TiO_2$. Hence, a good hydrogen production rate of $420\,\mu mol\,h^{-1}$ was achieved under visible light irradiation. Yamato *et al.* [260] synthesized a series of

AgGaS$_2$ photocatalysts doped with iron, cobalt and nickel, which extended the absorption edge to the near-IR region. Among all metal-cation-doped AgGaS$_2$, the nickel-doped AgGaS$_2$ realized hydrogen production even in the absence of any co-catalyst. Its performance could be further enhanced with rhodium as the co-catalyst, which gave out apparent quantum yields of about 1% at 540–620 nm and could utilize visible light up to 760 nm.

### 4.3.2.　CuGaS$_2$

CuGaS$_2$ is a p-type semiconductor with a bandgap of ∼2.4 eV [261]. Owing to its high absorption coefficient ($10^5$–$10^6$ cm$^{-1}$) and relatively good stability [262], CuGaS$_2$ is a visible-light-responsive material for photocatalysis, and a number of CuGaS$_2$ with various sizes, morphologies and phases have been reported [263–266]. CuGaS$_2$ alone has not been reported as an efficient photocatalyst for hydrogen evolution, which might be ascribed to the fast recombination of photoexcited electrons and the positive holes that widely exist in CuGaS$_2$ as it is a p-type semiconductor. Thus, the photoactivity of CuGaS$_2$ was dramatically undermined. A few studies have been carried out to improve the hydrogen evolution of CuGaS$_2$ by different methods [267–270]. Kaga *et al.* [267] prepared a series of Ag$^+$ substituted CuGaS$_2$ with a single chalcopyrite phase by controlling the amount of excessive Ga$_2$S$_3$ during sample preparation. Upon the incorporation of Ag$^+$, both the conduction and VBs shifted to a more positive position, and the mobility of the holes at the VB was improved compared to that of pristine CuGaS$_2$, which accounted for the enhanced photoactivity. Similarly, Kandiel *et al.* [268] reported that CuGaS$_2$ with Zn$^{2+}$ substitution could increase its photocatalytic activity from negligible to a reasonable rate. Moreover, Zhao *et al.* [269] synthesized CuGaS$_2$–ZnS p–n-type nanoheterostructures via a solution route (Fig. 18), where Cu$_{1.94}$S nanocrystals first grew on ZnS and further served as sacrificial seeds to form CuGaS$_2$. A hydrogen production rate of 131 $\mu$mol g$^{-1}$h$^{-1}$ was achieved by CuGaS$_2$-ZnS nanocomposites under visible light irradiation, which was 15-fold higher than the rate of pristine CuGaS$_2$ and was comparable to that of the reported CdS nanophase. The catalytic performance of

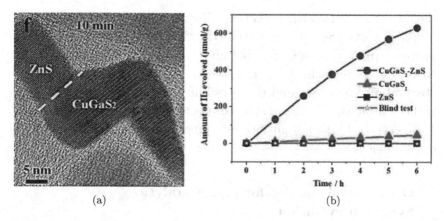

(a)                                    (b)

**Fig. 18.** (a) HRTEM image of $CuGaS_2$–ZnS composites with nanoheterostructures. (b) Photocatalytic hydrogen evolution of $CuGaS_2$ (triangle), ZnS (square) and $CuGaS_2$–ZnS (dot) under visible light irradiation, and $CuGaS_2$–ZnS under light-off environment (star).

*Source*: Reprinted with permission from Ref. [269].

$CuGaS_2$–ZnS was boosted by the efficient charge separation due to the formation of the p–n heterojunctions. Besides, noble metal-loaded $CuGaS_2$ has been successfully employed as a hydrogen-evolving photocatalyst in a Z-scheme system for efficient water splitting under simulated sunlight irradiation [270].

### 4.3.3. $CuGa_3S_5$

Tabata and his co-workers [271] have done pioneering work on the exploration of $CuGa_3S_5$ for photocatalytic hydrogen evolution. $CuGa_3S_5$ was prepared by a solid-state reaction and possessed a chalcopyrite-type structure with a bandgap of 2.4 eV. In the presence of $Na_2S$ and $Na_2SO_3$ as sacrificial electron donors, the prepared $CuGa_3S_5$ could generate hydrogen at a rate of $\sim$15 $\mu$mol h$^{-1}$ under visible light irradiation even without a co-catalyst. Its hydrogen evolution activity could be improved to 40 $\mu$mol h$^{-1}$ after loading an optimized amount of Rh. Moreover, a series of dispersed metal sulfides have been studied as the co-catalysts for $CuGa_3S_5$. Interestingly, with NiS suspended in the reaction solution, $CuGa_3S_5$ achieved a significantly improved hydrogen production rate, which was $\sim$3 times the rate of optimized Rh–CuGa$_3$S$_5$. The results indicated that for the prepared CuGa$_3$S$_5$ with dispersed NiS, the entire electron migration process (from CuGa$_3$S$_5$ to NiS and subsequent H$_2$ evolution on NiS) was more facile than a noble metal-loaded CuGa$_3$S$_5$. Such observations demonstrated the possibility of efficient photocatalysis without a physical junction structure between the photocatalyst and the co-catalyst, though an intimate junction was generally believed to be efficient.

## 5. Other metal sulfides for photocatalytic hydrogen evolution

### 5.1. *Quaternary metal sulfides*

Aside from the abovementioned binary and ternary metal sulfides, a new class of quaternary metal sulfide (QMS) photocatalysts (such as $X$Ga$_2$In$_3$S$_8$ ($X$ = Cu, Ag) [272–274], A$_2^{\mathrm{I}}$–Zn–A$^{\mathrm{IV}}$–S$_4$ (A$^{\mathrm{I}}$ = Cu and Ag; A$^{\mathrm{IV}}$ = Sn and Ge) [275–279]) has also been developed for photocatalytic hydrogen evolution in recent years. Similar to ternary metal sulfides, the bandgap of QMSs can be tuned by controlling their stoichiometry and composition [280, 281]. Thus, the optical and photocatalytic properties of QMS photocatalysts are significantly influenced by their stoichiometric compositions. For example, by the inclusion of indium into CuGa$_3$S$_5$, Kudo *et al.* [272] developed a novel microsized QMS CuGa$_2$In$_3$S$_8$ (CGIS) photocatalyst, which could absorb light up to 700 nm (which therefore covers almost 50% of the solar spectrum), whereas CuGa$_3$S$_5$ could absorb light only up to 516 nm. Kandiel *et al.* [282] recently reported the effect of the Ga-to-In ratio in CuGa$_x$In$_{5-x}$S$_8$ ($x$ = 0–5) on the photocatalytic activity for visible-light-driven hydrogen evolution, and found that CuGa$_2$In$_3$S$_8$ showed the highest activity and stability.

The morphological feature of QMS photocatalysts also plays an important role in their photocatalytic performance. However, the synthesis of QMS photocatalysts is still challenging because of the difficulty of controlling their stoichiometry and phase structure [280]. The typical preparation method for QMS is usually based on solid-state reactions, and when using such a conventional solid-state reaction route, QMS particles can only be prepared in the bulk form [272, 275]. Recently, several solution-based synthetic methods (such as solvothermal synthesis [278], the hot injection method [273, 274, 279, 282] and the sol–gel method [277]) have been reported to prepare nanosized QMS photocatalysts for photocatalytic hydrogen evolution. For example, Kandiel *et al.* [273] demonstrated the successful preparation of a nanosized $CuGa_2In_3S_8$ (CGIS) photocatalyst by using the hot injection method. The nanosized CGIS photocatalyst showed much higher photocatalytic activity for hydrogen production than microsized particles, which might be attributed to its higher surface area and good dispersion properties. Wang *et al.* [277] reported an interesting work, where ultrathin $Cu_2ZnSnS_4$ nanosheets were prepared via a facile template-free sol–gel approach. The prepared $Cu_2ZnSnS_4$ nanosheets exhibited not only a special unilamellar structure with a thickness of only 2–3 nm but also high photocatalytic activity and stability for $H_2$ evolution from aqueous solution containing $S^{2-}$ and $SO_3^{2-}$ even without loading any noble metals.

In addition, much effort has been also made to fabricate heterostructured QMS photocatalysts to further improve their photocatalytic activity. For example, the deposition of $CuGa_xIn_{5-x}S_8$ (CGIS) nanocrystals onto $TiO_2$ nanoparticles was prepared by a unique solvent-induced deposition process [282], and the CGIS–$TiO_2$ photocatalyst showed comparable photocatalytic activity for hydrogen production to that obtained using bare CGIS nanocrystals. The enhanced activities at low CGIS loadings observed in the presence of $TiO_2$ could be ascribed to the improved dispersion of the powder suspension and optical path in the photoreactor. In another interesting work [283], quasi-spherical $Cu_2ZnSnS_4$ (CZTS) nanoparticles with a narrow size distribution were prepared, and the CZTS–Au and CZTS–Pt heterostructured nanoparticles were then produced by a simple impregnation method (Fig. 19). These heterostructured nanoparticles exhibited significantly enhanced photocatalytic properties toward degradation of Rhodamine B and

*D. Chen and X. Li*

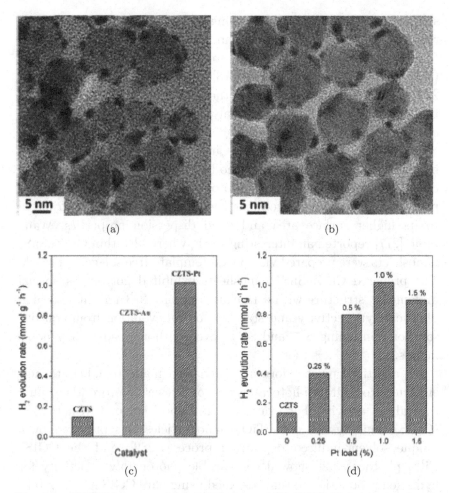

**Fig. 19.** HRTEM images of (a) CZTS–Au and (b) CZTS–Pt heterostructured nanoparticles; (c) hydrogen evolution rate of CZTS, CZTS–Au and CZTS–Pt nanomaterials during a 4-h test; (d) dependence of the hydrogen evolution rate on the Pt load on CZTS–Pt heterostructures.

*Source*: Reprinted with permission from Ref. [283].

hydrogen generation from water splitting when compared to pure CZTS. The enhanced photocatalytic activity could be attributed to the built intimate contact between the two materials, which would facilitate the efficient charge transfer of photogenerated carriers from the semiconductor to the metal co-catalyst.

## 5.2. Hybrid metal sulfides

### 5.2.1. Binary/binary metal sulfides

As mentioned above, CdS is an important photocatalyst for photocatalytic hydrogen evolution in terms of its wide light response range and high CB minimum, and its practical application is greatly restrained by its poor stability arising from photocorrosion during the photocatalytic reaction. To overcome this problem, the widely used strategy is loading co-catalysts or constructing semiconductor heterostructures with staggered band alignments, which would facilitate the separation of photogenerated electrons and holes and suppress the photocorrosion of CdS. To improve the photocatalytic activity for $H_2$ evolution, loading appropriate co-catalysts onto the surface of the photocatalysts can significantly facilitate the reactions by synergistic effects of enriching the active sites, suppressing the charge recombination and reverse reactions, as well as lowering the activation energy [284]. On the contrary, some binary metal sulphides, such as $MoS_2$, $WS_2$ and NiS, have been demonstrated as co-catalysts to replace noble metals, as discussed in the previous section. In this regard, the coupling of CdS with sulphide co-catalyst is expected to effectively improve the photocatalytic activity and stability of CdS. Till now, progress has been made on the development of hybrid sulphides consisting of CdS and sulphide co-catalysts, such as $CdS/MoS_2$ [285–289], $CdS/WS_2$ [290–292], $CdS/NiS$ [293, 294], $CdS/CuS$ [295] and $CdS/Ag_2S$ [296], which exhibited enhanced photocatalytic activities for hydrogen evolution in varying degrees compared with CdS, although the activities still need further improvement. For example, Liu *et al.* [287] reported the fabrication of stable $1T-MoS_2$ slabs grown *in situ* on CdS nanorods ($1T-MoS_2@CdS$) by using a solvothermal method (Figs. 20(a) and 20(b)). The heterostructure with an optimum loading of 0.2 wt.% $1T-MoS_2$ exhibited an almost 39-fold enhancement in the photocatalytic activity relative to that exhibited by bare CdS (Fig. 20(c)). The enhanced photocatalytic activity was attributed to the synergistic effects resulting from formation of the intimate nanojunction between the interfaces and effective electron transport in the metallic phase

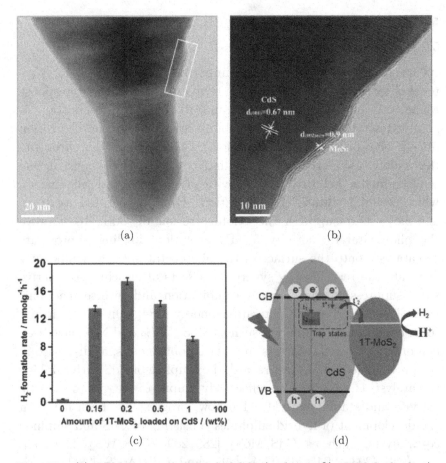

**Fig. 20.** (a) TEM image of 1T-MoS$_2$@CdS (with 1 wt.% 1T-MoS$_2$ loading); (b) magnification of the area marked by a box in panel a showing the folded edges of the 1T-MoS$_2$ layers on the CdS NRs; (c) the rate of H$_2$ evolution on 1T-MoS$_2$@CdS photocatalysts loaded with different amounts of 1T-MoS$_2$ under visible light irradiation; (d) schematic illustration of the 1T-MoS$_2$ co-catalyst strategy for accelerating electron transfer.

*Source*: Reprinted with permission from Ref. [287].

of 1T-MoS$_2$, as schematically illustrated in Fig. 20(d). This work represents a step toward the *in situ* realization of a 1T-phase MoS$_2$-based heterostructure as a promising co-catalyst with high performance and low cost. For another example, Zong *et al.* [290]

prepared $WS_2/CdS$ photocatalysts with different amounts of $WS_2$ co-catalyst, and found that the $WS_2$ co-catalyst played a crucial role in $H_2$ production. The highest $H_2$ production rate was increased by up to 28 times when CdS was loaded with only 1.0 wt.% $WS_2$. The enhanced $H_2$ evolution of $WS_2/CdS$ was largely attributed to the junctions formed between $WS_2$ and CdS as well as the excellent performance of $WS_2$ as a co-catalyst in catalyzing $H_2$ evolution.

In addition, an alternative way to prevent the photocorrosion of CdS is surface passivation since the photogenerated carriers of CdS are easily captured by its surface states [297]. In particular, a ZnS passivation layer has been widely used to passivate some surface states of CdS, which is further favorable for the photocatalytic reaction [298–301]. For instance, Xie *et al.* [299] prepared a CdS–ZnS composite with a mesoporous ZnS shell grown *in situ* on a CdS core by using a one-pot surfactant-free hydrothermal route. The obtained CdS–ZnS composite showed a significantly enhanced hydrogen evolution rate, which was 169 and 56 times higher than that of ZnS and CdS, respectively, under visible light. The enhanced photocatalytic activity was ascribed to the synergistic effect of ZnS as a passivation layer as well as the unique spatial distribution of the photoexcited charge carriers in the CdS–ZnS core–shell particles.

Similarly, other heterostructures consisting of two binary sulphides, such as CuS–ZnS [302, 303] and $Ag_2S$–ZnS [304], have also been proved to promote the generation and separation of photoinduced electrons and holes, thus improving the photocatalytic performance significantly for hydrogen evolution. For example, Zhang *et al.* [302] reported the preparation of CuS–ZnS porous nanosheet photocatalysts for visible light photocatalytic hydrogen production. The prepared CuS–ZnS porous nanosheets with 2 mol% loading content of CuS reached a high hydrogen production rate of $4147 \, \mu mol \, h^{-1} \, g^{-1}$ even without a Pt co-catalyst. Such high visible light photocatalytic hydrogen production activity was suggested to be due to the photoinduced interfacial charge transfer from the VB of ZnS to CuS arising from the heterostructure between ZnS and CuS.

### 5.2.2. Binary/ternary metal sulfides

As discussed above, ternary sulphides (such as $ZnIn_2S_4$ and $CuIn_2S_4$) are promising visible-light-driven photocatalysts for photocatalytic hydrogen evolution, owing to their suitable bandgaps, considerable photocatalytic activity and chemical stability as well as low toxicity. On the contrary, co-catalysts are always indispensable for visible-light-driven metal sulphide photocatalysts to obtain high photocatalytic hydrogen evolution activity. More importantly, several binary metal sulphides (such as $MoS_2$ and $WS_2$) have been demonstrated as efficient co-catalysts in the replacement of noble metals (such as Pt and Pd) for photocatalytic $H_2$ production in various photocatalytic systems. Thus, the loading of binary sulphide-based co-catalysts on the surface of ternary sulphides is expected to effectively inhibit the recombination of photogenerated electron and holes, lower the overpotential for hydrogen evolution and also provide redox reaction sites for $H_2$ evolution.

Recently, binary sulfides like $MoS_2$ and $NiS_2$ have been demonstrated to be excellent co-catalysts for ternary sulphides (such as $MoS_2$–$ZnIn_2S_4$ [305–307], $MoS_2$–$CuIn_2S_4$ [308] and $NiS_2$––$CdLa_2S_4$ [309]) in photocatalytic hydrogen evolution. For example, Wei et al. [305] prepared $MoS_2$–$ZnIn_2S_4$ nanocomposites for photocatalytic hydrogen evolution, and demonstrated that the photocatalytic hydrogen evolution activity of $ZnIn_2S_4$ was significantly increased by loading $MoS_2$ as a co-catalyst. The enhanced photocatalytic activity of $MoS_2$–$ZnIn_2S_4$ was ascribed to the improved charge separation at the interface of $MoS_2$ and $ZnIn_2S_4$ as well as the existence of many defect sites in $MoS_2$, which could act as adsorption sites for hydrogen atoms, thus leading to hydrogen evolution. Yuan et al. revealed that the loading of a $NiS_2$ co-catalyst could significantly enhance the hydrogen production rate over the $CdLa_2S_4$ photocatalyst under visible light irradiation, and the 2 wt.% $NiS_2$-loaded $CdLa_2S_4$ photocatalyst exhibited the highest $H_2$ production rate of $2.5\,\mathrm{mmol\,h^{-1}\,g^{-1}}$, which was over 3 times higher than that of the pristine $CdLa_2S_4$ photocatalyst. The promoted photocatalytic $H_2$ production by $NiS_2$-loading was attributed to the enhanced separation of photogenerated electrons and holes as well as the activation effect of $NiS_2$ for $H_2$ evolution.

In addition, considerable attention has been paid to the solid solution-type sulphide photocatalysts, since making a solid solution between wide and narrow bandgap semiconductors is a powerful strategy for bandgap control [310]. Recently, much effort has been expended on the solid solutions of the wide bandgap ZnS photocatalyst and narrow bandgap semiconductors, such as $ZnS$–$CuInS_2$ [311, 312] and $ZnS$–$AgInS_2$ [313], for photocatalytic hydrogen evolution from water splitting. For example, an often-studied $CuInS_2$-based solid solution is the $ZnS$–$CuInS_2$ composite, which is generally prepared by alloying ternary $CuInS_2$ with binary ZnS [312, 314, 315]. By varying the ratio of $CuInS_2$ to ZnS, the bandgap of the $ZnS$–$CuInS_2$ solid solution can be easily tuned to absorb visible light [314], thus making it very promising for visible-light-driven photocatalytic hydrogen evolution. Lin *et al.* [312] prepared a series of $(ZnS)_x(CuInS_2)_{1-x}$ hierarchical microspheres by using a facile one-pot solvothermal method. The obtained solid solution samples exhibited an excellent hydrogen evolution rate up to $91\,\mu\mathrm{mol}\,\mathrm{h}^{-1}\,\mathrm{g}^{-1}$ without noble metal loading. The excellent photocatalytic performance of these solid solutions was attributed to the synergetic effect of the tunable electronic structure and the unique hierarchical porous structure.

# 6. Summary

In summary, herein an up-to-date review of the recent progress regarding the studies of metal sulphide nanomaterials for photocatalytic hydrogen evolution is presented. Over the past two decades, metal sulphide-based photocatalysts have been studied extensively for their potential for photocatalytic hydrogen evolution. Undoubtedly, there are aspects of these sulphide-based photocatalysts that make them fascinating for applications in the area of photocatalysis, and remarkably rapid progress in this area has already been made. Needless to say, due to the tremendous research effort and space limitations, this review article is unable to cover all the exciting research achievements reported in this field.

The research involving metal sulphide-based photocatalysts for photocatalytic hydrogen evolution will continue to be exciting and highly rewarding. Nonetheless, the research in this field is still at an early stage. A number of challenges remain to be explored for their practical applications. Many of these challenges are related to the development of synthetic methods for facile synthesis of metal sulphides at a high quality and in large quantities. The remaining issues include but are not limited to reliable control of morphological feature, hierarchical structure, crystalline phase as well as a considerable number of defects. In addition, these issues still remain a grand challenge to the fundamental research. So far, we are still at a fairly early stage of understanding fundamental theories of various aspects of metal sulphide-based photocatalysts, including the electronic structure, charge transport, electrochemical properties and dependence of photocatalytic properties on their bandgap structures. In terms of metal sulphide-related photocatalytic hydrogen evolution, specific attention should be focused more on further exploration of the electronic structures and properties of metal sulphides, gaining a better understanding of the charge transfer, separation and recombination of photoinduced carriers in metal sulphide-based photocatalysts, and the interpretation of intrinsic mechanism for photocatalytic hydrogen evolution in different metal sulphide-based photocatalytic systems. Thus, the photocatalytic mechanism of metal sulphide-based photocatalysts for photocatalytic hydrogen evolution still remains an intriguing issue and needs to be explored further. Furthermore, for a given metal sulphide photocatalyst, the construction of a suitable photocatalytic reaction system is another challenging work. Despite a number of remaining challenges, the future of metal sulphide nanomaterials for photocatalytic hydrogen evolution should be very interesting and fantastic.

## Acknowledgments

This work is financially supported by the National Natural Science Foundation of China (No. 51872271, 51972294), and Zhejiang

Provincial Natural Science Foundation of China (No. LY19E020003, LQ20F040007).

## References

[1] Mills, S. LeHunte, An overview of semiconductor photocatalysis, *J. Photochem. Photobiol. A-Chem.*, 108 (1997) 1–35.

[2] D. Ravelli, D. Dondi, M. Fagnoni, A. Albini, Photocatalysis. A multifaceted concept for green chemistry, *Chem. Soc. Rev.*, 38 (2009) 1999–2011.

[3] P.V. Kamat, Meeting the clean energy demand: Nanostructure architectures for solar energy conversion, *J. Phys. Chem. C*, 111 (2007) 2834–2860.

[4] J.M. Herrmann, Heterogeneous photocatalysis: fundamentals and applications to the removal of various types of aqueous pollutants, *Catal. Today*, 53 (1999) 115–129.

[5] K. Nakata, A. Fujishima, $TiO_2$ photocatalysis: Design and applications, *J. Photochem. Photobiol. C*, 13 (2012) 169–189.

[6] M.D. Hernandez-Alonso, F. Fresno, S. Suarez, J.M. Coronado, Development of alternative photocatalysts to $TiO_2$: Challenges and opportunities, *Energy Environ. Sci.*, 2 (2009) 1231–1257.

[7] Y.Q. Qu, X.F. Duan, Progress, challenge and perspective of heterogeneous photocatalysts, *Chem. Soc. Rev.*, 42 (2013) 2568–2580.

[8] Z.F. Huang, L. Pan, J.J. Zou, X.W. Zhang, L. Wang, Nanostructured bismuth vanadate-based materials for solar-energy-driven water oxidation: a review on recent progress, *Nanoscale*, 6 (2014) 14044–14063.

[9] L.W. Zhang, Y.F. Zhu, A review of controllable synthesis and enhancement of performances of bismuth tungstate visible-light-driven photocatalysts, *Catal. Sci. Technol.*, 2 (2012) 694–706.

[10] X.P. Dong, F.X. Cheng, Recent development in exfoliated two-dimensional g-$C_3N_4$ nanosheets for photocatalytic applications, *J. Mater. Chem. A*, 3 (2015) 23642–23652.

[11] X.J. Chen, Y.Z. Dai, X.Y. Wang, Methods and mechanism for improvement of photocatalytic activity and stability of $Ag_3PO_4$: A review, *J. Alloy Compd.*, 649 (2015) 910–932.

[12] U. Shamraiz, R.A. Hussain, A. Badshah, B. Raza, Functional metal sulfides and selenides for the removal of hazardous dyes from water, *J. Photochem. Photobiol. B*, 159 (2016) 33–41.

[13] K. Chang, X. Hai, J.H. Ye, Transition metal disulfides as noble-metal-alternative co-catalysts for solar hydrogen production, *Adv. Energy Mater.*, 6 (2016) 1502555.

[14] A. Fujishima, K. Honda, Electrochemical photolysis of water at a semiconductor electrode, *Nature*, 238 (1972) 37–38.

[15] A. Kudo, Y. Miseki, Heterogeneous photocatalyst materials for water splitting, *Chem. Soc. Rev.*, 38 (2009) 253–278.

[16] R. Abe, Recent progress on photocatalytic and photoelectrochemical water splitting under visible light irradiation, *J. Photochem. Photobiol. C*, 11 (2010) 179–209.

[17] M. Ni, M.K.H. Leung, D.Y.C. Leung, K. Sumathy, A review and recent developments in photocatalytic water-splitting using $TiO_2$ for hydrogen production, *Renew. Sust. Energy Rev.*, 11 (2007) 401–425.

[18] Y.X. Li, H.G. Gou, J.J. Lu, C.Y. Wang, A two-step synthesis of $NaTaO_3$ microspheres for photocatalytic water splitting, *Int. J. Hydrogen Energy*, 39 (2014) 13481–13485.

[19] T. Takata, C.S. Pan, K. Domen, Recent progress in oxynitride photocatalysts for visible-light-driven water splitting, *Sci. Technol. Adv. Mater.*, 16 (2015) 033506.

[20] H.Y. Lin, T.H. Lee, C.Y. Sie, Photocatalytic hydrogen production with nickel oxide intercalated $K_4Nb_6O_{17}$ under visible light irradiation, *Int. J. Hydrogen Energy*, 33 (2008) 4055–4063.

[21] S.H. Kim, S. Park, C.W. Lee, B.S. Han, S.W. Seo, J.S. Kim, I.S. Cho, K.S. Hong, Photophysical and photocatalytic water splitting performance of stibiotantalite type-structure compounds, $SbMO_4$ (M = Nb, Ta), *Int. J. Hydrogen Energy*, 37 (2012) 16895–16902.

[22] H.H. Yang, X.R. Liu, Z.H. Zhou, L.J. Guo, Preparation of a novel $Cd_2Ta_2O_7$ photocatalyst and its photocatalytic activity in water splitting, *Catal. Commun.*, 31 (2013) 71–75.

[23] W.Q. Cui, S.S. Ma, L. Liu, Y.H. Liang, PbS-sensitized $K_2Ti_4O_9$ composite: Preparation and photocatalytic properties for hydrogen evolution under visible light irradiation, *Chem. Eng. J.*, 204 (2012) 1–7.

[24] K. Maeda, D.L. Lu, K. Domen, Solar-driven Z-scheme water splitting using modified $BaZrO_3$-$BaTaO_2N$ solid solutions as photocatalysts, *ACS Catal.*, 3 (2013) 1026–1033.

[25] K. Zhang, L.J. Guo, Metal sulphide semiconductors for photocatalytic hydrogen production, *Catal. Sci. Technol.*, 3 (2013) 1672–1690.

[26] N.Z. Bao, L.M. Shen, T. Takata, K. Domen, Self-templated synthesis of nanoporous CdS nanostructures for highly efficient photocatalytic hydrogen production under visible, *Chem. Mater.*, 20 (2008) 110–117.

[27] H. Zhang, Y.F. Zhu, Significant visible photoactivity and antiphotocorrosion performance of CdS photocatalysts after monolayer polyaniline hybridization, *J. Phys. Chem. C*, 114 (2010) 5822–5826.

[28] F. Fang, L. Chen, Y.B. Chen, L.M. Wu, Synthesis and photocatalysis of $ZnIn_2S_4$ nano/micropeony, *J. Phys. Chem. C*, 114 (2010) 2393–2397.

[29] J.J. Ding, M.H. Yan, S. Sun, J. Bao, C. Gao, Hydrothermal synthesis of $CaIn_2S_4$-reduced graphene oxide nanocomposites with increased photocatalytic performance, *ACS Appl. Mater. Interfaces*, 6 (2014) 12877–12884.

[30] J. Kolny-Olesiak, H. Weller, Synthesis and application of colloidal $CuInS_2$ semiconductor nanocrystals, *ACS Appl. Mater. Interfaces*, 5 (2013) 12221–12237.

[31] D. Moore, Z.L. Wang, Growth of anisotropic one-dimensional ZnS nanostructures, *J. Mater. Chem.*, 16 (2006) 3898–3905.

[32] K. Koci, L. Matejova, O. Kozak, L. Capek, V. Vales, M. Reli, P. Praus, K. Safarova, A. Kotarba, L. Obalova, ZnS/MMT nanocomposites: The effect of ZnS loading in MMT on the photocatalytic reduction of carbon dioxide, *Appl. Catal. B*, 158 (2014) 410–417.

[33] J.S. Hu, L.L. Ren, Y.G. Guo, H.P. Liang, A.M. Cao, L.J. Wan, C.L. Bai, Mass production and high photocatalytic activity of ZnS nanoporous nanoparticles, *Angew. Chem. Int. Ed.*, 44 (2005) 1269–1273.

[34] A. Kudo, M. Sekizawa, Photocatalytic $H_2$ evolution under visible light irradiation on Ni-doped ZnS photocatalyst, *Chem. Commun.*, 15 (2000) 1371–1372.

[35] A. Kudo, M. Sekizawa, Photocatalytic $H_2$ evolution under visible light irradiation on Zn1-x CuxS solid solution, *Catal. Lett.*, 58 (1999) 241–243.

[36] I. Tsuji, A. Kudo, $H_2$ evolution from aqueous sulfite solutions under visible-light irradiation over Pb and halogen-codoped ZnS photocatalysts, *J. Photochem. Photobiol. A: Chem.*, 156 (2003) 249–252.

[37] N. Li, L. Zhang, J. Zhou, D. Jing, Y. Sun, Localized nano-solid-solution induced by Cu doping in ZnS for efficient solar hydrogen generation, *Dalton Trans.*, 43 (2014) 11533–11541.

[38] C.J. Chang, K.W. Chu, M.H. Hsu, C.Y. Chen, Ni-doped ZnS decorated graphene composites with enhanced photocatalytic hydrogen-production performance, *Int. J. Hydrogen Energy*, 40 (2015) 14498–14506.

[39] M. Kimi, L. Yuliati, M. Shamsuddin, Preparation and characterization of In and Cu co-doped ZnS photocatalysts for hydrogen production under visible light irradiation, *J. Energy Chem.*, 25 (2016) 512–516.

[40] M. Muruganandham, Y. Kusumoto, Synthesis of N, C codoped hierarchical porous microsphere ZnS as a visible light-responsive photocatalyst, *J. Phys. Chem. C*, 113 (2009) 16144–16150.

[41] A.D. Mani, P. Ghosal, C. Subrahmanyam, Novel synthesis of C, N doped rice grain shaped ZnS nanomaterials-towards enhanced visible light photocatalytic activity for aqueous pollutant removal and $H_2$ production, *RSC Adv.*, 4 (2014) 23292–23298.

[42] Y. Zhou, G. Chen, Y. Yu, Y. Feng, Y. Zheng, F. He, Z. Han, An efficient method to enhance the stability of sulphide semiconductor photocatalysts: a case study of N-doped ZnS, *Phys. Chem. Chem. Phys.*, 17 (2015) 1870–1876.

[43] S. Shen, L. Zhao, L. Guo, $Zn_mIn_2S_{3+m}$ ($m = 1$–5, integer): A new series of visible-light-driven photocatalysts for splitting water to hydrogen, *Int. J. Hydrogen Energy*, 35 (2010) 10148–10154.

[44] P.W. Blom, V.D. Mihailetchi, L.J.A. Koster, D.E. Markov, Device physics of polymer: fullerene bulk heterojunction solar cells, *Adv. Mater.*, 19 (2007) 1551–1566.

[45] H.X. Sang, X.T. Wang, C.C. Fan, F. Wang, Enhanced photocatalytic $H_2$ production from glycerol solution over ZnO/ZnS core/shell nanorods

prepared by a low temperature route, *Int. J. Hydrogen Energy*, 37 (2012) 1348–1355.

[46] Z. Wang, S.W. Cao, S.C.J. Loo, C. Xue, Nanoparticle heterojunctions in ZnS-ZnO hybrid nanowires for visible-light-driven photocatalytic hydrogen generation, *CrystEngComm*, 15 (2013) 5688–5693.

[47] J.S. Jang, C.J. Yu, S.H. Choi, S.M. Ji, E.S. Kim, J.S. Lee, Topotactic synthesis of mesoporous ZnS and ZnO nanoplates and their photocatalytic activity, *J. Catal.*, 254 (2008) 144–155.

[48] D. Bao, P. Gao, X. Zhu, S. Sun, Y. Wang, X. Li, P. Yang, ZnO/ZnS Heterostructured nanorod arrays and their efficient photocatalytic hydrogen evolution, *Chem. Eur. J.*, 21 (2015) 12728–12734.

[49] A. Wu, L. Jing, J. Wang, Y. Qu, Y. Xie, B. Jiang, H. Fu, ZnO-dotted porous ZnS cluster microspheres for high efficient, Pt-free photocatalytic hydrogen evolution, *Sci. Rep.*, 5 (2015) 8858.

[50] F. Shi, L. Chen, C. Xing, D. Jiang, D. Li, M. Chen, ZnS microsphere/g-$C_3N_4$ nanocomposite photo-catalyst with greatly enhanced visible light performance for hydrogen evolution: synthesis and synergistic mechanism study, *RSC Adv.*, 4 (2014) 62223–62229.

[51] C.J. Chang, K.W. Chu, ZnS–polyaniline composites with improved dispersing stability and high photocatalytic hydrogen production activity, *Int. J. Hydrogen Energy*, 41 (2016) 21764–21773.

[52] J. Wang, Y.F. Lim, G.W. Ho, Carbon-ensemble-manipulated ZnS heterostructures for enhanced photocatalytic $H_2$ evolution, *Nanoscale*, 6 (2014) 9673–9680.

[53] C.J. Chang, Z. Lee, M. Wei, C.C. Chang, K.W. Chu, Photocatalytic hydrogen production by magnetically separable $Fe_3O_4$@ZnS and $NiCo_2O_4$@ZnS core-shell nanoparticles, *Int. J. Hydrogen Energy*, 40 (2015) 11436–11443.

[54] D. Meissner, R. Memming, B. Kastening, Photoelectrochemistry of cadmium sulfide. 1. Reanalysis of photocorrosion and flat-band potential, *J. Phys. Chem.*, 92 (1988) 3476–3483.

[55] J.F. Reber, M. Rusek, Photochemical hydrogen production with platinized suspensions of cadmium sulfide and cadmium zinc sulfide modified by silver sulfide, *J. Phys. Chem.*, 90 (1986) 824–834.

[56] Y. Nosaka, Y. Ishizuka, H. Miyama, Separation mechanism of a photoinduced electron-hole pair in metal-loaded semiconductor powders, *Ber. Bunsenges. Phys. Chem.*, 90 (1986) 1199–1204.

[57] M. Matsumura, T. Uchihara, K. Hanafusa, H. Tsubomura, Interfacial band structure of platinum-loaded CdS powder and its correlation with the photocatalytic activity, *J. Electrochem. Soc.*, 136 (1989) 1704–1709.

[58] Q. Wu, S. Xiong, P. Shen, S. Zhao, Y. Li, D. Su, A. Orlov, Exceptional activity of sub-nm Pt clusters on CdS for photocatalytic hydrogen production: a combined experimental and first-principles study, *Catal. Sci. Technol.*, 5 (2015) 2059–2064.

[59] L. Ma, K. Chen, F. Nan, J.H. Wang, D.J. Yang, L. Zhou, Q.Q. Wang, Improved hydrogen production of Au–Pt–CdS hetero-nanostructures by

efficient plasmon-induced multipathway electron transfer, *Adv. Funct. Mater.*, 26 (2016) 6076–6083.

[60] J. Feng, J. Liu, G. Wei, J. Zhang, S. Wang, Z. Wang, C. An, Solar-driven Pt modified hollow structured CdS photocatalyst for efficient hydrogen evolution, *RSC Adv.*, 4 (2014) 36665–36670.

[61] M. Luo, W. Yao, C. Huang, Q. Wu, Q. Xu, Shape effects of Pt nanoparticles on hydrogen production via Pt/CdS photocatalysts under visible light, *J. Mater. Chem. A*, 3 (2015) 13884–13891.

[62] Q. Wang, J. Li, N. An, Y. Bai, X. Lu, J. Li, W. Shangguan, Preparation of a novel recyclable cocatalyst wool-Pd for enhancement of photocatalytic $H_2$ evolution on CdS, *Int. J. Hydrogen Energy*, 38 (2013) 10761–10767.

[63] Q. Yue, Y. Wan, Z. Sun, X. Wu, Y. Yuan, P. Du, MoP is a novel, noble-metal-free cocatalyst for enhanced photocatalytic hydrogen production from water under visible light, *J. Mater. Chem. A*, 3 (2015) 16941–16947.

[64] B. Ma, Y. Liu, J. Li, K. Lin, W. Liu, H. Zhan, $Mo_2N$: An efficient non-noble metal cocatalyst on CdS for enhanced photocatalytic $H_2$ evolution under visible light irradiation, *Int. J. Hydrogen Energy*, 41 (2016) 22009–22016.

[65] S. Cao, Y. Chen, C.C. Hou, X.J. Lv, W.F. Fu, Cobalt phosphide as a highly active non-precious metal cocatalyst for photocatalytic hydrogen production under visible light irradiation, *J. Mater. Chem. A*, 3 (2015) 6096–6101.

[66] J.S. Jang, D.J. Ham, N. Lakshminarasimhan, W.Y. Choi, J.S. Lee, Role of platinum-like tungsten carbide as cocatalyst of CdS photocatalyst for hydrogen production under visible light irradiation, *Appl. Catal. A: Gen.*, 346 (2008) 149–154.

[67] M. Matsumura, S. Furukawa, Y. Saho, H. Tsubomura, Cadmium sulfide photocatalyzed hydrogen production from aqueous solutions of sulfite: effect of crystal structure and preparation method of the catalyst, *J. Phys. Chem.*, 89 (1985) 1327–1329.

[68] Y. Li, L. Tang, S. Peng, Z. Li, G. Lu, Phosphate-assisted hydrothermal synthesis of hexagonal CdS for efficient photocatalytic hydrogen evolution, *CrystEngComm*, 14 (2012) 6974–6982.

[69] Y. Xu, W. Zhao, R. Xu, Y. Shi, B. Zhang, Synthesis of ultrathin CdS nanosheets as efficient visible-light-driven water splitting photocatalysts for hydrogen evolution, *Chem. Commun.*, 49 (2013) 9803–9805.

[70] F. Vaquero, R.M. Navarro, J.L.G. Fierro, Evolution of the nanostructure of CdS using solvothermal synthesis at different temperature and its influence on the photoactivity for hydrogen production, *Int. J. Hydrogen Energy*, 41 (2016) 11558–11567.

[71] Y. Li, Y. Hu, S. Peng, G. Lu, S. Li, Synthesis of CdS nanorods by an ethylenediamine assisted hydrothermal method for photocatalytic hydrogen evolution, *J. Phys. Chem. C*, 113 (2009) 9352–9358.

[72] X. Zou, P.P. Wang, C. Li, J. Zhao, D. Wang, T. Asefa, G.D. Li, One-pot cation exchange synthesis of 1D porous CdS/ZnO heterostructures for visible-light-driven $H_2$ evolution, *J. Mater. Chem. A*, 2 (2014) 4682–4689.

[73] Z. Shen, G. Chen, Y. Yu, Q. Wang, C. Zhou, L. Hao, R. Mu, Sonochemistry synthesis of nanocrystals embedded in a $MoO_3$-CdS core-shell photocatalyst with enhanced hydrogen production and photodegradation, *J. Mater. Chem.*, 22 (2012) 19646–19651.

[74] Y. Huang, J. Chen, W. Zou, L. Zhang, L. Hu, M. He, X. Xing, $TiO_2$/CdS porous hollow microspheres rapidly synthesized by salt-assistant aerosol decomposition method for excellent photocatalytic hydrogen evolution performance, *Dalton Trans.*, 45 (2016) 1160–1165.

[75] X. Wang, G. Liu, G.Q. Lu, H.M. Cheng, Stable photocatalytic hydrogen evolution from water over ZnO-CdS core-shell nanorods, *Int. J. Hydrogen Energy*, 35 (2010) 8199–8205.

[76] X. Wang, G. Liu, L. Wang, Z. Chen, G.. Lu, Q.M. G., H.M. Cheng, ZnO-CdS/Cd heterostructure for effective photocatalytic hydrogen generation, *Adv. Energy Mater.*, 2 (2012) 42–46.

[77] X.H. Lu, S.L. Xie, T. Zhai, Y.F. Zhao, P. Zhang, Y.L. Zhang, Y.X. Tong, Monodisperse $CeO_2$/CdS heterostructured spheres: one-pot synthesis and enhanced photocatalytic hydrogen activity, *RSC Adv.*, 1 (2011) 1207–1210.

[78] Z. Wang, J. Hou, C. Yang, S. Jiao, H. Zhu, Three-dimensional $MoS_2$-CdS-$\gamma$-TaON hollow composites for enhanced visible-light-driven hydrogen evolution, *Chem. Commun.*, 50 (2014) 1731–1734.

[79] G. Yang, W. Yan, Q. Zhang, S. Shen, S. Ding, One-dimensional CdS/ZnO core/shell nanofibers via single-spinneret electrospinning: tunable morphology and efficient photocatalytic hydrogen production, *Nanoscale*, 5 (2013) 12432–12439.

[80] Y. Li, W. Zhang, L. Li, C. Yi, H. Lv, Q. Song, Litchi-like CdS/$CdTiO_3$-$TiO_2$ composite: synthesis and enhanced photocatalytic performance for crystal violet degradation and hydrogen production, *RSC Adv.*, 6 (2016) 51374–51386.

[81] S. Peng, Y. Huang, Y. Li, Rare earth doped $TiO_2$-CdS and $TiO_2$-CdS composites with improvement of photocatalytic hydrogen evolution under visible light irradiation, *Mat. Sci. Semicon. Proc.*, 16 (2013) 62–69.

[82] Q. Li, B. Guo, J. Yu, J. Ran, B. Zhang, H. Yan, J.R. Gong, Highly efficient visible-light-driven photocatalytic hydrogen production of CdS-cluster-decorated graphene nanosheets, *J. Am. Chem. Soc.*, 133 (2011) 10878–10884.

[83] A. Ye, W. Fan, Q. Zhang, W. Deng, Y. Wang, CdS-graphene and CdS-CNT nanocomposites as visible-light photocatalysts for hydrogen evolution and organic dye degradation, *Catal. Sci. Technol.*, 2 (2012) 969–978.

[84] L. Jia, D.H. Wang, Y.X. Huang, A.W. Xu, H.Q. Yu, Highly durable N-doped graphene/CdS nanocomposites with enhanced photocatalytic hydrogen evolution from water under visible light irradiation, *J. Phys. Chem. C*, 115 (2011) 11466–11473.

[85] M.E. Khan, M.M. Khan, M.H. Cho, CdS-graphene Nanocomposite for Efficient Visible-light-driven photocatalytic and photoelectrochemical applications, *J. Colloid Interf. Sci.*, 482 (2016) 221–232.

[86] Y.K. Kim, M. Kim, S.H. Hwang, S.K. Lim, H. Park, S. Kim, CdS-loaded flexible carbon nanofiber mats as a platform for solar hydrogen production, *Int. J. Hydrogen Energy*, 40 (2015) 136–145.

[87] Q.Z. Wang, J.H. Lian, Q. Ma, S.L. Zhang, J.J. He, J.B. Zhong, J.Z. Li, H.H. Huang, B. Su, Preparation of carbon spheres supported CdS photocatalyst for enhancement its photocatalytic $H_2$ evolution, *Catal. Today*, 281 (2017) 662–668.

[88] W. Li, C. Feng, S. Dai, J. Yue, F. Hua, H. Hou, Fabrication of sulfur-doped g-$C_3N_4$/Au/CdS Z-scheme photocatalyst to improve the photocatalytic performance under visible light, *Appl. Catal. B-Environ.*, 168 (2015) 465–471.

[89] H. Yu, F. Chen, F. Chen, X. Wang, *In situ* self-transformation synthesis of g-$C_3N_4$-modified CdS heterostructure with enhanced photocatalytic activity, *Appl. Surf. Sci.*, 358 (2015) 385–392.

[90] D. Zheng, X. Wang, Integrating CdS quantum dots on hollow graphitic carbon nitride nanospheres for hydrogen evolution photocatalysis, *Appl. Catal. B-Environ.*, 179 (2015) 479–488.

[91] F. Ma, G. Zhao, C. Li, T. Wang, Y. Wu, J. Lv, Y. Zhong, X. Hao, Fabrication of CdS/BNNSs nanocomposites with broadband solar absorption for efficient photocatalytic hydrogen evolution, *CrystEngComm*, 18 (2016) 631–637.

[92] Y. Peng, Z. Guo, D. Wang, N. Pan, W. Yuan, Heterogeneous nucleation of CdS to enhance visible-light photocatalytic hydrogen evolution of SiC/CdS composite, *Appl. Phys. Lett.*, 107 (2015) 012102.

[93] Q.H. Wang, K. Kalantar-Zadeh, A. Kis, J.N. Coleman, M.S. Strano, Electronics and optoelectronics of two-dimensional transition metal dichalcogenides, *Nat. Nanotechnol.*, 7 (2012) 699–712.

[94] B. Hinnemann, P.G. Moses, J. Bonde, K.P. Jørgensen, J.H. Nielsen, S. Horch, I. Chorkendorff, J.K. Nørskov, Biomimetic hydrogen evolution: $MoS_2$ nanoparticles as catalyst for hydrogen evolution, *J. Am. Chem. Soc.*, 127 (2005) 5308–5309.

[95] K.F. Mak, C. Lee, J. Hone, J. Shan, T.F. Heinz, Atomically thin $MoS_2$: a new direct-gap semiconductor, *Phys. Rev. Lett.*, 105 (2010) 136805.

[96] T.F. Jaramillo, K.P. Jørgensen, J. Bonde, J.H. Nielsen, S. Horch, I. Chorkendorff, Identification of active edge sites for electrochemical $H_2$ evolution from $MoS_2$ nanocatalysts, *Science*, 317 (2007) 100–102.

[97] D.Y. Chung, S.K. Park, Y.H. Chung, S.H. Yu, D.H. Lim, N. Jung, Y.E. Sung, Edge-exposed $MoS_2$ nano-assembled structures as efficient electrocatalysts for hydrogen evolution reaction, *Nanoscale*, 6 (2014) 2131–2136.

[98] Z. Wu, B. Fang, Z. Wang, C. Wang, Z. Liu, F. Liu, D.P. Wilkinson, $MoS_2$ nanosheets: a designed structure with high active site density for the hydrogen evolution reaction, *ACS Catal.*, 3 (2013) 2101-2107.

[99] D. Gopalakrishnan, D. Damien, M.M. Shaijumon, $MoS_2$ quantum dot-interspersed exfoliated $MoS_2$ nanosheets, *ACS Nano*, 8 (2014) 5297–5303.

[100] D. Wang, Z. Wang, C. Wang, P. Zhou, Z. Wu, Z. Liu, Distorted MoS$_2$ nanostructures: An efficient catalyst for the electrochemical hydrogen evolution reaction, *Electrochem. Commun.*, 34 (2013) 219–222.

[101] J. Xie, H. Zhang, S. Li, R. Wang, X. Sun, M. Zhou, Y. Xie, Defect–rich MoS$_2$ ultrathin nanosheets with additional active edge sites for enhanced electrocatalytic hydrogen evolution, *Adv. Mater.*, 25 (2013) 5807–5813.

[102] M.A. Lukowski, A.S. Daniel, F. Meng, A. Forticaux, L. Li, S. Jin, Enhanced hydrogen evolution catalysis from chemically exfoliated metallic MoS$_2$ nanosheets, *J. Am. Chem. Soc.*, 135 (2013) 10274–10277.

[103] D. Voiry, M. Salehi, R. Silva, T. Fujita, M. Chen, T. Asefa, M. Chhowalla, Conducting MoS$_2$ nanosheets as catalysts for hydrogen evolution reaction, *Nano Lett.*, 13 (2013) 6222–6227.

[104] Q. Liu, X. Li, Q. He, A. Khalil, D. Liu, T. Xiang, L. Song, Gram-scale aqueous synthesis of stable few-layered 1T-MoS$_2$: Applications for visible-light-driven photocatalytic hydrogen evolution, *Small*, 11 (2015) 5556–5564.

[105] Q. Ding, F. Meng, C.R. English, M. Cabán-Acevedo, M.J. Shearer, D. Liang, S. Jin, Efficient photoelectrochemical hydrogen generation using heterostructures of Si and chemically exfoliated metallic MoS$_2$, *J. Am. Chem. Soc.*, 136 (2014) 8504–8507.

[106] J. Shi, Y. Zhang, Y. Hu, X. Guan, Z. Zhou, L. Guo, NH$_3$-treated MoS$_2$ nanosheets as photocatalysts for enhanced H$_2$ evolution under visible-light irradiation, *J. alloy. compd.*, 688 (2016) 368–375.

[107] Q. Zhang, S.Z. Kang, D. Wang, X. Li, L. Qin, J. Mu, Multi-layered mesh-like MoS$_2$ hierarchical nanostructure fabricated on Ti foil: An efficient noble metal-free photocatalyst for visible-light-driven H$_2$ evolution from water, *Catal. Commun.*, 82 (2016) 7–10.

[108] R. Peng, L. Liang, Z.D. Hood, A. Boulesbaa, A.A. Puretzky, A. Ievlev, M. Chi, In-plane heterojunctions enable multiphasic two-dimensional (2D) MoS$_2$ nanosheets as efficient photocatalysts for hydrogen evolution from water reduction, *ACS Catal.*, 6 (2016) 6723–6729.

[109] X. Zong, Y. Na, F. Wen, G. Ma, J. Yang, D. Wang, C. Li, Visible light driven H$_2$ production in molecular systems employing colloidal MoS$_2$ nanoparticles as catalyst, *Chem. Commun.*, 30 (2009) 4536–4538.

[110] W. Zhou, Z. Yin, Y. Du, X. Huang, Z. Zeng, Z. Fan, H. Zhang, Synthesis of few-layer MoS$_2$ nanosheet-coated tio$_2$ nanobelt heterostructures for enhanced photocatalytic activities, *Small*, 9 (2013) 140–147.

[111] C. Liu, L. Wang, Y. Tang, S. Luo, Y. Liu, S. Zhang, Y. Xu, Vertical single or few-layer MoS$_2$ nanosheets rooting into TiO$_2$ nanofibers for highly efficient photocatalytic hydrogen evolution, *Appl. Catal. B-Environ.*, 164 (2015) 1–9.

[112] P. Zhang, T. Tachikawa, M. Fujitsuka, T. Majima, Efficient charge separation on 3D architectures of TiO$_2$ mesocrystals packed with a chemically exfoliated MoS$_2$ shell in synergetic hydrogen evolution, *Chem. Commun.*, 51 (2015) 7187–7190.

[113] Y. Yuan, H. Lu, Z. Ji, J. Zhong, M. Ding, D. Chen, Z. Zou, Enhanced visible-light-induced hydrogen evolution from water in a noble-metal-free system catalyzed by ZnTCPP-MoS$_2$/TiO$_2$ assembly, *Chem. Eng. J.*, 275 (2015) 8–16.

[114] Y. Zhu, Q. Ling, Y. Liu, H. Wang, Y. Zhu, Photocatalytic H$_2$ evolution on MoS$_2$-TiO$_2$ catalysts synthesized via mechanochemistry, *Phys. Chem. Chem. Phys.*, 17 (2015) 933–940.

[115] Y. Tian, L. Ge, K. Wang, Y. Chai, Synthesis of novel MoS$_2$/g-C$_3$N$_4$ heterojunction photocatalysts with enhanced hydrogen evolution activity, *Mater. Charact.*, 87 (2014) 70–73.

[116] J. Xu, Y. Li, S. Peng, Photocatalytic hydrogen evolution over Erythrosin B-sensitized graphitic carbon nitride with *in situ* grown molybdenum sulfide cocatalyst, *Int. J. Hydrogen Energy*, 40 (2015) 353–362.

[117] M. Li, L. Zhang, X. Fan, M. Wu, Y. Du, M. Wang, J. Shi, Dual synergetic effects in MoS$_2$/pyridine-modified g-C$_3$N$_4$ composite for highly active and stable photocatalytic hydrogen evolution under visible light, *Appl. Catal. B-Environ.*, 190 (2016) 36–43.

[118] D. Zheng, G. Zhang, Y. Hou, X. Wang, Layering MoS$_2$ on soft hollow g-C$_3$N$_4$ nanostructures for photocatalytic hydrogen evolution, *Appl. Catal. A-Gen.*, 521 (2016) 2–8.

[119] X. Zong, H. Yan, G. Wu, G. Ma, F. Wen, L. Wang, C. Li, Enhancement of photocatalytic H$_2$ evolution on CdS by loading MoS$_2$ as cocatalyst under visible light irradiation, *J. Am. Chem. Soc.*, 130 (2008) 7176–7177.

[120] X. Zong, G. Wu, H. Yan, G. Ma, J. Shi, F. Wen, C. Li, Photocatalytic H$_2$ evolution on MoS$_2$/CdS catalysts under visible light irradiation, *J. Phys. Chem. C*, 114 (2010) 1963–1968.

[121] Y.J. Yuan, F. Wang, B. Hu, H.W. Lu, Z.T. Yu, Z.G. Zou, Significant enhancement in photocatalytic hydrogen evolution from water using a MoS$_2$ nanosheet-coated ZnO heterostructure photocatalyst, *Dalton Trans.*, 44 (2015) 10997–11003.

[122] Y.J. Yuan, J.R. Tu, Z.J. Ye, H.W. Lu, Z.G. Ji, B. Hu, Z.G. Zou, Visible-light-driven hydrogen production from water in a noble-metal-free system catalyzed by zinc porphyrin sensitized MoS$_2$/ZnO, *Dyes Pigments*, 123 (2015) 285–292.

[123] Y. Dong, Y. Chen, P. Jiang, G. Wang, X. Wu, R. Wu, C. Zhang, Efficient and stable MoS$_2$/CdSe/NiO photocathode for photoelectro-chemical hydrogen generation from Water, *Chem. Asian J.*, 10 (2015) 1660–1667.

[124] Y.J. Yuan, Z.T. Yu, Y.H. Li, H.W. Lu, X. Chen, W.G. Tu, Z.G. Zou, A MoS$_{2/6}$, 13-pentacenequinone composite catalyst for visible-light-induced hydrogen evolution in water, *Appl. Catal. B-Environ.*, 181 (2016) 16–23.

[125] D. Merki, X. Hu, Recent developments of molybdenum and tungsten sulfides as hydrogen evolution catalysts, *Energy Environ. Sci.*, 4 (2011) 3878–3888.

[126] J.W. Seo, Y.W. Jun, S.W. Park, H. Nah, T. Moon, B. Park, J. Cheon, Two-dimensional nanosheet crystals, *Angew. Chem. Int. Ed.*, 46 (2007) 8828–8831.

[127] V.G. Pol, S.V. Pol, A. Gedanken, Micro to nano conversion: A one-step, environmentally friendly, solid state, bulk fabrication of $WS_2$ and $MoS_2$ nanoplates, *Cryst. Growth Des.*, 8 (2008) 1126–1132.

[128] L. Cheng, W. Huang, Q. Gong, C. Liu, Z. Liu, Y. Li, H. Dai, Ultrathin $WS_2$ nanoflakes as a high-performance electrocatalyst for the hydrogen evolution reaction, *Angew. Chem. Int. Ed.*, 53 (2014) 7860–7863.

[129] X. Zhao, X. Ma, J. Sun, D. Li, X. Yang, Enhanced catalytic activities of surfactant-assisted exfoliated $WS_2$ nanodots for hydrogen evolution, *ACS Nano*, 10 (2016) 2159–2166.

[130] M.A. Lukowski, A.S. Daniel, C.R. English, F. Meng, A. Forticaux, R.J. Hamers, S. Jin, Highly active hydrogen evolution catalysis from metallic $WS_2$ nanosheets, *Energy Environ. Sci.*, 7 (2014) 2608–2613.

[131] Y. Zhang, J. Shi, G. Han, M. Li, Q. Ji, D. Ma, Z. Liu, Chemical vapor deposition of monolayer $WS_2$ nanosheets on Au foils toward direct application in hydrogen evolution, *Nano Res.*, 8 (2015) 2881–2890.

[132] X. Xiao, C. Engelbrekt, Z. Li, P. Si, Hydrogen evolution at nanoporous gold/tungsten sulfide composite film and its optimization, *Electrochim. Acta*, 173 (2015) 393–398.

[133] Z. Huang, C. Wang, Z. Chen, H. Meng, C. Lv, Z. Chen, C. Zhang, Tungsten sulfide enhancing solar-driven hydrogen production from silicon nanowires, *ACS Appl. Mater. Interfaces*, 6 (2014) 10408–10414.

[134] A. Di Paola, L. Palmisano, V. Augugliaro, Photocatalytic behavior of mixed $WO_3/WS_2$ powders, *Catal. Today*, 58 (2000) 141–149.

[135] A. Di Paola, L. Palmisano, M. Derrigo, V. Augugliaro, Preparation and characterization of tungsten chalcogenide photocatalysts, *J. Phys. Chem. B*, 101 (1997) 876–883.

[136] Y. Sang, Z. Zhao, M. Zhao, P. Hao, Y. Leng, H. Liu, From UV to near-infrared, $WS_2$ nanosheet: A novel photocatalyst for full solar light spectrum photodegradation, *Adv. Mater.*, 27 (2015) 363–369.

[137] D. Jing, L. Guo, $WS_2$ sensitized mesoporous $TiO_2$ for efficient photocatalytic hydrogen production from water under visible light irradiation, *Catal. Commun.*, 8 (2007) 795–799.

[138] Y. Hou, Y. Zhu, Y. Xu, X. Wang, Photocatalytic hydrogen production over carbon nitride loaded with $WS_2$ as cocatalyst under visible light, *Appl. Catal. B-Environ.*, 156 (2014) 122–127.

[139] M.S. Akple, J. Low, S. Wageh, A.A. Al-Ghamdi, J. Yu, J. Zhang, Enhanced visible light photocatalytic $H_2$-production of g-$C_3N_4/WS_2$ composite heterostructures, *Appl. Surf. Sci.*, 358 (2015) 196–203.

[140] Z. Liu, J. Liang, S. Li, S. Peng, Y. Qian, Synthesis and growth mechanism of $Bi_2S_3$ nanoribbons, *Chem. Eur. J.*, 10 (2004) 634–640.

[141] H. Zhou, S. Xiong, L. Wei, B. Xi, Y. Zhu, Y. Qian, Acetylacetone-directed controllable synthesis of $Bi_2S_3$ nanostructures with tunable morphology, *Cryst. Growth Des.*, 9 (2009) 3862–3867.

[142]  B. Zhang, X. Ye, W. Hou, Y. Zhao, Y. Xie, Biomolecule-assisted synthesis and electrochemical hydrogen storage of $Bi_2S_3$ flowerlike patterns with well-aligned nanorods, *J. Phys. Chem. B*, 110 (2006) 8978–8985.

[143]  K. Yao, W.W. Gong, Y.F. Hu, X.L. Liang, Q. Chen, L.M. Peng, Individual $Bi_2S_3$ nanowire-based room-temperature $H_2$ sensor, *J. Phys. Chem. C*, 112 (2008) 8721–8724.

[144]  Y. Bessekhouad, M. Mohammedi, M. Trari, Hydrogen photoproduction from hydrogen sulfide on $Bi_2S_3$ catalyst, *Sol. Energ. Mat. Sol. Cells*, 73 (2002) 339–350.

[145]  A. Abdi, A. Denoyelle, N. Commenges-Bernole, M. Trari, Photocatalytic hydrogen evolution on new mesoporous material $Bi_2S_3$/Y-zeolite, *Int. J. Hydrogen Energy*, 38 (2013) 2070–2078.

[146]  S.R. Kadam, R.P. Panmand, R.S. Sonawane, S.W. Gosavi, B.B. Kale, A stable $Bi_2S_3$ quantum dot-glass nanosystem: size tuneable photocatalytic hydrogen production under solar light, *RSC Adv.*, 5 (2015) 58485–58490.

[147]  C. García-Mendoza, S. Oros-Ruiz, A. Hernández-Gordillo, R. López, G. Jácome-Acatitla, H.A. Calderón, R. Gómez, Suitable preparation of $Bi_2S_3$ nanorods-$TiO_2$ heterojunction semiconductors with improved photocatalytic hydrogen production from water/methanol decomposition, *J. Chem. Technol. Biotechnol.*, 91 (2016) 2198–2204.

[148]  J. Kim, M. Kang, High photocatalytic hydrogen production over the bandgap-tuned urchin-like $Bi_2S_3$-loaded $TiO_2$ composites system, *Int. J. Hydrogen Energy*, 37 (2012) 8249–8256.

[149]  R. Brahimi, Y. Bessekhouad, A. Bouguelia, M. Trari, Visible light induced hydrogen evolution over the heterosystem $Bi_2S_3$/$TiO_2$, *Catal. Today*, 122 (2007) 62–65.

[150]  J. Santos Cruz, S.A. Mayén Hernández, F. Paraguay Delgado, O. Zelaya Angel, R. Castanedo Pérez, G. Torres Delgado, Optical and electrical properties of thin films of CuS nanodisks ensembles annealed in a vacuum and their photocatalytic activity, *Int. J. Photoenergy*, 2013 (2013) 195–203.

[151]  Y. Wang, X. Zhang, P. Chen, H. Liao, S. Cheng, *In situ* preparation of CuS cathode with unique stability and high rate performance for lithium ion batteries, *Electrochim. Acta*, 80 (2012) 264–268.

[152]  L. Isac, A. Duta, A. Kriza, S. Manolache, M. Nanu, Copper sulfides obtained by spray pyrolysis-possible absorbers in solid-state solar cells, *Thin Solid Films*, 515 (2007) 5755–5758.

[153]  X.L. Yu, Y. Wang, H.L.W. Chan, C.B. Cao, Novel gas sensing materials based on CuS hollow spheres, *Micropor. Mesopor. Mater.*, 118 (2009) 423–426.

[154]  P. Gomathisankar, K. Hachisuka, H. Katsumata, T. Suzuki, K. Funasaka, S. Kaneco, Photocatalytic hydrogen production with CuS/ZnO from aqueous $Na_2S + Na_2SO_3$ solution, *Int. J. Hydrogen Energy*, 38 (2013) 8625–8630.

[155]  Q. Wang, N. An, Y. Bai, H. Hang, J. Li, X. Lu, Z. Lei, High photocatalytic hydrogen production from methanol aqueous solution using the photocatalysts CuS/$TiO_2$, *Int. J. Hydrogen Energy*, 38 (2013) 10739–10745.

[156] Q. Wang, G. Yun, Y. Bai, N. An, Y. Chen, R. Wang, W. Shangguan, CuS, NiS as co-catalyst for enhanced photocatalytic hydrogen evolution over $TiO_2$, *Int. J. Hydrogen Energy*, 39 (2014) 13421–13428.

[157] L. Zhang, Y.N. Liu, M. Zhou, J. Yan, Improving photocatalytic hydrogen evolution over $CuO/Al_2O_3$ by platinum-depositing and CuS-loading, *Appl. Surf. Sci.*, 282 (2013) 531–537.

[158] P. Kumari, P. Chandran, S.S. Khan, Synthesis and characterization of silver sulfide nanoparticles for photocatalytic and antimicrobial applications, *J. Photochem. Photobiol. B*, 141 (2014) 235–240.

[159] D. Wang, C. Hao, W. Zheng, Q. Peng, T. Wang, Z. Liao, Y. Li, Ultralong single-crystalline $Ag_2S$ nanowires: promising candidates for photoswitches and roomtemperature oxygen sensors, *Adv. Mater.*, 20 (2008) 2628–2632.

[160] Z.D. Meng, T. Ghosh, L. Zhu, J.G. Choi, C.Y. Park, W.C. Oh, Synthesis of fullerene modified with $Ag_2S$ with high photocatalytic activity under visible light, *J. Mat. Chem.*, 22 (2012) 16127–16135.

[161] K. Nagasuna, T. Akita, M. Fujishima, H. Tada, Photodeposition of $Ag_2S$ quantum dots and application to photoelectrochemical cells for hydrogen production under simulated sunlight, *Langmuir*, 27 (2011) 7294–7300.

[162] R. Vogel, P. Hoyer, H. Weller, Quantum-sized PbS, CdS, $Ag_2S$, $Sb_2S_3$, and $Bi_2S_3$ particles as sensitizers for various nanoporous wide-bandgap semiconductors, *J. Phys. Chem.*, 98 (1994) 3183–3188.

[163] I. Hwang, K. Yong, Environmentally benign and efficient $Ag_2S$-ZnO nanowires as photoanodes for solar cells: comparison with CdS-ZnO nanowires, *ChemPhysChem*, 14 (2013) 364–368.

[164] D. Jiang, L. Chen, J. Xie, M. Chen, $Ag_2S/g$-$C_3N_4$ composite photocatalysts for efficient Pt-free hydrogen production. The co-catalyst function of $Ag/Ag_2S$ formed by simultaneous photodeposition, *Dalton Trans.*, 43 (2014) 4878–4885.

[165] S. Liu, X. Wang, W. Zhao, K. Wang, H. Sang, Z. He, Synthesis, characterization and enhanced photocatalytic performance of $Ag_2S$-coupled ZnO/ZnS core/shell nanorods, *J. Alloy. Compd.*, 568 (2013) 84–91.

[166] X. Liu, Z. Liu, J. Lu, X. Wu, W. Chu, Silver sulfide nanoparticles sensitized titanium dioxide nanotube arrays synthesized by *in situ* sulfurization for photocatalytic hydrogen production, *J. Colloid Interf. Sci.*, 413 (2014) 17–23.

[167] M. Gholami, M. Qorbani, O. Moradlou, N. Naseri, A.Z. Moshfegh, Optimal $Ag_2S$ nanoparticle incorporated $TiO_2$ nanotube array for visible water splitting, *RSC Adv.*, 4 (2014) 7838–7844.

[168] W.L. Ong, G.W. Ho, Enhanced photocatalytic performance of $TiO_2$ hierarchical spheres decorated with $Ag_2S$ nanoparticles, *Procedia Engineering*, 141 (2016) 7–14.

[169] P. Pistor, J.M. Merino Alvarez, M. Leon, M. di Michiel, S. Schorr, R. Klenk, S. Lehmann, Structure reinvestigation of $\alpha$-, $\beta$- and $\gamma$-$In_2S_3$, *Acta Crystallogr. B*, 72 (2016) 410–415.

[170] W.T. Kim, C.D. Kim, Optical energy gaps of $\beta$-In$_2$S$_3$ thin films grown by spray pyrolysis, *J. Appl. Phys.*, 60 (1986) 2631–2633.

[171] W. Qiu, M. Xu, X. Yang, F. Chen, Y. Nan, J. Zhang, H. Chen, Biomolecule-assisted hydrothermal synthesis of In$_2$S$_3$ porous films and enhanced photocatalytic properties, *J. Mater. Chem.*, 21 (2011) 13327–13333.

[172] Y. Xing, H. Zhang, S. Song, J. Feng, Y. Lei, L. Zhao, M. Li, Hydrothermal synthesis and photoluminescent properties of stacked indium sulfide superstructures, *Chem. Commun.*, 12 (2008) 1476–1478.

[173] Y.H. Kim, J.H. Lee, D.W. Shin, S.M. Park, J.S. Moon, J.G. Nam, J.B. Yoo, Synthesis of shape-controlled $\beta$-In$_2$S$_3$ nanotubes through oriented attachment of nanoparticles, *Chem. Commun.*, 46 (2010) 2292–2294.

[174] R. Sumi, A.R. Warrier, C. Vijayan, Visible-light driven photocatalytic activity of $\beta$-indium sulfide (In$_2$S$_3$) quantum dots embedded in Nafion matrix, *J. Phys. D-Appl. Phys.*, 47 (2014) 105103.

[175] Y. Li, S. Luo, Z. Wei, D. Meng, M. Ding, C. Liu, Electrodeposition technique-dependent photoelectrochemical and photocatalytic properties of an In$_2$S$_3$/TiO$_2$ nanotube array, *Phys. Chem. Chem. Phys.*, 16 (2014) 4361–4368.

[176] C. Xing, Z. Wu, D. Jiang, M. Chen, Hydrothermal synthesis of In$_2$S$_3$/g-C$_3$N$_4$ heterojunctions with enhanced photocatalytic activity, *J. Colloid Interf. Sci.*, 433 (2014) 9–15.

[177] X. Zhang, C. Shao, X. Li, N. Lu, K. Wang, F. Miao, Y. Liu, In$_2$S$_3$/carbon nanofibers/Au ternary synergetic system: Hierarchical assembly and enhanced visible-light photocatalytic activity, *J. Hazard. Mater.*, 283 (2015) 599–607.

[178] F. Wang, W. Li, S. Gu, H. Li, H. Zhou, X. Wu, Novel In$_2$S$_3$/ZnWO$_4$ heterojunction photocatalysts: facile synthesis and high-efficiency visible-light-driven photocatalytic activity, *RSC Adv.*, 5 (2015) 89940–89950.

[179] X. Zhang, X. Li, C. Shao, J. Li, M. Zhang, P. Zhang, Y. Liu, One-dimensional hierarchical heterostructures of In$_2$S$_3$ nanosheets on electrospun TiO$_2$ nanofibers with enhanced visible photocatalytic activity, *J. Hazard. Mater.*, 260 (2013) 892–900.

[180] B. Chai, T. Peng, P. Zeng, J. Mao, Synthesis of floriated In$_2$S$_3$ decorated with TiO$_2$ nanoparticles for efficient photocatalytic hydrogen production under visible light, *J. Mater. Chem.*, 21 (2011) 14587–14593.

[181] X. Yang, J. Xu, T. Wong, Q. Yang, C.S. Lee, Synthesis of In$_2$O$_3$-In$_2$S$_3$ core-shell nanorods with inverted type-I structure for photocatalytic H$_2$ generation, *Phys. Chem. Chem. Phys.*, 15 (2013) 12688–12693.

[182] S. Shen, L. Guo, Structural, textural and photocatalytic properties of quantum-sized In$_2$S$_3$-sensitized Ti-MCM-41 prepared by ion-exchange and sulfidation methods, *J. Solid State Chem.*, 179 (2006) 2629–2635.

[183] C. Gao, J. Li, Z. Shan, F. Huang, H. Shen, Preparation and visible-light photocatalytic activity of In$_2$S$_3$/TiO$_2$ composite, *Mater. Chem. Phys.*, 122 (2010) 183–187.

[184] F. Wang, Z. Jin, Y. Jiang, E.H. Backus, M. Bonn, S.N. Lou, R. Amal, Probing the charge separation process on $In_2S_3/Pt$-$TiO_2$ nanocomposites for boosted visible-light photocatalytic hydrogen production, *Appl. Catal. B-Environ.*, 198 (2016) 25–31.

[185] J. Liu, T. Ding, Z. Li, J. Zhao, S. Li, J. Liu, Photocatalytic hydrogen production over $In_2S_3$-$Pt$-$Na_2Ti_3O_7$ nanotube films under visible light irradiation, *Ceram. Int.*, 39 (2013) 8059–8063.

[186] X. Fu, X. Wang, Z. Chen, Z. Zhang, Z. Li, D.Y. Leung, X. Fu, Photocatalytic performance of tetragonal and cubic $\beta$-$In_2S_3$ for the water splitting under visible light irradiation, *Appl. Catal. B-Environ.*, 95 (2010) 393–399.

[187] S.D. Sartale, C.D. Lokhande, Preparation and characterization of nickel sulphide thin films using successive ionic layer adsorption and reaction (SILAR) method, *Mater. Chem. Phys.*, 72 (2001) 101–104.

[188] J. Hong, Y. Wang, Y. Wang, W. Zhang, R. Xu, Noble-metal-free $NiS/C_3N_4$ for efficient photocatalytic hydrogen evolution from water, *ChemSusChem*, 6 (2013) 2263–2268.

[189] Y. Zhong, J. Yuan, J. Wen, X. Li, Y. Xu, W. Liu, Y. Fang, Earth-abundant NiS co-catalyst modified metal-free mpg-$C_3N_4$/CNT nanocomposites for highly efficient visible-light photocatalytic $H_2$ evolution, *Dalton Trans.*, 44 (2015) 18260–18269.

[190] J. Yuan, J. Wen, Y. Zhong, X. Li, Y. Fang, S. Zhang, W. Liu, Enhanced photocatalytic $H_2$ evolution over noble-metal-free NiS cocatalyst modified CdS nanorods/g-$C_3N_4$ heterojunctions, *J. Mater. Chem. A*, 3 (2015) 18244–18255.

[191] J. Wen, X. Li, H. Li, S. Ma, K. He, Y. Xu, Q. Gao, Enhanced visible-light $H_2$ evolution of g-$C_3N_4$ photocatalysts via the synergetic effect of amorphous NiS and cheap metal-free carbon black nanoparticles as co-catalysts, *Appl. Surf. Sci.*, 358 (2015) 204–212.

[192] Y. Lu, D. Chu, M. Zhu, Y. Du, P. Yang, Exfoliated carbon nitride nanosheets decorated with NiS as an efficient noble-metal-free visible-light-driven photocatalyst for hydrogen evolution, *Phys. Chem. Chem. Phys.*, 17 (2015) 17355–17361.

[193] Y. Liu, C. Tang, Enhancement of photocatalytic $H_2$ evolution over $TiO_2$ nano-sheet films by surface loading NiS nanoparticles, *Russ. J. Phys. Chem. A*, 90 (2016) 1042–1048.

[194] J. Huang, Z. Shi, X. Dong, Nickel sulfide modified $TiO_2$ nanotubes with highly efficient photocatalytic $H_2$ evolution activity, *J. Energy Chem.*, 25 (2016) 136–140.

[195] M. Mollavali, C. Falamaki, S. Rohani, High performance NiS-nanoparticles sensitized $TiO_2$ nanotube arrays for water reduction, *Int. J. Hydrogen Energy*, 41 (2016) 5887–5901.

[196] A.W. Peters, Z. Li, O.K. Farha, J.T. Hupp, Toward inexpensive photocatalytic hydrogen evolution: a nickel sulfide catalyst supported on a high-stability metal-organic framework, *ACS Appl. Mater. Interfaces*, 8 (2016) 20675–20681.

[197] S.I. Radautsan, F.G. Donika, G.A. Kyosse, I.G. Mustya, Polytypism of ternary phases in the system ZnInS, *Phys. Status Solidi (b)*, 37 (1970) K123–K127.

[198] M.A. Sriram, P.H. McMichael, A. Waghray, P.N. Kumta, S. Misture, X.L. Wang, Chemical synthesis of the high-pressure cubic-spinel phase of ZnIn$_2$S$_4$, *J. Mater. Sci.*, 33 (1998) 4333–4339.

[199] S. Shen, P. Guo, L. Zhao, Y. Du, L. Guo, Insights into photoluminescence property and photocatalytic activity of cubic and rhombohedral ZnIn$_2$S$_4$, *J. Solid State Chem.*, 184 (2011) 2250–2256.

[200] X. Gou, F. Cheng, Y. Shi, L. Zhang, S. Peng, J. Chen, P. Shen, Shape-controlled synthesis of ternary chalcogenide ZnIn$_2$S$_4$ and CuIn(S, Se)$_2$ nano-/microstructures via facile solution route, *J. Am. Chem. Soc.*, 128 (2006) 7222–7229.

[201] W. Cai, Y. Zhao, J. Hu, J. Zhong, W. Xiang, Solvothermal synthesis and characterization of zinc indium sulfide microspheres, *J. Mater. Sci. Technol.*, 27 (2011) 559–562.

[202] F. Tian, R. Zhu, K. Song, M. Niu, F. Ouyang, G. Cao, The effects of hydrothermal temperature on the photocatalytic performance of ZnIn$_2$S$_4$ for hydrogen generation under visible light irradiation, *Mater. Res. Bull.*, 70 (2015) 645–650.

[203] N.S. Chaudhari, S.S. Warule, B.B. Kale, Architecture of rose and hollow marigold-like ZnIn$_2$S$_4$ flowers: structural, optical and photocatalytic study, *RSC Adv.*, 4 (2014) 12182–12187.

[204] N.S. Chaudhari, A.P. Bhirud, R.S. Sonawane, L.K. Nikam, S.S. Warule, V.H. Rane, B.B. Kale, Ecofriendly hydrogen production from abundant hydrogen sulfide using solar light-driven hierarchical nanostructured ZnIn$_2$S$_4$ photocatalyst, *Green Chem.*, 13 (2011) 2500–2506.

[205] G. Wang, G. Chen, Y. Yu, X. Zhou, Y. Teng, Mixed solvothermal synthesis of hierarchical ZnIn$_2$S$_4$ spheres: specific facet-induced photocatalytic activity enhancement and a DFT elucidation, *RSC Adv.*, 3 (2013) 18579–18586.

[206] J. Shen, J. Zai, Y. Yuan, X. Qian, 3D hierarchical ZnIn$_2$S$_4$: the preparation and photocatalytic properties on water splitting, *Int. J. Hydrogen Energy*, 37 (2012) 16986–16993.

[207] Z. Xu, Y. Li, S. Peng, G. Lu, S. Li, NaCl-assisted low temperature synthesis of layered Zn-In-S photocatalyst with high visible-light activity for hydrogen evolution, *RSC Adv.*, 2 (2012) 3458–3466.

[208] S. Shen, L. Zhao, X. Guan, L. Guo, Improving visible-light photocatalytic activity for hydrogen evolution over ZnIn$_2$S$_4$: a case study of alkaline-earth metal doping, *J. Phys. Chem. Solids*, 73 (2012) 79–83.

[209] F. Li, J. Luo, G. Chen, Y. Fan, Q. Huang, Y. Luo, Q. Meng, Hydrothermal synthesis of zinc indium sulfide microspheres with Ag$^+$ doping for enhanced H$_2$ production by photocatalytic water splitting under visible light, *Catal. Sci. Technol.*, 4 (2014) 1144–1150.

[210] S. Shen, L. Zhao, Z. Zhou, L. Guo, Enhanced photocatalytic hydrogen evolution over Cu-doped ZnIn$_2$S$_4$ under visible light irradiation, *J. Phys. Chem. C*, 112 (2008) 16148–16155.

[211] D. Jing, M. Liu, L. Guo, Enhanced hydrogen production from water over Ni doped $ZnIn_2S_4$ microsphere photocatalysts, *Catal. Lett.*, 140 (2010) 167–171.

[212] F. Tian, R. Zhu, Y. He, F. Ouyang, Improving photocatalytic activity for hydrogen evolution over $ZnIn_2S_4$ under visible-light: A case study of rare earth modification, *Int. J. Hydrogen Energy*, 39 (2014) 6335–6344.

[213] W. Yang, L. Zhang, J. Xie, X. Zhang, Q. Liu, T. Yao, Y. Xie, Enhanced photoexcited carrier separation in oxygen-doped $ZnIn_2S_4$ nanosheets for hydrogen evolution, *Angew. Chem. Int. Ed.*, 55 (2016) 6716–6720.

[214] Q. Liu, F. Wu, F. Cao, L. Chen, X. Xie, W. Wang, L. Li, A multijunction of $ZnIn_2S_4$ nanosheet/$TiO_2$ film/Si nanowire for significant performance enhancement of water splitting, *Nano Res.*, 8 (2015) 3524–3534.

[215] W. Cui, D. Guo, L. Liu, J. Hu, D. Rana, Y. Liang, Preparation of $ZnIn_2S_4/K_2La_2Ti_3O_{10}$ composites and their photocatalytic $H_2$ evolution from aqueous $Na_2S/Na_2SO_3$ under visible light irradiation, *Catal. Commun.*, 48 (2014) 55–59.

[216] H. Liu, Z. Jin, Z. Xu, Z. Zhang, D. Ao, Fabrication of $ZnIn_2S_4$–g-$C_3N_4$ sheet-on-sheet nanocomposites for efficient visible-light photocatalytic $H_2$-evolution and degradation of organic pollutants, *RSC Adv.*, 5 (2015) 97951–97961.

[217] W. Chen, T.Y. Liu, T. Huang, X.H. Liu, J.W. Zhu, G.R. Duan, X.J. Yang, One-pot hydrothermal route to synthesize the $ZnIn_2S_4/g$-$C_3N_4$ composites with enhanced photocatalytic activity, *J. Mater. Sci.*, 50 (2015) 8142–8152.

[218] L. Ye, J. Fu, Z. Xu, R. Yuan, Z. Li, Facile one-pot solvothermal method to synthesize sheet-on-sheet reduced graphene oxide (RGO)/$ZnIn_2S_4$ nanocomposites with superior photocatalytic performance, *ACS Appl. Mater. Interfaces*, 6 (2014) 3483–3490.

[219] F. Tian, R. Zhu, J. Zhong, P. Wang, F. Ouyang, G. Cao, An efficient preparation method of RGO/$ZnIn_2S_4$ for photocatalytic hydrogen generation under visible light, *Int. J. Hydrogen Energy*, 41 (2016) 20156–20171.

[220] S.B. Kale, R.S. Kalubarme, M.A. Mahadadalkar, H.S. Jadhav, A.P. Bhirud, J.D. Ambekar, B.B. Kale, Hierarchical 3D $ZnIn_2S_4$/graphene nanoheterostructures: their *in situ* fabrication with dual functionality in solar hydrogen production and as anodes for lithium ion batteries, *Phys. Chem. Chem. Phys.*, 17 (2015) 31850–31861.

[221] Y. Chen, G. Tian, Z. Ren, K. Pan, Y. Shi, J. Wang, H. Fu, Hierarchical Core–shell carbon nanofiber@$ZnIn_2S_4$ composites for enhanced hydrogen evolution performance, *ACS Appl. Mater. Interfaces*, 6 (2014) 13841–13849.

[222] J. Zhou, G. Tian, Y. Chen, X. Meng, Y. Shi, X. Cao, H. Fu, *In situ* controlled growth of $ZnIn_2S_4$ nanosheets on reduced graphene oxide for enhanced photocatalytic hydrogen production performance, *Chem. Commun.*, 49 (2013) 2237–2239.

[223] B. Chai, T. Peng, P. Zeng, X. Zhang, Preparation of a MWCNTs/$ZnIn_2S_4$ composite and its enhanced photocatalytic hydrogen production under visible-light irradiation, *Dalton Trans.*, 41 (2012) 1179–1186.

[224] L. Ye, Z. Li, Rapid microwave-assisted syntheses of reduced graphene oxide (RGO)/ZnIn$_2$S$_4$ microspheres as superior noble-metal-free photocatalyst for hydrogen evolutions under visible light, *Appl. Catal. B-Environ.*, 160 (2014) 552–557.

[225] Q. Li, C. Cui, H. Meng, J. Yu, Visible-light photocatalytic hydrogen production activity of ZnIn$_2$S$_4$ microspheres using carbon quantum dots and platinum as dual co-catalysts, *Chem. Asian J.*, 9 (2014) 1766–1770.

[226] C.K. Lowe-Ma, Powder diffraction data for ZnGa$_2$S$_4$, *Powder Diffract.*, 5 (1990) 223–224.

[227] H.G. Kim, W.T. Kim, Optical absorption of ZnGa$_2$S$_4$ and ZnGa$_2$S$_4$: Co$^{2+}$ crystals, *Phys. Rev. B*, 41 (1990) 8541–8544.

[228] X. Jiang, W.R. Lambrecht, Electronic band structure of ordered vacancy defect chalcopyrite compounds with formula II-III2-VI4, *Phys. Rev. B*, 69 (2004) 1129–1133.

[229] D. Peng, Z. Min, X. Zhonglei, C. Lihong, Synthesis of the ZnGa$_2$S$_4$ nanocrystals and their visible-light photocatalytic degradation property, *J. Nanomater.*, 2015 (2015) 1–7.

[230] H. Kaga, A. Kudo, Cosubstituting effects of copper (I) and gallium (III) for ZnGa$_2$S$_4$ with defect chalcopyrite structure on photocatalytic activity for hydrogen evolution, *J. Catal.*, 310 (2014) 31–36.

[231] J. Yang, H. Fu, D. Yang, W. Gao, R. Cong, T. Yang, ZnGa$_{2-x}$In$_x$S$_4$ ($0 \leq x \leq 0.4$) and Zn$_{1-2y}$(CuGa)$_y$Ga$_{1.7}$In$_{0.3}$S$_4$ ($0.1 \leq y \leq 0.2$): Optimize visible light photocatalytic H$_2$ evolution by fine modulation of band structures, *Inorg. Chem.*, 54 (2015) 2467–2473.

[232] J.J.M. Binsma, L.J. Giling, J. Bloem, Phase relations in the system Cu$_2$S-In$_2$S$_3$, *J. Cryst. Growth*, 50 (1980) 429–436.

[233] W. Yue, S. Han, R. Peng, W. Shen, H. Geng, F. Wu, M. Wang, CuInS$_2$ quantum dots synthesized by a solvothermal route and their application as effective electron acceptors for hybrid solar cells, *J. Mater. Chem.*, 20 (2010) 7570–7578.

[234] T. Todorov, E. Cordoncillo, J.F. Sanchez-Royo, J. Carda, P. Escribano, CuInS$_2$ films for photovoltaic applications deposited by a low-cost method, *Chem. Mater.*, 18 (2006) 3145–3150.

[235] X. Wang, D. Pan, D. Weng, C.Y. Low, L. Rice, J. Han, Y. Lu, A General Synthesis of Cu-In-S based multicomponent solid-solution nanocrystals with tunable bandgap, size, and structure, *J. Phys. Chem. C*, 114 (2010) 17293–17297.

[236] E. Garskaite, G.T. Pan, T.C.K. Yang, S.T. Huang, A. Kareiva, The study of preparation and photoelectrical properties of chemical bath deposited Zn, Sb and Ni-doped CuInS$_2$ films for hydrogen production, *Sol. Energy*, 86 (2012) 2584–2591.

[237] K. Guo, Z. Liu, J. Han, Z. Liu, Y. Li, B. Wang, C. Zhou, Hierarchical TiO$_2$-CuInS$_2$ core-shell nanoarrays for photoelectrochemical water splitting, *Phys. Chem. Chem. Phys.*, 16 (2014) 16204–16213.

[238] C. Li, Z. Xi, W. Fang, M. Xing, J. Zhang, Enhanced photocatalytic hydrogen evolution activity of $CuInS_2$ loaded $TiO_2$ under solar light irradiation, *J. Solid State Chem.*, 226 (2015) 94–100.

[239] M. Kruszynska, H. Borchert, J. Parisi, J. Kolny-Olesiak, Synthesis and shape control of $CuInS_2$ nanoparticles, *J. Am. Chem. Soc.*, 132 (2010) 15976–15986.

[240] H. Zhong, S.S. Lo, T. Mirkovic, Y. Li, Y. Ding, Y. Li, G.D. Scholes, Noninjection gram-scale synthesis of monodisperse pyramidal $CuInS_2$ nanocrystals and their size-dependent properties, *ACS Nano*, 4 (2010) 5253–5262.

[241] W. Yue, S. Han, R. Peng, W. Shen, H. Geng, F. Wu, M. Wang, $CuInS_2$ quantum dots synthesized by a solvothermal route and their application as effective electron acceptors for hybrid solar cells, *J. Mater. Chem.*, 20 (2010) 7570–7578.

[242] Y. Luo, G. Chang, W. Lu, X. Sun, Synthesis and characterization of $CuInS_2$ nanoflowers, *Colloid J.*, 72 (2010) 282–285.

[243] X. Sheng, L. Wang, Y. Luo, D. Yang, Synthesis of hexagonal structured wurtzite and chalcopyrite $CuInS_2$ via a simple solution route, *Nanoscale Res. Lett.*, 6 (2011) 562.

[244] L. Zheng, Y. Xu, Y. Song, C. Wu, M. Zhang, Y. Xie, Nearly monodisperse $CuInS_2$ hierarchical microarchitectures for photocatalytic $H_2$ evolution under visible light, *Inorg. Chem.*, 48 (2009) 4003–4009.

[245] K. Kobayakawa, A. Teranishi, T. Tsurumaki, Y. Sato, A. Fujishima, Photocatalytic activity of $CuInS_2$ and $CuIn_5S_8$, *Electrochim. Acta*, 37 (1992) 465–467.

[246] G. Delgado, A.J. Mora, C. Pineda, T. Tinoco, Simultaneous Rietveld refinement of three phases in the Ag-In-S semiconducting system from X-ray powder diffraction, *Mater. Res. Bull.*, 36 (2001) 2507–2517.

[247] M.A. Aguilera, J.A. Hernández, M.G. Trujillo, M.O. López, G.C. Puente, Photoluminescence studies of chalcopyrite and orthorhombic $AgInS_2$ thin films deposited by spray pyrolysis technique, *Thin Solid Films*, 515 (2007) 6272–6275.

[248] J. Liu, E. Hua, Electronic structure and absolute band edge position of tetragonal $AgInS_2$ photocatalyst: A hybrid density functional study, *Mat. Sci. Semiconduct. Proc.*, 40 (2015) 446–452.

[249] J. Liu, S. Chen, Q. Liu, Y. Zhu, Y. Lu, Density functional theory study on electronic and photocatalytic properties of orthorhombic $AgInS_2$, *Comp. Mater. Sci.*, 91 (2014) 159–164.

[250] Z. Liu, K. Guo, J. Han, Y. Li, T. Cui, B. Wang, C. Zhou, Dendritic $TiO_2/In_2S_3/AgInS_2$ trilaminar core-shell branched nanoarrays and the enhanced activity for photoelectrochemical water splitting, *Small*, 10 (2014) 3153–3161.

[251] I. Tsuji, H. Kato, A. Kudo, Visible-light-induced $H_2$ evolution from an aqueous solution containing sulfide and sulfite over a $ZnS$-$CuInS_2$-$AgInS_2$ solid-solution photocatalyst, *Angew. Chem. Int. Ed.*, 44 (2005) 3565–3568.

[252] I. Tsuji, H. Kato, A. Kudo, Photocatalytic hydrogen evolution on ZnS-CuInS$_2$-AgInS$_2$ solid solution photocatalysts with wide visible light absorption bands, *Chem. Mater.*, 18 (2006) 1969–1975.

[253] T. Kameyama, T. Takahashi, T. Machida, Y. Kamiya, T. Yamamoto, S. Kuwabata, T. Torimoto, Controlling the electronic energy structure of ZnS-AgInS$_2$ solid solution nanocrystals for photoluminescence and photocatalytic hydrogen evolution, *J. Phys. Chem. C*, 119 (2015) 24740–24749.

[254] M. Jagadeeswararao, S. Dey, A. Nag, C.N.R. Rao, Visible light-induced hydrogen generation using colloidal $(ZnS)_{0.4}(AgInS_2)_{0.6}$ nanocrystals capped by S$_2$-ions, *J. Mater. Chem. A*, 3 (2015) 8276–8279.

[255] S.C. Abrahams, J.L. Bernstein, Crystal structure of piezoelectric nonlinear-optic AgGaS$_2$, *J. Chem. Phys.*, 59 (1973) 1625–1629.

[256] C.M. Fan, M.D. Regulacio, C. Ye, S.H. Lim, Y. Zheng, Q.H. Xu, M.Y. Han, Colloidal synthesis and photocatalytic properties of orthorhombic AgGaS$_2$ nanocrystals, *Chem. Commun.*, 50 (2014) 7128–7131.

[257] J.S. Jang, S.H. Choi, N. Shin, C. Yu, J.S. Lee, AgGaS$_2$-type photocatalysts for hydrogen production under visible light: Effects of post-synthetic H$_2$S treatment, *J. Solid State Chem.*, 180 (2007) 1110–1118.

[258] A. Kudo, Development of photocatalyst materials for water splitting, *Int. J. Hydrogen Energy*, 31 (2006) 197–202.

[259] J.S. Jang, S.J. Hong, J.Y. Kim, J.S. Lee, Heterojunction photocatalyst TiO$_2$/AgGaS$_2$ for hydrogen production from water under visible light, *Chem. Phys. Lett.*, 475 (2009) 78–81.

[260] K. Yamato, A. Iwase, A. Kudo, Photocatalysis using a wide range of the visible light spectrum: hydrogen evolution from doped AgGaS$_2$, *ChemSusChem*, 8 (2015) 2902–2906.

[261] J.L. Shay, J.H. Wernick, Ternary Chalcopyrite Semiconductors: Growth, Electronic Properties, and Applications: *International Series of Monographs in The Science of The Solid State* (Vol. 7), Elsevier, (2013).

[262] G. Boyd, H. Kasper, J. McFee, Linear and nonlinear optical properties of AgGaS$_2$, CuGaS$_2$, and CuInS$_2$, and theory of the wedge technique for the measurement of nonlinear coefficients, *IEEE J. Quantum Electron.*, 7 (1971) 563–573.

[263] M.D. Regulacio, C. Ye, S.H. Lim, Y. Zheng, Q.H. Xu, M.Y. Han, Facile noninjection synthesis and photocatalytic properties of wurtzite-phase CuGaS$_2$ nanocrystals with elongated morphologies, *CrystEngComm*, 15 (2013) 5214–5217.

[264] S.H. Chang, B.C. Chiu, T.L. Gao, S.L. Jheng, H.Y. Tuan, Selective synthesis of copper gallium sulfide (CuGaS$_2$) nanostructures of different sizes, crystal phases, and morphologies, *CrystEngComm*, 16 (2014) 3323–3330.

[265] Q. Zhou, S.Z. Kang, X. Li, L. Wang, L. Qin, J. Mu, One-pot hydrothermal preparation of wurtzite CuGaS$_2$ and its application as a photoluminescent probe for trace detection of l-noradrenaline, *Colloid Surface A*, 465 (2015) 124–129.

[266] M. Sabet, M. Ramezani, K. Motevalli, M. Salavati-Niasari, O. Amiri, Synthesis and characterization of different morphologies $CuGaS_2/CuS$ nanostructures with a simple sonochemical method, *J. Mater. Sci.*, 28 (2017) 2427–2434.

[267] H. Kaga, Y. Tsutsui, A. Nagane, A. Iwase, A. Kudo, An effect of Ag (i)-substitution at Cu sites in $CuGaS_2$ on photocatalytic and photoelectrochemical properties for solar hydrogen evolution, *J. Mater. Chem. A*, 3 (2015) 21815–21823.

[268] T.A. Kandiel, D.H. Anjum, P. Sautet, T. Le Bahers, K. Takanabe, Electronic structure and photocatalytic activity of wurtzite Cu-Ga-S nanocrystals and their Zn substitution, *J. Mater. Chem. A*, 3 (2015) 8896–8904.

[269] M. Zhao, F. Huang, H. Lin, J. Zhou, J. Xu, Q. Wu, Y. Wang, $CuGaS_2$-ZnS p–n nanoheterostructures: a promising visible light photo-catalyst for water-splitting hydrogen production, *Nanoscale*, 8 (2016) 16670–16676.

[270] K. Iwashina, A. Iwase, Y.H. Ng, R. Amal, A. Kudo, Z-schematic water splitting into $H_2$ and $O_2$ using metal sulfide as a hydrogen-evolving photocatalyst and reduced graphene oxide as a solid-state electron mediator, *J. Am. Chem. Soc.*, 137 (2015) 604–607.

[271] M. Tabata, K. Maeda, T. Ishihara, T. Minegishi, T. Takata, K. Domen, Photocatalytic hydrogen evolution from water using copper gallium sulfide under visible-light irradiation, *J. Phys. Chem. C*, 114 (2010) 11215–11220.

[272] H. Kaga, K. Saito, A. Kudo, Solar hydrogen production over novel metal sulfide photocatalysts of $AGa_{(2)}In_{(3)}S_{(8)}$ (A = Cu or Ag) with layered structures, *Chem. Commun.*, 46 (2010) 3779–3781.

[273] T.A. Kandiel, D.H. Anjum, K. Takanabe, Nano-sized quaternary $CuGa_2In_3S_8$ as an efficient photocatalyst for solar hydrogen production, *ChemSusChem*, 7 (2014) 3112–3121.

[274] T.A. Kandiel, G.A.M. Hutton, E. Reisner, Visible light driven hydrogen evolution with a noble metal free $CuGa_2In_3S_8$ nanoparticle system in water, *Catal. Sci. Technol.*, 6 (2016) 6536–6541.

[275] I. Tsuji, Y. Shimodaira, H. Kato, H. Kobayashi, A. Kudo, Novel Stannite-type complex sulfide photocatalysts $A_2^I$-Zn-$A^{IV}$-$S_4$ ($A^I$ = Cu and Ag; $A^{IV}$ = Sn and Ge) for hydrogen evolution under visible-light irradiation, *Chem. Mater.*, 22 (2010) 1402–1409.

[276] S. Ikeda, T. Nakamura, T. Harada, M. Matsumura, Multicomponent sulfides as narrow gap hydrogen evolution photocatalysts, *Phys. Chem. Chem. Phys.*, 12 (2010) 13943–13949.

[277] L. Wang, W.Z. Wang, S.M. Sun, A simple template-free synthesis of ultrathin $Cu_2ZnSnS_4$ nanosheets for highly stable photocatalytic $H_2$ evolution, *J. Mater. Chem.*, 22 (2012) 6553–6555.

[278] K. Li, B. Chai, T.Y. Peng, J. Mao, L. Zan, Synthesis of multicomponent sulfide $Ag_2ZnSnS_4$ as an efficient photocatalyst for $H_2$ production under visible light irradiation, *RSC Adv.*, 3 (2013) 253–258.

[279] C.M. Fan, M.D. Regulacio, C. Ye, S.H. Lim, S.K. Lua, Q.H. Xu, Z.L. Dong, A.W. Xu, M.Y. Han, Colloidal nanocrystals of orthorhombic $Cu_2ZnGeS_4$: phase-controlled synthesis, formation mechanism and photocatalytic behavior, *Nanoscale*, 7 (2015) 3247–3253.

[280] C. Sun, J.S. Gardner, G. Long, C. Bajracharya, A. Thurber, A. Punnoose, R.G. Rodriguez, J.J. Pak, Controlled stoichiometry for quaternary $CuIn_xGa_{1-x}S_2$ chalcopyrite nanoparticles from single-source precursors via microwave irradiation, *Chem. Mater.*, 22 (2010) 2699–2701.

[281] K.L. Ou, J.C. Fan, J.K. Chen, C.C. Huang, L.Y. Chen, J.H. Ho, J.Y. Chang, Hot-injection synthesis of monodispersed $Cu_2ZnSn(S_xSe_{1-x})_4$ nanocrystals: tunable composition and optical properties, *J. Mater. Chem.*, 22 (2012) 14667–14673.

[282] T.A. Kandiel, K. Takanabe, Solvent-induced deposition of Cu-Ga-In-S nanocrystals onto a titanium dioxide surface for visible-light-driven photocatalytic hydrogen production, *Appl. Catal. B-Environ.*, 184 (2016) 264–269.

[283] X.L. Yu, A. Shavel, X.Q. An, Z.S. Luo, M. Ibanez, A. Cabot, $Cu_2ZnSnS_4$-Pt and $Cu_2ZnSnS_4$-Au heterostructured nanoparticles for photocatalytic water splitting and pollutant degradation, *J. Am. Chem. Soc.*, 136 (2014) 9236–9239.

[284] J.H. Yang, D.E. Wang, H.X. Han, C. Li, Roles of cocatalysts in photocatalysis and photoelectrocatalysis, *Acc. Chem. Res.*, 46 (2013) 1900–1909.

[285] G.P. Chen, D.M. Li, F. Li, Y.Z. Fan, H.F. Zhao, Y.H. Luo, R.C. Yu, Q.B. Meng, Ball-milling combined calcination synthesis of $MoS_2/CdS$ photocatalysts for high photocatalytic $H_2$ evolution activity under visible light irradiation, *Appl. Catal. A-Gen.*, 443–444 (2012) 138–144.

[286] J. Xu, X.J. Gao, Characterization and mechanism of $MoS_2/CdS$ composite photocatalyst used for hydrogen production from water splitting under visible light, *Chem. Eng. J.*, 260 (2015) 642–648.

[287] Q. Liu, Q.C. Shang, A. Khalil, Q. Fang, S.M. Chen, Q. He, T. Xiang, D.B. Liu, Q. Zhang, Y. Luo, L. Song, *In situ* integration of a metallic 1T-$MoS_2$/CdS heterostructure as a means to promote visible-light-driven photocatalytic hydrogen evolution, *ChemCatChem*, 8 (2016) 2614–2619.

[288] X.L. Yin, L.L. Li, W.J. Jiang, Y. Zhang, X. Zhang, L.J. Wan, J.S. Hu, $MoS_2$/CdS nanosheets-on-nanorod heterostructure for highly efficient photocatalytic $H_2$ generation under visible light irradiation, *ACS Appl. Mater. Interfaces*, 8 (2016) 15258–15266.

[289] B. Han, S.Q. Liu, N. Zhang, Y.J. Xu, Z.R. Tang, One-dimensional CdS@$MoS_2$ core-shell nanowires for boosted photocatalytic hydrogen evolution under visible light, *Appl. Catal. B-Environ.*, 202 (2017) 298–304.

[290] X. Zong, J.F. Han, G.J. Ma, H.J. Yan, G.P. Wu, C. Li, Photocatalytic $H_2$ evolution on CdS loaded with $WS_2$ as cocatalyst under visible light irradiation, *J. Phys. Chem. C*, 115 (2011) 12202–12208.

[291] J. He, L. Chen, Z.Q. Yi, C.T. Au, S.F. Yin, CdS nanorods coupled with WS$_2$ nanosheets for enhanced photocatalytic hydrogen evolution activity, *Ind. Eng. Chem. Res.*, 55 (2016) 8327–8333.

[292] Y.Y. Zhong, G. Zhao, F.K. Ma, Y.Z. Wu, X.P. Hao, Utilizing photocorrosion-recrystallization to prepare a highly stable and efficient CdS/WS$_2$ nanocomposite photocatalyst for hydrogen evolution, *Appl. Catal. B-Environ.*, 199 (2016) 466–472.

[293] W. Zhang, Y.B. Wang, Z. Wang, Z.Y. Zhong, R. Xu, Highly efficient and noble metal-free NiS/CdS photocatalysts for H$_2$ evolution from lactic acid sacrificial solution under visible light, *Chem. Commun.*, 46 (2010) 7631–7633.

[294] J. Zhang, S.Z. Qiao, L.F. Qi, J.G. Yu, Fabrication of NiS modified CdS nanorod p–n junction photocatalysts with enhanced visible-light photocatalytic H$_2$-production activity, *Phys. Chem. Chem. Phys.*, 15 (2013) 12088–12094.

[295] F.Y. Cheng, Q.J. Xiang, A solid-state approach to fabricate CdS/CuS nano-heterojunction with promoted visible-light photocatalytic H$_2$-evolution activity, *RSC Adv.*, 6 (2016) 76269–76272.

[296] S.H. Shen, L.J. Guo, X.B. Chen, F. Ren, S.S. Mao, Effect of Ag$_2$S on solar-driven photocatalytic hydrogen evolution of nanostructured CdS, *Int. J. Hydrogen Energy*, 35 (2010) 7110–7115.

[297] G.P. Chen, F. Li, Y.Z. Fan, Y.H. Luo, D.M. Li, Q.B. Meng, A novel noble metal-free ZnS-WS$_2$/CdS composite photocatalyst for H$_2$ evolution under visible light irradiation, *Catal. Commun.*, 40 (2013) 51–54.

[298] X. Wang, W.C. Peng, X.Y. Li, Photocatalytic hydrogen generation with simultaneous organic degradation by composite CdS-ZnS nanoparticles under visible light, *Int. J. Hydrogen Energy*, 39 (2014) 13454–13461.

[299] Y.P. Xie, Z.B. Yu, G. Liu, X.L. Ma, H.M. Cheng, CdS-mesoporous ZnS core-shell particles for efficient and stable photocatalytic hydrogen evolution under visible light, *Energy Environ. Sci.*, 7 (2014) 1895–1901.

[300] X.J. Xu, L.F. Hu, N. Gao, S.X. Liu, S. Wageh, A.A. Al-Ghamdi, A. Alshahrie, X.S. Fang, Controlled growth from ZnS nanoparticles to ZnS-CdS nanoparticle hybrids with enhanced photoactivity, *Adv. Funct. Mater.*, 25 (2015) 445–454.

[301] D.C. Jiang, Z.J. Sun, H.X. Jia, D.P. Lu, P.W. Du, A Cocatalyst-free CdS nanorods/ZnS nanoparticles composite for high-performance visible-light-driven hydrogen production from water, *J. Mater. Chem. A*, 4 (2016) 675–683.

[302] J. Zhang, J.G. Yu, Y.M. Zhang, Q. Li, J.R. Gong, Visible light photocatalytic H$_2$-production activity of CuS/ZnS porous nanosheets based on photoinduced interfacial charge transfer, *Nano Lett.*, 11 (2011) 4774–4779.

[303] L. Xiao, H. Chen, J.H. Huang, Visible light-driven photocatalytic H$_2$-generation activity of CuS/ZnS composite particles, *Mater. Res. Bull.*, 64 (2015) 370–374.

[304] X. Yang, H.T. Xue, J. Xu, X. Huang, J. Zhang, Y.B. Tang, T.W. Ng, H.L. Kwong, X.M. Meng, C.S. Lee, Synthesis of porous ZnS:Ag$_2$S nanosheets

by ion exchange for photocatalytic $H_2$ generation, *ACS Appl. Mater. Interfaces*, 6 (2014) 9078–9084.

[305] L. Wei, Y.J. Chen, Y.P. Lin, H.S. Wu, R.S. Yuan, Z.H. Li, $MoS_2$ as non-noble-metal co-catalyst for photocatalytic hydrogen evolution over hexagonal $ZnIn_2S_4$ under visible light irradiations, *Appl. Catal. B-Environ.*, 144 (2014) 521–527.

[306] G.P. Chen, N. Ding, F. Li, Y.Z. Fan, Y.H. Luo, D.M. Li, Q.B. Meng, Enhancement of photocatalytic $H_2$ evolution on $ZnIn_2S_4$ loaded with *in-situ* photo-deposited $MoS_2$ under visible light irradiation, *Appl. Catal. B-Environ.*, 160 (2014) 614–620.

[307] G.H. Tian, Y.J. Chen, Z.Y. Ren, C.G. Tian, K. Pan, W. Zhou, J.Q. Wang, H.G. Fu, Enhanced photocatalytic hydrogen evolution over hierarchical composites of $ZnIn_2S_4$ nanosheets grown on $MoS_2$ slices, *Chem. Asia. J.*, 9 (2014) 1291–1297.

[308] Y.J. Yuan, D.Q. Chen, Y.W. Huang, Z.T. Yu, J.S. Zhong, T.T. Chen, W.G. Tu, Z.J. Guan, D.P. Cao, Z.G. Zou, $MoS_2$ nanosheet-modified $CuInS_2$ photocatalyst for visible-light-driven hydrogen production from water, *ChemSusChem*, 9 (2016) 1003–1009.

[309] Y.P. Yuan, S.W. Cao, L.S. Yin, L. Xu, C. Xue, $NiS_2$ Co-catalyst decoration on $CdLa_2S_4$ nanocrystals for efficient photocatalytic hydrogen generation under visible light irradiation, *Int. J. Hydrogen Energy*, 38 (2013) 7218–7223.

[310] I. Tsuji, H. Kato, A. Kudo, Photocatalytic hydrogen evolution on $ZnS$-$CuInS_2$-$AgInS_2$ solid solution photocatalysts with wide visible light absorption bands, *Chem. Mater.*, 18 (2016) 1969–1975.

[311] Y.H. Lin, F. Zhang, D.C. Pan, A facile route to $(ZnS)_x(CuInS_2)_{1-x}$ hierarchical microspheres with excellent water-splitting ability, *J. Mater. Chem.*, 22 (2012) 22619–22623.

[312] C. Ye, M.D. Regulacio, S.H. Lim, S. Li, Q.H. Xu, M.Y. Han, Alloyed $ZnS$-$CuInS_2$ semiconductor nanorods and their nanoscale heterostructures for visible-light-driven photocatalytic hydrogen generation, *Chem. Eur. J.*, 21 (2015) 9514–9519.

[313] T. Kameyama, T. Takahashi, T. Machida, Y. Kamiya, T. Yamamoto, S. Kuwabata, T. Torimoto, Controlling the electronic energy structure of $ZnS$-$AgInS_2$ solid solution nanocrystals for photoluminescence and photocatalytic hydrogen evolution, *J. Phys. Chem. C*, 119 (2015) 24740–24749.

[314] D.C. Pan, D. Weng, X.L. Wang, Q.F. Xiao, W. Chen, C.L. Xu, Z.Z. Yang, Y.F. Lu, Alloyed semiconductor nanocrystals with broad tunable bandgaps, *Chem. Commun.*, (2009) 4221–4223.

[315] Z. Tan, Y. Zhang, C. Xie, H.P. Su, J. Liu, C.F. Zhang, N. Dellas, S.E. Mohney, Y.Q. Wang, J.K. Wang, J. Xu, Near-band-edge electroluminescence from heavy-metal-free colloidal quantum dots, *Adv. Mater.*, 23 (2011) 3553–3558.

# Chapter 3

# Graphene-based Nanocomposites for Photocatalysis

Xinjuan Liu*,§, Taiqiang Chen*, Hengchao Sun†, Yan Guo†
and Likun Pan‡

*Institute of Optoelectronic Materials and Devices,
College of Optical and Electronic Technology,
China Jiliang University,
Hangzhou 310018, China
†Beijing Smart-Chip Microelectronics Technology Co., Ltd.,
Beijing, 100192, China
‡Shanghai Key Laboratory of Magnetic Resonance,
School of Physics and Electronic Science,
East China Normal University,
Shanghai 200062, China
§lxj669635@126.com

## 1. Introduction

Graphene was first prepared in 2004 by peeling a single layer of graphene using sticky tape and a pencil [1]. As an emerging carbon material with a unique 2D conjugated chemical structure, graphene has attracted a great deal of attention from both theoretical and experimental scientists owing to its excellent electronic, capacitive and mechanical properties, superior chemical stability and high specific surface area arising from its unique structure [2–4]. As a low-dimensional material, graphene has a large theoretical specific surface area of $2{,}630\,\mathrm{m^2 g^{-1}}$, high intrinsic mobility of $2{,}000{,}000\,\mathrm{cm^2\,v^{-1}\,s^{-1}}$, excellent conductivity of $7{,}200\,\mathrm{S\,m^{-1}}$, high Young's modulus of $1.0\,\mathrm{TPa}$, thermal conductivity of $5{,}000\,\mathrm{W m^{-1}\,K^{-1}}$ and optical transmittance of $97.7\%$. These intriguing properties of graphene have offered some promising applications in many fields such as photocatalysis [5], energy storage [6, 7], batteries [8], ultraviolet (UV) sensors, transparent electrodes [9], supercapacitors [10, 11], dye-sensitized solar cells, [12] organic solar cells [13] and field emission [14] Various methods for producing graphene have been developed, such as physical/mechanical or chemical exfoliation [15, 16] epitaxial growth via chemical vapor deposition [17] the unzipping of CNTs (via electrochemical, chemical or physical methods) and the reduction of sugars (such as glucose or sucrose) [18]. The recent dramatic improvement in chemical methodology for synthesizing graphene has brought it close to large-scale preparation.

Currently, one attractive challenge is to hybridize these 2D carbon nanostructures with other materials to form new hybrid materials with potential applications. Attempts to combine other materials and graphene have been reported in efforts to obtain hybrid materials with superior optical or electrical properties, which act as an excellent electron acceptor/transport materials for photocatalysis [19]. Here, we discuss recent advances in the field of graphene-based nanocomposites from the stand point of photocatalysis, such as photocatalytic reduction of $Cr(VI)$ and $CO_2$, photocatalytic selective oxidation/reduction and photocatalytic degradation of organic pollutions. The chapter is organized into a few major sections: (i) a brief introduction of graphene; (ii) a brief introduction of the basic working principle of photocatalysis; (iii) the application of various graphene-based semiconductor composites in photocatalytic reduction of $Cr(VI)$ and $CO_2$, photocatalytic selective oxidation/reduction and photocatalytic degradation of organic pollutions; (iv) an introduction of the mechanism for the enhanced photocatalytic activity; and (v) finally, the major challenges for the future development of graphene-based semiconductor composites are identified.

## 2. Photocatalysis

### 2.1. *Background*

Currently, considerable attention has been paid to semiconductor photocatalysis for water treatment due to its good destruction ability of pollutants and broad compound applicability, ever since the research of photocatalytic water splitting on $TiO_2$ electrodes was conducted in 1972 [20–24]. It has drawn considerable academic interest as a very attractive, non-selective, room-temperature process for the degradation of organic pollutants and reduction of metal ions. In contrast with other semiconductors (i.e., $WO_3$, $ZnS$, $Fe_2O_3$, $CdS$ and $SrTiO_3$), $ZnO$ and $TiO_2$ have been proven to be the promising photocatalysts for widespread environmental applications due to their intriguing optical and electric properties, low cost and ease of availability [25]. However, their universal use is restricted to UV light due to the intrinsically wide bang gap (3.2 eV in the anatase phase), and the quick recombination of photoinduced electron–hole pairs has significantly decreased the photocatalytic activity. Currently, a particularly attractive option is to design and develop hybrid materials based on pure semiconductors to solve this problem, which involves anion and cation element doping [26], semiconductor compounding [27, 28], dye sensitization [29], loading of noble metals [30], and graphene hybrid [31]. The strategies of modification are as follows: (i) inhibiting recombination by increasing the charge separation; (ii) increasing the wavelength response range; and (iii) changing the selectivity or yield of a particular product.

### 2.2. *Basic working principle*

A photocatalytic reaction can be defined as a reaction induced by photoabsorption of a solid material, the photocatalyst, which remains unchanged during the reaction. Photocatalysis is normally based on the light absorption of semiconductor oxide photocatalysts to excite the electrons from the valence band (VB) to the conduction band (CB) and create electron–hole pairs. The energy of the incident

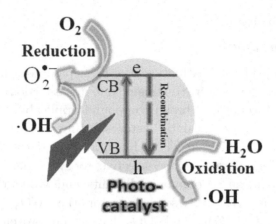

**Fig. 1.**    Schematic illustration of photocatalytic mechanism.

light must be greater than the bandgap of the photocatalyst. The activation equation can be written as follows:

$$\text{photocatalyst} + \text{h}\nu \rightarrow \text{h}^+ + \text{e}^-.$$

These electrons and holes can migrate and initiate redox reactions with water and oxygen, and then degrade organic molecules and reduction metal ions absorbed on the surface of the photocatalyst. For reduction metal ions, the CB level should be more negative than the metal ion level, while the VB should be more positive than the water oxidation level for efficient production from water by photocatalysis. The photogenerated electrons and holes can also recombine in bulk or on the surface of the semiconductor within a very short time, releasing energy in the form of heat or photons. As revealed by time-resolved spectroscopic studies, electron–hole trapping or recombination rates are extremely fast, about the order of $10^{-6}$ to $10^{-15}$ s, and can substantially lower the photocatalytic activity. Attempts to combine photocatalysts and graphene have been reported in efforts to obtain hybrid materials acting as excellent electron acceptor/transport materials for photocatalysis, which exhibit an enhanced photocatalytic activity due to the increased light absorption intensity and range, and the reduction of photoelectron–hole pair recombination with the introduction of grapheme [32].

## 3. Various graphene-based composites for photocatalysis

Several investigations have been carried out to produce graphene composite materials via reduction of exfoliated graphite oxide (GO) such as chemical reduction using highly toxic hydrazine and hydrazine derivatives [33], hydrothermal reduction [34, 35], electrochemical reduction [36], solvothermal reduction [37–39], sonolytic reduction [40], and high-temperature annealing reduction [41]. As an inexpensive, quick, versatile technique, microwave can heat the reactant to a high temperature in a short time by transferring energy selectively to microwave-absorbing polar solvents with a simultaneous increase in self-generated pressure inside the sealed reaction vessel. Thus, it can facilitate mass production in a short time with little energy cost and form an intimate contact between other materials and graphene, which is crucial for the formation of electronic interaction and interelectron transfer at the interface. Furthermore, as an environmentally friendly and efficient approach, UV-assisted photocatalytic reduction of GO has been successfully applied to the synthesis of graphene-based composites materials. This one-step strategy can be used to fabricate high-quality graphene-based nanocomposites without using any stabilizing reagent.

Graphene-based composites have stimulated interest in different applications, such as sensors, energy storage devices, bio-applications and particularly graphene-based photocatalysts with improved solar-to-fuel conversion efficiency. This effect is attributed mainly to the following effects: (i) decrease in the bandgap energy of the composite catalyst; (ii) increase in the adsorptive and active sites toward the reagents; (iii) acting as electron acceptor and transport channel to inhibit the recombination of photoinduced electron–hole pairs; and (iv) support material for enhanced structure stability.

### 3.1. *Photocatalytic reduction of Cr(VI)*

The possible photocatalytic reduction of the Cr(VI) mechanism using graphene-based hybrid composites is proposed. In the photocatalytic reduction of the Cr(VI) process, the semiconductor is excited by UV or visible light, leading to the generation of electron–hole pairs.

The work function of graphene is $-0.08\,V$ (vs. NHE) [32]. The CB level of the semiconductor is negative compared to the work function of grapheme; thus, the photoinduced electrons can be efficiently transferred from semiconductor CB to graphene, which could hinder the recombination of photoinduced charge carries. Moreover, the $Cr(VI)/Cr(III)$ potential is $0.51\,V$ vs. NHE [42], which is positive compared to the CB of the semiconductor and the work function of graphene. Therefore, the photoinduced electrons can reduce the $Cr(VI)$ to $Cr(III)$, and the hole can oxidize the water to form oxygen in the photocatalytic reaction process, which has been confirmed by the gas chromatograph [43]. Meanwhile, the photogenerated electrons are found to govern the catalytic reaction. The photocatalytic reduction of $Cr(VI)$ procedures is summarized as follows:

$$\text{photocatalyst} + h\nu \rightarrow h^+ + e^-.$$

$$Cr_2O_7^{2-} + 14H^+ + 6e^- \rightarrow 2Cr^{3+} + 7H_2O$$

$$2H_2O + 4h^+ \rightarrow O_2 + 4H^+$$

During the past decade, graphene has shown great ability to enhance the photocatalytic reduction $Cr(VI)$ activity of semiconductor oxide photocatalysts. ZnO-reduced graphene oxide (RGO) composites are successfully synthesized via UV-assisted photocatalytic reduction of graphite oxide by ZnO nanoparticles (NPs) in ethanol [44]. As shown in Fig. 2, RGO nanosheets are decorated densely by ZnO NPs to form a good combination between RGO and ZnO. ZnO–RGO composites exhibit an enhanced photocatalytic performance in reduction of $Cr(VI)$ with a maximum removal rate of 96% under UV light irradiation as compared with pure ZnO (67%) due to the increased light absorption intensity and range as well as the reduction of electron–hole pair recombination in ZnO with the introduction of RGO. However, when the RGO content is further increased above its optimum value, the photocatalytic performance deteriorates. This is ascribed to the following reasons: (i) RGO may absorb some UV light and thus there exists a light-harvesting competition between ZnO and RGO with the increase of RGO content, which leads to the decrease of the photocatalytic performance [33]; (ii) the excessive RGO can act as a kind of recombination center instead of providing

**Fig. 2.** FESEM images of (a) ZnO NPs and (b) ZnO–RGO composites by measurement; (c) low-magnification and (d) high-magnification HRTEM images of ZnO–RGO composites [44].

an electron pathway and promote the recombination of electron–hole pair in RGO [45].

TiO$_2$–graphene composites show also excellent photocatalytic activity in the reduction of Cr(VI) under visible light irradiation [46] TiO$_2$–graphene composites with 5 wt.% graphene decoration exhibited the best photocatalytic reduction of Cr(VI), which was mainly ascribed to the extended wavelength region of TiO$_2$, higher specific surface area and the quick transfer of electrons from excited TiO$_2$ particles to graphene nanosheets in the composite. Moreover, 3D TiO$_2$–graphene hydrogel structure also exhibited superb adsorption–photocatalysis performance for removing Cr(VI) from aqueous solutions [47]. 100% Cr(VI) from a solution containing (5 mg/L) could be removed within 30 min by the synergy performance of adsorption and photocatalysis, which is attributed to the non-porous surface adsorption and $\pi$–$\pi$ interaction adsorption for graphene, and the combination between graphene and TiO$_2$ nanospheres promoted photoinduced charge transport and separation, thereby facilitating photocatalytic reduction of Cr(VI).

Zero-dimensional quantum dots (QDs) have also been anchored onto graphene to synthetize graphene–hybrid photocatalysts with excellent photocatalytic activity. New composite materials consisting of $SnS_2$ QDs grown on RGO were also prepared by a simple and cost-effective one-step hydrothermal process for photocatalytic reduction of Cr(VI) under visible light irradiation [48]. $SnS_2$ QDs/RGO photocatalysts exhibited significantly enhanced photocatalytic activities for Cr(VI) reduction and the 1.5% $SnS_2$ QDs/RGO photocatalyst achieved the highest photoreduction efficiency of 95.3% under visible light irradiation, which was mainly due to excellent electron transportation ability of graphene that impedes the recombination of electron–hole pairs. $SnS_2$ QDs/RGO photocatalysts are visible light responsive, environmentally friendly, highly active and durable photocatalysts for photocatalytic reduction of Cr(VI). Carbon quantum dots/graphene aerogel (CQDs/GA) composites were synthesized by a facile one-step hydrothermal method for photoreduction of Cr(VI) [49]. 3D graphene aerogel with interconnected networks exhibits efficient spatial separation and transportation of photoexcited charge carriers, thereby leading to the improved photocatalytic activity of the CQDs/GA composites. The optimal photocatalytic activity of the CQDs/GA composites has been demonstrated to be 2.59 times as high as that of the blank CQDs.

Compared with other solution-phase synthetic methods by conventional heating, microwave hydrothermal or solvothermal synthesis has been accepted as a promising method for rapid heating, higher reaction rate and selectivity, lower reaction temperature, less reaction time, homogeneous thermal transmission and phase purity with better yield [50]. The microwave-assisted method can fulfill *in situ* formation of semiconductor photocatalysts on GO nanosheets and simultaneous reduction of GO to RGO without any toxic chemicals, and form an intimate contact between photocatalysts and RGO, which is crucial for the formation of electronic interaction and interelectron transfer at the interface.

In our previous work, $Ni_3S_2$/CdS/ZnO/$TiO_2$–RGO composites were successfully synthesized via the microwave-assisted reduction of graphite oxide using a microwave synthesis system (Explorer-48, CEM, USA) for photocatalytic reduction of Cr(VI). The ZnO–RGO composite (1.0 wt.% RGO) exhibited an enhanced photocatalytic performance with a maximum reduction rate of 98% under UV light irradiation for 4 h as compared with pure ZnO (58%) and P25 (70%) [51]. The $TiO_2$–RGO composite with 0.8 wt.% RGO exhibited an enhanced photocatalytic performance in the reduction of Cr(VI) (10 mg/L) with a maximum removal rate of 91% under UV light irradiation (500-W high-pressure Hg lamp with the main wave crest at 365 nm) for 4 h as compared with pure $TiO_2$ (83%) and commercial $TiO_2$ P25 (70%) [52]. The CdS–RGO composite (1.5 wt.% RGO) exhibited an enhanced photocatalytic performance in the reduction of Cr(VI) with a maximum removal rate of 92% as compared with pure CdS (79%) [53]. $Ni_3S_2$–RGO hybrids show excellent visible light photocatalytic activity in the reduction of Cr(VI) compared to pure $Ni_3S_2$ [54]. The as-synthesized $Ni_3S_2$–RGO samples with 0.5, 1 and 2 wt.% RGO were named NG-0.5, NG-1 and NG-2, respectively. The morphologies of $Ni_3S_2$–RGO hybrid composites are similar to that of pure $Ni_3S_2$. The ultrathin curled RGO nanosheets uniformly grew on the surface of $Ni_3S_2$ sheets to form a sheet-on-sheet structure (Fig. 3). The sheet-on-sheet structure in $Ni_3S_2$–RGO hybrid composites is beneficial to the separation of photoinduced charge carriers and ultimately contributes to the photocatalytic activity. As shown in Fig. 4, the Cr(VI) reduction rate of higher than 90% has been achieved with 1 wt.% reduced graphene oxide under visible light irradiation at 180 min, which is attributed to its efficient charge separation and more active sites due to the integrative effect and good interfacial contact between $Ni_3S_2$ and reduced graphene oxide. The RGO as co-catalyst can promote the transfer and inhibit the recombination of photoinduced charge carriers due to the matched energy levels and good interfacial contact between $Ni_3S_2$ and RGO, as shown in Fig. 5.

The photocatalytic reduction Cr(VI) activity of those graphene-based binary photocatalysts can be improved by introducing an additional component to form graphene-based ternary composite photocatalysts, as reported for ZnO–$TiO_2$–RGO composites. ZnO–$TiO_2$–RGO composites were successfully synthesized by

118                            X. Liu et al.

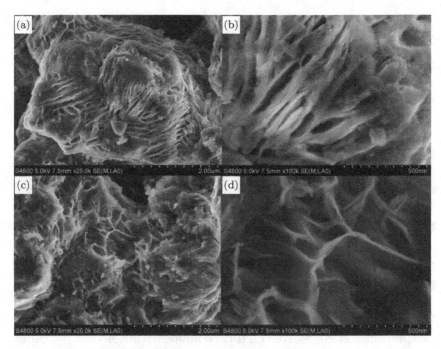

**Fig. 3.** FESEM images of (a) and (b) pure Ni$_3$S$_2$, and (c) and (d) Ni$_3$S$_2$–RGO hybrid [54].

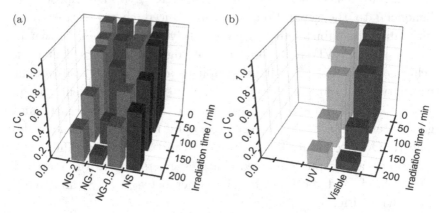

**Fig. 4.** (a) Photocatalytic reduction of Cr(VI) by (a) pure NS, NG-0.5, NG-1 and NG-2 under visible light irradiation, and (b) NG-1 under UV and visible light irradiation [54].

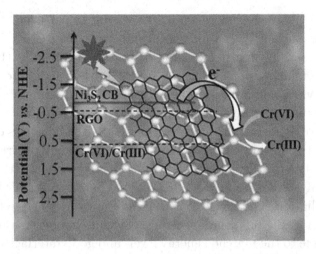

**Fig. 5.** Proposed photocatalytic mechanism for Ni$_3$S$_2$–RGO hybrid composites [54].

microwave-assisted reduction of graphite oxide in ZnO precursor solution with TiO$_2$ suspension using a microwave synthesis system [55]. According to the data reported on the CB and VB of ZnO ($-4.05$ and $-7.25$ eV vs. vacuum) [56], TiO$_2$ ($-4.2$ and $-7.4$ eV vs. vacuum) [57] and the work function of graphene ($-4.42$ eV vs. vacuum) [58], the energy levels of the three materials follow the order ZnO(CB) $>$ TiO$_2$(CB) $>$ graphene. Such energy levels are beneficial for photoinduced electrons to transfer from ZnO CB via TiO$_2$ CB to graphene, which can efficiently separate the photoinduced electrons and hinder the charge recombination in the electron transfer processes, thus enhancing the photocatalytic performance. Therefore, ZnO–TiO$_2$–RGO composites exhibit a better photocatalytic performance than pure ZnO and ZnO–RGO composite. The ZnO–TiO$_2$–RGO composite with 10 wt.% TiO$_2$ achieves the highest Cr(VI) removal rate of 99.4%.

In semiconductor photocatalysis, it is considered an effective strategy to use sunlight, but it is still a great challenge to fully utilize it, ranging from UV to visible and even the near infrared (NIR) wavelength. It is well known that solar light is composed of 5% UV light, 45% visible light and about 50% NIR. Nevertheless,

NIR light, possessing more than 50% of the whole solar spectrum, cannot be efficiently used for photocatalysis.

Wu *et al.* [59] first showed the discovery of full solar (UV–Vis–NIR) spectrum-responsive CuS–RGO (CR) hybrid photocatalysts prepared via a microwave-assisted method. Their catalytic activities were systematically investigated in the reduction of Cr(VI) under the radiation of light over the full spectrum. CuS–RGO displays excellent light absorption and catalytic activity in full solar light irradiation due to its synergistic effect. Only 1 wt.% RGO emersion in the hybrid photocatalysts can reduce 90% of Cr(VI) under full UV–Vis–NIR light irradiation. The excellent activity is mainly ascribed to the good absorption in the full solar spectrum, efficient charge separation and transfer due to the matched energy levels and good interfacial contact between CuS and RGO.

Ren *et al.* [60] also showed the composite ZnS–RGO prepared via a microwave-assisted method for efficient Cr(VI) reduction over the full solar spectrum for the first time. The as-synthesized ZnS–RGO samples with 0.5, 1 and 1.5 wt.% RGO were named ZR-0.5, ZR-1 and ZR-1.5, respectively. As shown in Fig. 6, ZnS NPs are embedded into the RGO nanosheets to form a good contact, which promotes the transport and the separation of charge carriers between the adjacent components, consequently improving the catalytic activity. The ZnS–RGO displays high absorption and catalytic activity in the UV, visible and even NIR light regions (Fig. 7). The enhanced mechanism of ZnS–RGO hybrid photocatalysts is proposed, as shown in Fig. 8. The introduction of RGO can increase the light absorption, provide more active sites as well as separate and inhibit the recombination of charge carriers, consequently improving the catalytic activity. Although the introduction of RGO narrows the bandgap of ZnS–RGO hybrid photocatalysts, they still cannot be excited by visible and NIR light. In contrast, the ZnS–RGO hybrid photocatalyst with 1 wt.% RGO shows the best catalytic activity in the reduction of Cr(VI) under visible and NIR light irradiation. In fact, the RGO plays an importance role in the visible and NIR light catalytic processes. In the visible light photocatalytic process, the RGO can be regarded as a visible light photosensitizer to ZnS in the ZR hybrid

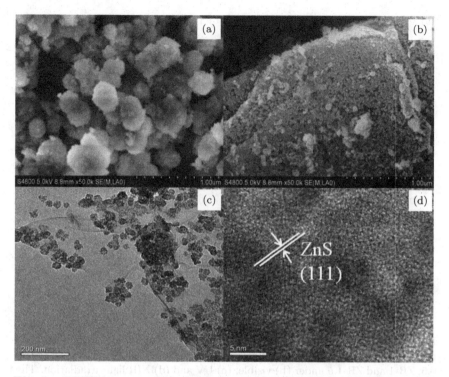

**Fig. 6.**   FESEM images of (a) pure ZnS and (b) ZnS–RGO hybrid. (c) Low- and (d) high-magnification HRTEM images of ZnS–RGO hybrid [60].

photocatalysts, which is similar to the dye self-sensitized degradation process [61]. The RGO is excited under visible light irradiation to generate the photoinduced electrons, which are transformed to the CB of ZnS and then participate in redox reactions. The improved catalytic activity of ZnS–RGO hybrid photocatalysts under visible light irradiation is ascribed to the increased light absorption in the visible light range, more active sites and the reduction of electron–hole pair recombination in ZnS with the introduction of RGO. In the NIR light photocatalytic process, the excellent catalytic activity of ZnS–RGO hybrid photocatalysts is mainly due to the high photothermal effect of RGO in the NIR region. RGO is a promising photothermal agent that possesses high absorption intensity in the NIR region and excellent heat generation efficiency, which has been used in cancer therapy [62] Furthermore, under NIR light

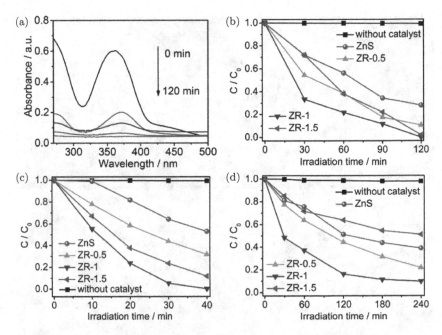

**Fig. 7.** (a) UV–Vis absorbance of Cr(VI) with the variation of irradiation time under visible light irradiation. Photocatalytic reduction of Cr(VI) by ZnS, ZR-0.5, ZR-1 and ZR-1.5 under (b) visible, (c) UV and (d) NIR light irradiation. The concentration of catalysts is $1.0 \, \text{g L}^{-1}$. The initial concentration and pH value of Cr(VI) are $80 \, \text{mg L}^{-1}$ and 7, respectively [60].

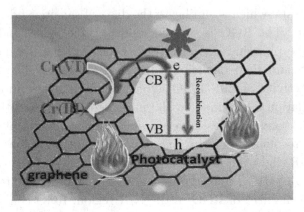

**Fig. 8.** Proposed photocatalytic mechanism for ZR hybrid photocatalysts [60].

irradiation, the photocatalytic reaction temperature reaches $\sim45°C$ in the reduction of the Cr(VI) process after 240 min irradiation. For comparison, the catalytic reduction of Cr(VI) by pure ZnS and ZnS–RGO hybrid photocatalysts was examined at various temperatures without NIR irradiation to clarify the temperature effect. It can be found that the Cr(VI) reduction rate increases with the increase of the reaction temperature. When the reaction temperature reaches 45°C, the Cr(VI) reduction rates are 20% and 71% for pure ZnS and ZnS–RGO hybrid photocatalysts, respectively. The result confirms that the reaction temperature can affect the photocatalytic activity in the reduction of Cr(VI) under NIR light irradiation, as also observed for other RGO-based composites [63–65].

## 3.2. *Photocatalytic reduction of* $CO_2$

Photocatalytic conversion of $CO_2$ into valuable fuels has drawn increasing attention because it can not only decrease the concentration of greenhouse gases but can also alleviate the energy shortage. Since Inoue *et al.* [66] first demonstrated the photoelectrocatalytic reduction of carbon dioxide into hydrocarbon fuels, many research studies have been performed to develop highly efficient photocatalysts. According to thermodynamic calculation, it takes 1,135 kJ/mol of energy to reduce $CO_2$ to $CH_4$, and at least 2.14 eV of reduction potential to reduce $CO_2$ to anionic radicals. The $CO_2$ reduction products include carbon monoxide (CO, $-0.53$ eV), formic acid (HCOOH, $-0.61$ eV), formaldehyde (HCHO, $-0.48$ eV), $CH_3OH$ ($-0.38$ eV), $CH_4$ ($-0.24$ eV) and other hydrocarbons [67–69], as shown in Fig. 9. To achieve the $CO_2$ photoreduction by water, the CB level of a given semiconductor must be much higher than the proton-assisted multielectron reduction potentials of $CO_2$, whereas the VB edge must be much more positive than the four-electron water oxidation potentials. As compared to the two-electron water reduction, the multielectron and multiproton reduction reactions of $CO_2$ are more favorable because of the substantially lower thermodynamic barriers [70]. In principle, all the photocatalysts used in photocatalytic $H_2$ evolution are capable of performing the reactions of photocatalytic $CO_2$ reduction into methanol and methane, which is due to the

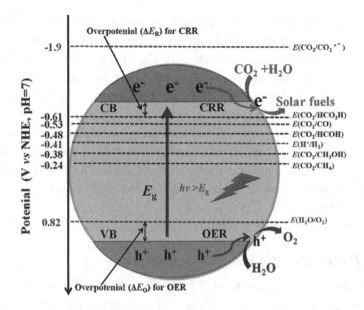

**Fig. 9.** Schematic illustration of $CO_2$ photoreduction on a semiconductor [70].

fact that their reduction potentials are even lower than that of $H_2$ generation from water splitting. However, it is especially challenging to get the target product with high selectivity due to very similar thermodynamic potentials [70]. Therefore, how to construct high-efficiency photocatalysts and realize the directional conversion of $CO_2$ are vital and challenging in photocatalytic $CO_2$ reduction. Even though photocatalytic transformation of $CO_2$ using solar energy is the most suitable way for the conversion of $CO_2$, some obstacles are involved in photocatalytic $CO_2$ reduction: (i) fast recombination of photoinduced electron and hole recombination; (ii) slow charge reduction during redox reaction; and (iii) lowly light employment.

Graphene-based hybrid photocatalysts exhibit excellent photo-catalytic activity in reduction of $CO_2$. For example, Zou et al. [71] successfully fabricated novel hollow spheres consisting of molecular-scale alternating $Ti_{0.91}O_2$ nanosheets and graphene nanosheets via a combination of the layer-by-layer assembly technique and a microwave irradiation technique for high-efficiency photocatalytic $CO_2$ conversion. The highly crystalline titania nanosheets are of extremely high 2D anisotropy with lateral dimensions of 0.1–1 $\mu m$ and a thickness of $\approx 0.75\,nm$. The 9 times increase of the photocatalytic activity of graphene–$Ti_{0.91}O_2$ hollow spheres relative to commercial P25 $TiO_2$ is confirmed with photoreduction of $CO_2$ into renewable fuels (CO and $CH_4$). G–$Ti_{0.91}O_2$ hollow spheres prove to be an efficient photocatalyst for $CO_2$ conversion in the presence of water vapor. With regard to generation of different products, i.e., CO and $CH_4$, via two-electron and eight-electron transfer processes, respectively, the generation of $CH_4$ is a major product over the titania catalyst through the reaction route $CO_2 \rightarrow CO \rightarrow C\cdot \rightarrow CH_2 \rightarrow CH_4$ [71].

Yu *et al.* [72] found that the CdS–RGO nanorod composites via a one-step microwave hydrothermal method exhibited obviously improved photocatalytic activity in reduction of $CO_2$ to $CH_4$ even without a noble metal Pt co-catalyst under visible light irradiation. 0.5 wt.% RGO–CdS composites exhibited a higher $CH_4$ production rate of $2.51\,mmol\,h^{-1}\,g^{-1}$ compared to blank CdS $(0.21\,mmol\,h^{-1}\,g^{-1})$ and 0.5% Pt–CdS nanorod composite $(1.52\,mmol\,h^{-1}\,g^{-1})$ under identical reaction conditions. The enhanced photoactivity of $CO_2$ reduction should be ascribed to the deposition of CdS onto RGO sheets, which promotes the separation and transfer of the photogenerated charge carriers and enhances the adsorption and activation of $CO_2$ molecules.

$Cu_2O$–graphene composites were also used for $CO_2$ photoconversion [73–76]. Tang *et al.* [77] synthesized the $Cu_2O$–RGO composites via a microwave-assisted hydrothermal method for photocatalytic reduction of $CO_2$ to CO in water. The results indicate that 0.5% RGO–$Cu_2O$ composites exhibit an improved photocatalytic $CO_2$ conversion rate of $50\,ppm\,g^{-1}\,h^{-1}$ for 20 h without the need for a noble metal co-catalyst. Stability is an issue for a $Cu_2O$ photocatalyst. The incorporation of RGO into $Cu_2O$ improves the photocatalyst's stability remarkably, which shows a great potential for $CO_2$ conversion in a sustainable manner. The superior photocatalytic activity and stability of $Cu_2O$–RGO composites are ascribed to the retarded electron–hole recombination, efficient charge transfer and protective function of RGO. Liu *et al.* [78] reported the effects of crystal facets and oxidation states of $Cu_2O$–graphene composites on the $CO_2$ reduction. The cubic, octahedral and rhombic dodecahedral $Cu_2O$ crystals with respective facets of (100), (111) and (110) decorated with graphene sheets were synthesized using a simple solution chemical route. $Cu_2O$–graphene with a rhombic dodecahedral structure exhibits the highest photocatalytic activity of $CO_2$ reduction as compared to cubic and octahedral structures, which is possibly due to the lower band bending of CB and VB of rhombic dodecahedral $Cu_2O$, resulting in the lower barrier and obstruction for the transfer of photogenerated electrons to the surface.

As shown in Fig. 10 [79], g-$C_3N_4$ has a suitable bandgap width (2.77 eV in bulk, 2.97 eV in nanosheets), which can effectively reduce $CO_2$ [80] The photogenerated electrons in g-$C_3N_4$ have a large thermodynamic driving force to reduce various small molecules such as $H_2O$, $CO_2$ and $O_2$. An effective strategy to obtain an intimate and large contact interface is to construct 2D–2D layered junctions to provide abundant surface-active sites and achieve efficient interfacial charge transfer. 2D–2D electron channels based on g-$C_3N_4$ at the interface significantly reduce the recombination of photogenerated electron–hole pairs, resulting in the enhanced photocatalytic performance [81] 2D–2D RGO–protonated g-$C_3N_4$ nanostructures with effective interfacial contact were synthesized by novel combined ultrasonic dispersion and electrostatic self-assembly

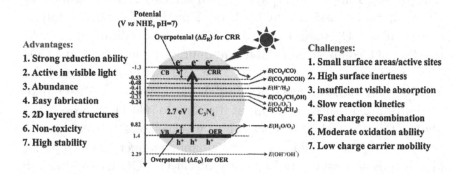

**Fig. 10.**  The redox potential of the relevant reactions of g-$C_3N_4$ band edges at pH = 7 [79].

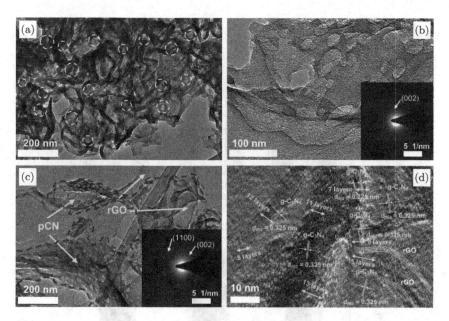

**Fig. 11.** TEM images of (a) bulk g-$C_3N_4$; (b) exfoliated protonated g-$C_3N_4$; (c) and (d) RGO–protonated g-$C_3N_4$ nanostructures [82].

strategy followed by a $NaBH_4$ reduction process [82]. As shown in Fig. 11 [82], exfoliation of protonated g-$C_3N_4$ successfully coupled with the negatively charged GO via electrostatic attraction, forming a more compact and well-dispersed sheet-on-sheet structure. The RGO–protonated g-$C_3N_4$ hybrid photocatalysts well-maintained the original 2D structure without a distinct change in the morphology while simultaneously establishing a good interfacial contact. The synergistic effect between g-$C_3N_4$ and RGO not only significantly enhanced the 2D–2D layered heterointerface region but also promoted the separation of electron–hole pairs, thereby leading to improved photocatalytic activity. The reduction rate of $CO_2$ for 15 wt.% RGO–g-$C_3N_4$ nanostructures reached a maximum of 13.93 $\mu$mol g$^{-1}$, which was a 5.4-fold enhancement over protonated g-$C_3N_4$ (Fig. 12). RGO–g-$C_3N_4$ nanostructures also showed good stability after three successive cycles with no obvious change in the production of $CH_4$ from reduction (Fig. 12). 2D–2D ultrathin SiC–RGO nanosheet heterojunction was also demonstrated to allow

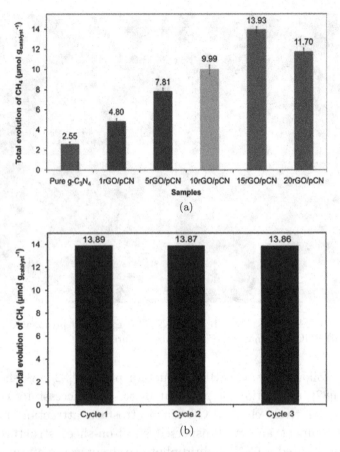

**Fig. 12.** (a) Total evolution of $CH_4$ over the pure g-$C_3N_4$ and a series of RGO–protonated g-$C_3N_4$ photocatalysts with different RGO contents under visible light irradiation for 10 h; (b) recycling runs for the photocatalytic reduction of $CO_2$ using 15 wt.% RGO–protonated g-$C_3N_4$ sample [82].

fast transfer of energetic electrons from SiC to RGO [83]. The $CO_2$ reduction rate is $58.17 \, \mu\mathrm{mol} \, \mathrm{g}^{-1} \, \mathrm{h}^{-1}$, 2.7-fold that of pure SiC ($20.25 \, \mu\mathrm{mol} \, \mathrm{g}^{-1} \, \mathrm{h}^{-1}$).

The role of graphene and its defects in the photocatalytic reduction of $CO_2$ was investigated by Hersam et al. [84]. The two major solution-based pathways for producing graphene, oxidation reduction and solvent exfoliation, result in nanoplatelets with different defect densities. Results show that nanocomposites based on the less defective solvent-exfoliated graphene (SEG) exhibit a significantly larger enhancement for the reduction of $CO_2$ to $CH_4$, with up to a 7-fold improvement compared to pure $TiO_2$ under visible illumination due to their superior electrical mobility, which facilitates the diffusion of photoexcited electrons to reactive sites. 0.55 wt.% SEG-P25 nanocomposites show the highest photoreductive $CO_2$ activity. Therefore, graphene is able to act as an efficient co-catalyst for improving the photocatalytic activity of $TiO_2$ for reduction of $CO_2$. The defect density and sheet resistance of graphene have a significant influence on the photoactivity enhancement of the resulting graphene semiconductor composite photocatalysts.

Ternary graphene-based photocatalysts with enhanced photocatalytic activity have also been fabricated for catalytic reduction of $CO_2$ in solar fuels and chemicals [85, 86]. Metal–organic frameworks (MOFs), composed of organic ligands and inorganic clusters, are a new class of porous inorganic–organic hybrid materials. MOFs have been exploited as photocatalysts for pollutant degradation, water splitting, $CO_2$ reduction and organic transformation owing to their unique properties, such as high specific surface area, tailorable pore size, designed frame structure and easy functional group.

Liu *et al.* [87] constructed a novel three-component Z-scheme system, which utilizes UiO–66–NH₂ and O–ZnO as the photocatalysts, and RGO as the electron mediator for $CO_2$ photoreduction. UiO–66–NH₂ and O–ZnO are evenly dispersed on RGO nanosheets (Fig. 13). In the Z-scheme photocatalytic system, the photoinduced electrons in the CB of O–ZnO transferred to RGO, and then recombined with the holes in the VB of UiO–66–NH₂, leading to enhanced charge separation efficiency. O–ZnO/RGO/UiO–66–NH₂ heterostructures could efficiently reduce $CO_2$ to $CH_3OH$ and HCOOH, and its activity was significantly superior to that of O–ZnO/ UiO–66–NH₂ and ZnO/RGO/UiO–66–NH₂ due to effective spatial separation of photogenerated electrons and holes via a Z-scheme charge transfer. The yield of $CH_3OH$ and HCOOH reaches 34.83 and $6.41\,\mu\mathrm{mol\,g^{-1}\,h^{-1}}$, respectively, under visible light irradiation, as shown in Fig. 14. This study offers an insight into the design and fabrication of novel MOF-based Z-scheme photocatalytic systems.

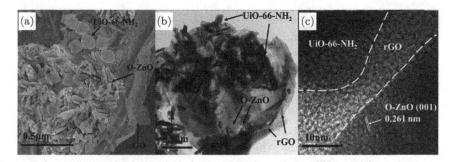

**Fig. 13.** (a) SEM images, (b) TEM and (c) HRTEM images of O–ZnO/
RGO/UiO–66-NH$_2$ heterostructures [87].

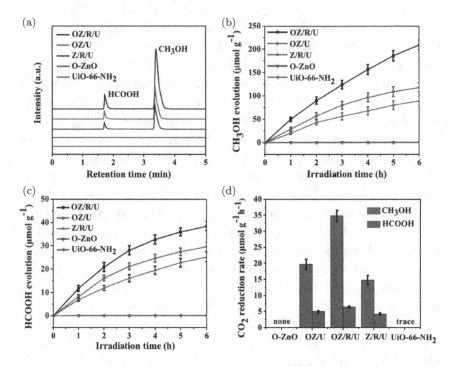

**Fig. 14.** (a) Original chromatograms for the as-synthesized samples after 6 h
of irradiation, (b) CH$_3$OH evolution amount vs illumination time, (c) HCOOH
evolution amount vs illumination time and (d) CO$_2$ reduction rate of the
as-prepared samples [87].

Transition metal dichalcogenides (TMDs) have attracted great interest due to their intriguing properties and potential applications in hydrogen evolution [88], lithium/sodium batteries [89] and photocatalysis [90–92]. Among these TMDs, $MX_2$ (M = Mo/W; X = S/Se) has emerged as a potential candidate due to its unique properties [93–95] $MX_2$-based composites demonstrated excellent photocatalytic activity under visible light irradiation [96–98]. A hierarchical porous structure with mesoporous $TiO_2$ and a few layers of $MoS_2$ on macroporous 3D graphene was synthesized via a simple one-pot hydrothermal method for $CO_2$ photoreduction [99]. The hierarchical structure increased the surface area, offering abundant reaction sites and efficient mass flow pathways at the randomly arranged porous backbone structure. The electron flow from $TiO_2$ via graphene into the few-layered $MoS_2$ could effectively lower the charge recombination rate and increase the potential for $CO_2$ reduction. A high CO selectivity of 97% and CO production yield (93.22 $\mu$mol g$^{-1}$ h$^{-1}$) was reached, which is 14.5 times that of bare $TiO_2$.

Oh *et al.* [100] prepared $WSe_2$–graphene–$TiO_2$ ternary nanocomposites using ultrasonication techniques for the photocatalytic reduction of $CO_2$ under UV–V is light irradiation. $WSe_2$ NPs support the $TiO_2$ NPs on the graphene sheet, and provide a bridge between $TiO_2$ and the graphene nanosheets, which is beneficial to the photocatalytic process. The ternary nanocomposites have good photocatalytic activity for $CO_2$ reduction to $CH_3OH$ under UV–V is light irradiation. $WSe_2$–graphene–$TiO_2$ ternary nanocomposites with an optimum loading of graphene 8 wt.% showed the highest photocatalytic activity with the $CH_3OH$ yield of 6.3262 $\mu$mol g$^{-1}$ h$^{-1}$ after 48 h.

Noble metal NPs as co-catalysts can suppress the recombination of photogenerated charges and improve the utilization efficiency of photogenerated electrons, which is beneficial to improve photogenerated charge separation and photocatalytic activity. In addition, metal NPs can also act as activity sites to improve the photocatalytic $CO_2$ reduction and selective conversion rate. There are some reports on the co-catalysts of metal NPs (Ag, Rh, Ni and Pt) and graphene for photocatalytic $CO_2$ reduction.

Core–shell-structured photocatalysts (Pt/TiO$_2$)@RGO of Pt–TiO$_2$ nanocrystals wrapped by RGO sheets were synthesized by a self-assembly method of a simple and pure chemical bonding reactions [101]. The (Pt/TiO$_2$)@RGO structure is not only favorable to the vectorial electron transfer of TiO$_2 \rightarrow$ Pt $\rightarrow$ RGO and enhances the separation efficiency of photogenerated electrons and holes but also the surface residual hydroxyl and extended $\pi$ bond of wrapping RGO sheets can improve the adsorption and activation capabilities for CO$_2$ reactant. Therefore, the (Pt–TiO$_2$)@RGO structure exhibited the highest photocatalytic activity and selectivity for CO$_2$ conversion with H$_2$O to CH$_4$ under simulated solar irradiation. The CH$_4$ formation rate with of 41.3 $\mu$mol g$^{-1}$ h$^{-1}$ is about 31-fold that of commercial P25, and the selectivity of CO$_2$ conversion to CH$_4$ product is close to 100%.

Graphene–CdS–Ag NPs [102], graphene–Rh$_2$O$_3$–Rh NPs [103] and graphene–NiO–Ni NPs [104] photocatalysts also exhibited enhanced photocatalytic activity for the conversion of CO$_2$ into CO–CH$_4$. The study demonstrates a semiconductor-based hybrid for efficient photocatalytic CO$_2$ reduction by virtue of the combinational effect of metal NPs and graphene.

### 3.3. Photocatalytic selective oxidation–reduction

CdS as another important semiconductor has been coupled with graphene for photocatalytic selective oxidation and reduction [105].

CdS–graphene nanocomposites were synthesized via a facile hydrothermal approach for selective oxidation of a range of alcohols and selective reduction of nitro organics to corresponding amino organics under visible light irradiation [106]. CdS–graphene nanocomposites exhibit enhanced photocatalytic performance for both the oxidation and reduction processes due to the increased adsorptivity, the improved lifetime and transfer of charge carriers (particularly for photogenerated electrons), and the addition of ammonium formate as quencher for photogenerated holes. CdS–graphene nanocomposites show excellent reusability due to the synergetic effect of the introduction of GR into the matrix of CdS

NSPs and the addition of ammonium formate as quencher for photogenerated holes. In addition, it is found that decreasing the defect density of graphene by using solvent-exfoliated graphene instead of graphene oxide as the precursor of graphene can efficiently enhance the photocatalytic activity of CdS–graphene nanocomposites due to its improved electron conductivity as compared to reduced graphene oxide [107].

In order to further improve the photocatalytic activity of CdS–graphene nanocomposites, ZnO or $TiO_2$ is introduced into CdS–graphene nanocomposites to form ternary $ZnO$–$TiO_2$–CdS–graphene nanocomposites [108] as efficient visible-light-driven photocatalysts for selective organic transformations. It is found that the ternary $ZnO$–$TiO_2$–CdS–graphene hybrid exhibits enhanced photocatalytic activity compared to its foundation matrix binary CdS–graphene due to the combined interaction of the longer lifetime of photogenerated electron–hole pairs, faster interfacial charge transfer rate and larger surface area.

Zhang *et al.* [109] reported a simple and general approach to boost the lifetime and transfer efficiency of photogenerated charge carriers across the interface between graphene and semiconductor CdS by introducing a small amount of metal ions ($Ca^{2+}$, $Cr^{3+}$, $Mn^{2+}$, $Fe^{2+}$, $Co^{2+}$, $Ni^{2+}$, $Cu^{2+}$ and $Zn^{2+}$) as the "mediator" into the interfacial layer matrix, significantly enhancing the photoactivity of graphene semiconductor composites in the selective oxidation of alcohol and anaerobic reduction of nitro-compound (Fig. 15). As shown in Fig. 16, the key role of metal ions is to act as "interfacial mediator" to optimize the transfer pathway of photogenerated charge carriers and improve their lifetime and transfer efficiency across the interface in the composite photocatalysts. The work highlights that the significant issue for improving the photoactivity of the graphene semiconductor via strengthening interfacial contact is not just a simple issue of tighter connection between graphene and the semiconductor but it is also the optimization of the atomic charge carrier transfer pathway across the interface between graphene and the semiconductor.

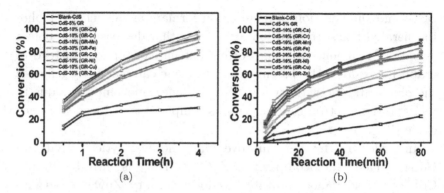

**Fig. 15.** Time-online profiles of conversion of (a) benzyl alcohol and (b) 4-nitroaniline over blank CdS, GR–CdS and CdS–(GR–M) (M = $Ca^{2+}$, $Cr^{3+}$, $Mn^{2+}$, $Fe^{2+}$, $Co^{2+}$, $Ni^{2+}$, $Cu^{2+}$ and $Zn^{2+}$) nanocomposites under visible light irradiation ($\lambda > 420$ nm) [109].

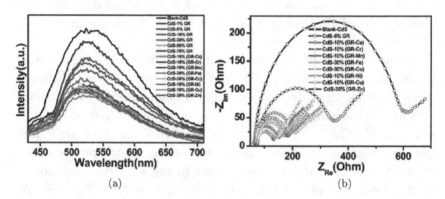

**Fig. 16.** Photoluminescence spectra and electrochemical impedance spectroscopy Nyquist diagrams of CdS–(GR–M) (M = $Ca^{2+}$, $Cr^{3+}$, $Mn^{2+}$, $Fe^{2+}$, $Co^{2+}$, $Ni^{2+}$, $Cu^{2+}$ and $Zn^{2+}$) nanocomposites [109].

Except for CdS, indium sulfide ($In_2S_3$), an n-type semiconductor with a bandgap of 2.0–2.3 eV, has attracted substantial attention as a potential semiconductor nanomaterial photocatalyst due to its high photosensitivity and photoconductivity, stable chemical and physical characteristics, and low toxicity. Li *et al.* [110] reported a simple *in situ* wet chemistry method to fabricate $In_2S_3$–graphene nanocomposites in which hierarchical $In_2S_3$ "petals" spread over the surface of graphene sheets in virtue of the "structure directing" role of GO as the precursor of graphene in a solution phase. $In_2S_3$–graphene nanocomposites show higher photoactivity than pure $In_2S_3$ toward selective oxidation of benzyl alcohol to benzaldehyde under visible light irradiation. $In_2S_3$–1% graphene exhibits the highest photocatalytic activity with high conversion of benzyl alcohol (55%) and selectivity (99%) after the visible light irradiation for 2 h. The enhanced photocatalytic activity is ascribed to the integrative effect of the increased separation and transfer efficiency of photogenerated electron–hole pairs and the larger surface area. Furthermore, $In_2S_3$–graphene nanocomposites synthesized by an efficient and simple self-assembly approach display much higher photoactivity toward selective reduction of nitroaromatic compounds to the corresponding amines than the photocatalysts of $In_2S_3$–graphene that is obtained from the simple "hard" integration of graphene nanosheets with solid $In_2S_3$ particles without modification of surface charge [111]. The more efficient interfacial contact due to the spontaneous attraction of electrostatic forces between positively charged $In_2S_3$ and negatively charged GO is beneficial for harnessing the electron conductivity of graphene, which contributes to improving the life span and transfer of photogenerated charge carriers and thus results in improvement of the photoactivity of $In_2S_3$–graphene more significantly.

Noble metal (Au, Ag, Pd) partially reduced graphene oxide (PRGO) nanocomposites were realized by a simple redox reaction between the noble metal precursors and graphene oxide (GO) in an aqueous solution for selective reduction of nitroaromatic compounds into the corresponding amino compounds under ambient conditions [112]. Results show that Au, Ag and Pd–PRGO nanocomposites also exhibit a high catalytic performance, and the reaction rates follow the order of Pd–PRGO > Ag–PRGO > Au–PRGO. The smallest noble metal particle size affords the highest catalytic activity.

### 3.4. *Photocatalytic degradation of organic pollutions*

The working mechanism of photocatalysis in the degradation of pollutants normally involves several key processes [113–116]. In brief,

*X. Liu et al.*

the electrons, excited from the VB to the conduction band CB of photocatalysts by absorbing light (Eq. (1)), are scavenged by molecular oxygen ($O_2$) to yield the superoxide radical anion ($O_2^{\bullet -}$) (Eq. (2)) and hydrogen peroxide ($H_2O_2$) (Eq. (3)):

$$\text{Photocatalyst} + h\nu \rightarrow \text{photocatalyst}(h^+) + e^- \tag{1}$$

$$e^- + O_2 \rightarrow O_2^{\bullet -} \tag{2}$$

$$2e^- + O_2 + 2H^+ \rightarrow H_2O_2 \tag{3}$$

These new intermediates or the holes will interact to produce the hydroxyl radical $\cdot OH$ (Eqs. (4) and (5)).

$$H_2O_2 + O_2^{\bullet -} \rightarrow \cdot OH + OH^- + O_2 \tag{4}$$

$$\text{photocatalyst}(h^+) + H_2O \rightarrow H^+ + \cdot OH \tag{5}$$

The $\cdot OH$ radical is a powerful oxidizing agent, which has sufficient capacity to degrade most pollutants (Eq. (6)) [117, 118]. The photocatalytic destruction of organic matter by photocatalysts proceeds mainly through these hydroxyl radicals [119]. The holes and reactive oxygen species may also contribute to the oxidative pathways for the degradation of pollutants (Eqs. (7) and (8)) [117, 118].

$$\cdot OH + \text{pollutants} \rightarrow H_2O + CO_2 \tag{6}$$

$$O_2^{\bullet -} + \text{pollutants} \rightarrow H_2O + CO_2 \tag{7}$$

$$h^+ + \text{pollutants} \rightarrow H_2O + CO_2 \tag{8}$$

Wang *et al.* [120] reported a facile and efficient strategy for preparing RGO nanosheets-ZnS composites assisted by microwave irradiation (400 W) using a MAS-I microwave oven (Sineo, China) at 85°C for 20 min. The process involved the reaction of GO nanosheets (300 mg) as the dispersant and 2D growth template for ZnS nanoballs, $Zn(CH_3COO)_2 \cdot 2H_2O$ (0.04 mol) as the zinc source and TAA (0.06 mol) as a sulfide source as well as reducing agent in aqueous solution (100 mL). ZnS nanoballs composed of many small self-assembled ZnS nanocrystals were formed *in situ* on GO and simultaneous reduction of GO to RGO nanosheets. By comparing the TEM images of ZnS and ZnS–RGO composites

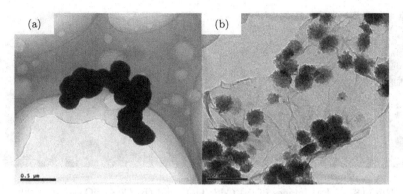

**Fig. 17.** (a) TEM images of ZnS nanoballs and (b) ZnS–RGO composites [120].

in Figs. 17(a) and 17(b), when adding RGO, the agglomeration of ZnS can be effectively avoided and the ZnS nanoballs disperse well in the RGO matrix. A possible mechanism was proposed as follows [120]: the positive $Zn^{2+}$ ions were easily anchored onto the surface of the GO. With TAA added to the mixture, the $Zn^{2+}$ ions *in situ* reacted with $S^{2-}$ uniformly released from TAA under microwave heating to form the initial ZnS nuclei. Afterward, the ZnS nuclei grew to form ZnS NPs, which were then self-assembled into nanoballs induced by continuous microwave radiation, while GO was reduced to RGO to form the ZnS–RGO composite. The as-prepared ZnS–RGO composites showed an excellent photocatalytic activity with a degradation rate of ~99% toward the photodegradation of MB (15 mg/L) under UV light irradiation using a high-pressure Hg lamp (500 W) for 30 min as compared with pure ZnS (~59%) after 150 min irradiation.

Pan *et al.* [121] applied the microwave irradiation in GO dispersion with $Zn(NO_3)_2$ (20 mL) at 100°C for 30 min under magnetic stirring to synthesize ZnO particle–RGO hybrid composites. ZnO displays a hexagonal nanorod structure and the RGO sheets are decorated densely by the ZnO nanorods. The incorporation of RGO could enhance the photocatalytic activity and photostability of ZnO, and the ZnO–RGO composite with 1.1 wt.% RGO achieved a maximum degradation rate of 88% at pH of 7% and 91.7% at pH of 9 as compared with pure ZnO (68%).

**Fig. 18.**  SEM images of (a) CdS–TiO$_2$ and (b) CdS–TiO$_2$–RGO [122].

Pan *et al.* [122] reported a rapid synthesis of CdS–TiO$_2$–RGO composite using microwave-assisted reduction of GO in CdS precursor solution with TiO$_2$ suspension. As shown in Fig. 18, RGO sheets are decorated densely by CdS–TiO$_2$, which displays a good combination between them. Due to the distribution of carboxylic acid groups on GO, TiO$_2$ NPs dispersed on the carbon support were eager to accumulate along the wrinkles and edge of RGO. Meanwhile, positively charged Cd$^{2+}$ ions form complexes with TiO$_2$ NPs and S$^{2-}$ from Na$_2$S could move quickly to react with Cd$^{2+}$ producing uniform CdS nuclei under microwave irradiation. The CdS nuclei grew, crystallized and stabilized on TiO$_2$ under the microwave irradiation, leading to a good contact between CdS and TiO$_2$. The photocatalytic experiments of the CdS–TiO$_2$–RGO in the degradation of MO (10 mg/L) showed that the RGO addition could enhance the photocatalytic performance of CdS–TiO$_2$ composites with a maximum degradation rate of 99.5% under visible light irradiation (400-W metal halogen lamp with cut off filter, $\lambda > 400$ nm) for 5 h as compared with the pure TiO$_2$ (43%) and the CdS–TiO$_2$ (79.9%).

## 4. Mechanism for the enhanced photocatalytic activity

Graphene can also act as an excellent electron acceptor/transport material, and attempts to combine semiconductor photocatalyst and graphene have been reported in efforts to obtain hybrid materials with superior performance. In the meantime, such an attachment of semiconductor particles onto RGO can also prevent the restacking and agglomeration of RGO sheets during the reduction process due to Van der Waals interactions between them. Due to the lower energy level of the work function of RGO than the CB of most photocatalysts, the stepwise structure of energy levels constructed in the RGO-photocatalyst composite can facilitate the electron transfer and separation, which enhances the photocatalytic performance of photocatalyst [123–132]. In addition, the introduction of RGO can enhance light absorption intensity and range [18, 133, 134] as well as increase the adsorption of pollutants [135]. However, it should be noticed that excessive RGO can act as a kind of recombination center instead of providing an electron pathway and light filter to decrease the number of charge carriers in the composite to be photogenerated, which will decrease the photocatalytic performance. As a result, the incorporation of suitable amount of RGO is necessary for optimum performance of the composite.

To affirm that graphene is able to manifest its truly unique and outstanding advantage over its forebears ($C_{60}$ and CNT) in enhancing the photocatalytic performance of semiconductors, a series of CdS–carbon ($C_{60}$, CNT and graphene) nanocomposites with intimate interfacial contact were synthesized via a simple solvothermal method for photocatalytic selective oxidation of alcohols under visible light irradiation [136]. The results reveal that graphene cannot manifest its unique advantage over its carbon allotropes in enhancing the photoactivity of semiconductor CdS. The primary roles among these carbon allotropes ($C_{60}$, CNT and graphene) in the nanocomposites carbon–CdS are quite similar, acting as electron reservoirs to accept or shuttle photogenerated electrons from semiconductors, improving the adsorption capacity and increasing the light absorption in the visible light region, which is strongly corroborated by the current systematic comparison of CdS–carbon in a reasonable benchmark framework. Furthermore, similar results are also observed in $TiO_2$–carbon (graphene, CNT and $C_{60}$) nanocomposites for photocatalytic oxidation of benzyl alcohol to benzaldehyde [137]. A significant difference of graphene on improving the photoactivity of $TiO_2$ as compared to CNT and $C_{60}$ cannot be found.

Furthermore, Zhang *et al.* [61] proved that graphene can act as an organic dye-like macromolecular "photosensitizer" instead of an electron reservoir in the photocatalytic process. ZnS–graphene nanocomposites synthesized by a facile two-step wet chemistry process exhibited visible light photoactivity toward selective oxidation of alcohols and alkenes under ambient conditions [61]. Graphene in the ZnS–graphene nanocomposites acts as an organic dye-like macromolecular "photosensitizer" for ZnS instead of an electron reservoir, with which the photogenerated electrons from graphene upon visible light irradiation can transfer to the conduction band of ZnS, while ZnS by itself is not bandgap photoexcited. The photocatalytic mechanism is remarkably different from previous research works for all other graphene semiconductor photocatalysts, for which graphene is always proposed to behave as an electron reservoir to capture or shuttle photogenerated electrons from the semiconductor. A similar role of graphene was also observed for the ZnO–graphene nanocomposites in the reduction of Cr(VI) in aqueous solution under ambient conditions [138].

## 5. Conclusions and outlook

Graphene-based hybrid photocatalysts are summarized in the chapter, which have opened up new pathways to design the novel hybrid photocatalysts with high catalytic activity and good stability. Although considerable progress has been achieved, a great challenge still exists in this field. First, the large-scale production of graphene-based photocatalysts with controlled morphologies and surface structure should be improved. Large-scale production methods for preparing graphene-based photocatalysts are still highly desirable in the future. Second, the enhanced photocatalytic mechanism of graphene-based photocatalysts should be clarified in detail through combing the theoretical calculations and experimental evidence. Third, the photocatalytic reduction $CO_2$ activity is very challenging and further improvement is required. Optimizing the charge carrier transfer pathway by introducing metal ions and metal NPs can inhibit the recombination of charge carrier, thus enhancing the photocatalytic activity. Considerable breakthroughs in the practical applications of graphene-based composite photocatalysts are expected to occur in the near future.

## Acknowledgments

The National Natural Science Foundation of China (No. 51902301) is gratefully acknowledged.

## References

[1] K.S. Novoselov, A.K. Geim, S.V. Morozov, D. Jiang, Y. Zhang, S.V. Dubonos, I.V. Grigorieva, A.A. Firsov, Electric field effect in atomically thin carbon films, *Science*, 306 (2004) 666–669.

[2] T.N. Lambert, C.A. Chavez, B. Hernandez-Sanchez, P. Lu, N.S. Bell, A. Ambrosini, T. Friedman, T.J. Boyle, D.R. Wheeler, D.L. Huber, Synthesis and characterization of titania-graphene nanocomposites, *J. Phys. Chem. C*, 113 (2009) 19812–19823.

[3] H.B. Li, T. Lu, L.K. Pan, Y.P. Zhang, Z. Sun, Electrosorption behavior of graphene in NaCl solutions, *J. Mater. Chem.*, 19 (2009) 6773–6779.

[4] L.L. Zhang, R. Zhou, X.S. Zhao, Graphene-based materials as supercapacitor electrodes, *J. Mater. Chem.*, 20 (2010) 5983–5992.

[5] K.F. Zhou, Y.H. Zhu, X.L. Yang, X. Jiang, C.Z. Li, Preparation of graphene-TiO$_2$ composites with enhanced photocatalytic activity, *New J. Chem.*, 35 (2011) 353–359.

[6] M. Pumera, Graphene-based nanomaterials for energy storage, *Energy Environ. Sci.*, 4 (2011) 668–674.

[7] A.V. Murugan, T. Muraliganth, A. Manthiram, Rapid, facile microwave-solvothermal synthesis of graphene nanosheets and their polyaniline nanocomposites for energy strorage, *Chem. Mater.*, 21 (2009) 5004–5006.

[8] J.K. Lee, K.B. Smith, C.M. Hayner, H.H. Kung, Silicon nanoparticles-graphene paper composites for Li ion battery anodes, *Chem. Commun.*, 46 (2010) 2025–2027.

[9] K.S. Kim, Y. Zhao, H. Jang, S.Y. Lee, J.M. Kim, K.S. Kim, J.H. Ahn, P. Kim, J.Y. Choi, B.H. Hong, Large-scale pattern growth of graphene films for stretchable transparent electrodes, *Nature*, 457 (2009) 706–710.

[10] Y.L. Chen, Z.A. Hu, Y.Q. Chang, H.W. Wang, Z.Y. Zhang, Y.Y. Yang, H.Y. Wu, Zinc oxide/reduced graphene oxide composites and electrochemical capacitance enhanced by homogeneous incorporation of reduced

graphene oxide sheets in zinc oxide matrix, *J. Phys. Chem. C*, 115 (2011) 2563–2571.

[11] T. Lu, Y.P. Zhang, H.B. Li, L.K. Pan, Y.L. Li, Z. Sun, Electrochemical behaviors of graphene-ZnO and graphene-$SnO_2$ composite films for supercapacitors, *Electrochim. Acta*, 55 (2010) 4170–4173.

[12] Y.B. Tang, C.S. Lee, J. Xu, Z.T. Liu, Z.H. Chen, Z.B. He, Y.L. Cao, G.D. Yuan, H.S. Song, L.M. Chen, L.B. Luo, H.M. Cheng, W.J. Zhang, I. Bello, S.T. Lee, Incorporation of graphenes in nanostructured $TiO_2$ films via molecular grafting for dye-sensitized solar cell application, *ACS Nano*, 4 (2010) 3482–3488.

[13] M. Cox, A. Gorodetsky, B. Kim, K.S. Kim, Z. Jia, P. Kim, C. Nuckolls, I. Kymissis, Single-layer graphene cathodes for organic photovoltaics, *Appl. Phys. Lett.*, 98 (2011) 123303.

[14] W.T. Zheng, Y.M. Ho, H.W. Tian, M. Wen, J.L. Qi, Y.A. Li, Field emission from a composite of graphene sheets and ZnO nanowires, *J. Phys. Chem. C*, 113 (2009) 9164–9168.

[15] D. Pan, S. Wang, B. Zhao, M. Wu, H. Zhang, Y. Wang, Z. Jiao, Li storage properties of disordered graphene nanosheets, *Chem. Mater.*, 21 (2009) 3136–3142.

[16] H.L. Guo, X.F. Wang, Q.Y. Qian, F.B. Wang, X.H. Xia, A green approach to the synthesis of graphene nanosheets, *ACS Nano*, 3 (2009) 2653–2659.

[17] X. Li, Y. Zhu, W. Cai, M. Borysiak, B. Han, D. Chen, R.D. Piner, L. Colombo, R.S. Ruoff, Transfer of large-area graphene films for high-performance transparent conductive electrodes, *Nano Lett.*, 9 (2009) 4359–4363.

[18] Y.J. Wang, R. Shi, J. Lin, Y.F. Zhu, Significant photocatalytic enhancement in methylene blue degradation of $TiO_2$ photocatalysts via graphene-like carbon *in situ* hybridization, *Appl. Catal. B: Environ.*, 100 (2010) 179–183.

[19] L. Jia, D.H. Wang, Y.X. Huang, A.W. Xu, H.Q. Yu, Highly durable n-doped graphene/CdS nanocomposites with enhanced photocatalytic hydrogen evolution from water under visible light irradiation, *J. Phys. Chem. C*, 115 (2011) 11466–11473.

[20] Z.X. Ren, X.J. Liu, Z.H. Zhuge, Y.Y. Gong, C.Q. Sun, $MoSe_2$/ZnO/ZnSe hybrids for efficient Cr(VI) reduction under visible light irradiation, *Chin. J. Catal.*, 41 (2020) 180–187.

[21] X.J. Liu, B.B. Liu, L. Li, Z.H. Zhuge, P.B. Chen, C. Li, Y.Y. Gong, L.Y. Niu, J.Y. Liu, L. Lei, C.Q. Sun, $Cu_2In_2ZnS_5$/$Gd_2O_2S$ : Tb for full solar spectrum photoreduction of Cr(VI) and $CO_2$ from UV/vis to near-infrared light, *Appl. Catal. B: Environ.*, 249 (2019) 82–90.

[22] B.B. Liu, X. Liu, L. Li, Z. Zhuge, Y.Q. Li, C. Li, Y.Y. Gong, L.Y. Niu, S.Q. Xu, C.Q. Sun, $CaIn_2S_4$ decorated $WS_2$ hybrid for efficient Cr(VI) reduction, *Appl. Surf. Sci.*, 484 (2019) 300–306.

[23] B.B. Liu, X.J. Liu, J.Y. Liu, C.J. Feng, Z. Li, C. Li, Y.Y. Gong, L.K. Pan, S.Q. Xu, C.Q. Sun, Efficient charge separation between UiO-66 and $ZnIn_2S_4$

flower like 3D microspheres for photoelectronchemical properties, *Appl. Catal. B: Environ.*, 226 (2018) 234–241.

[24] B.B. Liu, X.J. Liu, L. Li, J.W. Li, C. Li, Y.Y. Gong, L.Y. Niu, X.S. Zhao, C.Q. Sun, ZnIn$_2$S$_4$ flower like microspheres embedded with carbon quantum dots for efficient photocatalytic reduction of Cr(VI), *Chin. J. Catal.*, 39 (2018) 1901–1909.

[25] R. Long, N.J. English, Electronic structure of cation-codoped TiO$_2$ for visible-light photocatalyst applications from hybrid density functional theory calculations, *Appl. Phys. Lett.*, 98 (2011) 142103.

[26] T. Ochiai, K. Nakata, T. Murakami, A. Fujishima, Y. Yao, D.A. Tryk, Y. Kubota, Development of solar-driven electrochemical and photocatalytic water treatment system using a boron-doped diamond electrode and TiO$_2$ photocatalyst, *Water Res.*, 44 (2010) 904–910.

[27] K. Lalitha, G. Sadanandam, V.D. Kumari, M. Subrahmanyam, B. Sreedhar, N.Y. Hebalkar, Highly stabilized and finely dispersed Cu$_2$O/TiO$_2$: A promising visible sensitive photocatalyst for continuous production of hydrogen from glycerol: Water mixtures, *J. Phys. Chem. C*, 114 (2010) 22181–22189.

[28] Q. Li, T. Kako, J. Ye, PbS/CdS nanocrystal-sensitized titanate network films: Enhanced photocatalytic activities and super-amphiphilicity, *J. Mater. Chem.*, 20 (2010) 10187–10192.

[29] S. Fuldner, R. Mild, H.I. Siegmund, J.A. Schroeder, M. Gruber, B. Konig, Green-light photocatalytic reduction using dye-sensitized TiO$_2$ and transition metal nanoparticles, *Green Chem.*, 12 (2010) 400–406.

[30] J.J. Wu, C.H. Tseng, Photocatalytic properties of nc-Au/ZnO nanorod composites, *Appl. Catal. B: Environ.*, 66 (2006) 51–57.

[31] Y.Y. Liang, H.L. Wang, H. Sanchez Casalongue, Z. Chen, H.J. Dai, TiO$_2$ nanocrystals grown on graphene as advanced photocatalytic hybrid materials, *Nano Res.*, 3 (2010) 701–705.

[32] Q.J. Xiang, J.G. Yu, M. Jaroniec, Enhanced photocatalytic H$_2$-production activity of graphene-modified titania nanosheets, *Nanoscale*, 3 (2011) 3670–3678.

[33] T.G. Xu, L.W. Zhang, H.Y. Cheng, Y.F. Zhu, Significantly enhanced photocatalytic performance of ZnO via graphene hybridization and the mechanism study, *Appl. Catal. B: Environ.*, 101 (2011) 382–387.

[34] Y.H. Zhang, Z.R. Tang, X.Z. Fu, Y.J. Xu, TiO$_2$-graphene nanocomposites for gas-phase photocatalytic degradation of volatile aromatic pollutant: Is TiO$_2$–Graphene truly different from other TiO$_2$–Carbon composite materials?, *ACS Nano*, 2 (2010) 7303–7314.

[35] J.F. Shen, B. Yan, M. Shi, H.W. Ma, N. Li, M.X. Ye, One step hydrothermal synthesis of TiO$_2$-reduced graphene oxide sheets, *J. Mater. Chem.*, 21 (2011) 3415–3421.

[36] S. Liu, J. Wang, J. Zeng, J. Ou, Z. Li, X. Liu, S. Yang, "Green" electrochemical synthesis of Pt/graphene sheet nanocomposite film and its electrocatalytic property, *J. Power Sources*, 195 (2010) 4628–4633.

[37] J.L. Wu, X.P. Shen, L.J. Jiang, K. Wang, K.M. Chen, Solvothermal synthesis and characterization of sandwich-like graphene/ZnO nanocomposites, *Appl. Surf. Sci.*, 256 (2010) 2826–2830.

[38] W.B. Zou, J.W. Zhu, Y.X. Sun, X. Wang, Depositing ZnO nanoparticles onto graphene in a polyol system, *Mater. Chem. Phys.*, 125 (2011) 617–620.

[39] X.D. Huang, X.F. Zhou, L. Zhou, K. Qian, Y.H. Wang, Z.P. Liu, C.Z. Yu, A Facile one-step solvothermal synthesis of $SnO_2$/graphene nanocomposite and its application as an anode material for lithium-ion batteries, *ChemPhysChem*, 12 (2011) 278–281.

[40] K. Vinodgopal, B. Neppolian, I.V. Lightcap, F. Grieser, M. Ashokkumar, P.V. Kamat, Sonolytic design of graphene-Au nanocomposites. simultaneous and sequential reduction of graphene oxide and Au (III), *J. Phys. Chem. Lett.*, 1 (2010) 1987–1993.

[41] T.V. Cuong, V.H. Pham, Q.T. Tran, J.S. Chung, E.W. Shin, J.S. Kim, E.J. Kim, Optoelectronic properties of graphene thin films prepared by thermal reduction of graphene oxide, *Mater. Lett.*, 64 (2010) 765–767.

[42] X.L. Wang, S.O. Pehkonen, A.K. Ray, Removal of aqueous Cr (VI) by a combination of photocatalytic reduction and coprecipitation, *Ind. Eng. Chem. Res.*, 43 (2004) 1665–1672.

[43] H.P. Chu, X.J. Liu, B.B. Liu, G. Zhu, W.Y. Lei, H.G. Du, J.Y. Liu, J.W. Li, C. Li, C.Q. Sun, Hexagonal 2H-$MoSe_2$ broad spectrum active photocatalyst for Cr(VI) reduction, *Sci. Rep.*, 6 (2016) 35304.

[44] X.J. Liu, L.K. Pan, Q.F. Zhao, T. Lv, G. Zhu, T.Q. Chen, T. Lu, Z. Sun, C.Q. Sun, UV-assisted photocatalytic synthesis of ZnO-reduced graphene oxide composites with enhanced photocatalytic activity in reduction of Cr (VI), *Chem. Eng. J.*, 183 (2012) 238.

[45] G. Zhu, T. Xu, T. Lv, L.K. Pan, Q.F. Zhao, Z. Sun, Graphene-incorporated nanocrystalline $TiO_2$ films for CdS quantum dot-sensitized solar cells, *J. Electroanal. Chem.*, 650 (2011) 248–251.

[46] L. Liu, C. Luo, J. Xiong, Z.X. Yang, Y.B. Zhang, Y.X. Cai, H.S. Gu, Reduced graphene oxide (rGO) decorated $TiO_2$ microspheres for visible-light photocatalytic reduction of Cr(VI), *J. Alloys Compd.*, 690 (2017) 771–776.

[47] Y. Li, W.Q. Cui, L. Liu, R.L. Zong, W.Q. Yao, Y.H. Liang, Y.F. Zhu, Removal of Cr(VI) by 3D $TiO_2$–graphene hydrogel via adsorption enriched with photocatalytic reduction, *Appl. Catal. B: Environ.*, 199 (2016) 412–423.

[48] Y.J. Yuan, D.Q. Chen, X.F. Shi, J.R. Tu, B. Hu, L.X. Yang, Z.T. Yu, Z.G. Zou, Facile fabrication of "green" $SnS_2$ quantum dots/reduced graphene oxide composites with enhanced photocatalytic performance, *Chem. Eng. J.*, 313 (2017) 1438–1446.

[49] R. Wang, K.Q. Lu, F. Zhang, Z.R. Tang, Y.J. Xu, 3D carbon quantum dots/graphene aerogel as a metal-free catalyst for enhanced photosensitization efficiency, *Appl. Catal. B: Environ.*, 233 (2018) 11–18.

[50] L.K. Pan, X.J. Liu, Z. Sun, C.Q. Sun, Nanophotocatalysts via microwave-assisted solution-phase synthesis for efficient photocatalysis, *J. Mater. Chem. A*, 1 (2013) 8299–8326.

[51] X.J. Liu, L.K. Pan, T. Lv, T. Lu, G. Zhu, Z. Sun, C.Q. Sun, Microwave-assisted synthesis of ZnO-graphene nanocomposite for photocatalytic reduction of Cr(VI), *Catal. Sci. Technol.*, 1 (2011) 1189–1193.

[52] X.J. Liu, L.K. Pan, T. Lv, G. Zhu, T. Lu, Z. Sun, C.Q. Sun, Microwave-assisted synthesis of TiO$_2$-reduced graphene oxide composites for photocatalytic reduction of Cr(VI), *RSC Adv.*, 1 (2011) 1245–1249.

[53] X.J. Liu, L.K. Pan, T. Lv, G. Zhu, Z. Sun, C.Q. Sun, Microwave-assisted synthesis of CdS-reduced graphene oxide composites for photocatalytic reduction of Cr(VI), *Chem. Commun.*, 47 (2011) 11984–11986.

[54] P. Hu, X. Liu, B. Liu, L. Li, W. Qin, H. Yu, S. Zhong, Y. Li, Z. Ren, M. Wang, Hierarchical layered Ni$_3$S$_2$-graphene hybrid composites for efficient photocatalytic reduction of Cr(VI), *J. Colloid Interface Sci.*, 496 (2017) 254–260.

[55] X.J. Liu, L.K. Pan, T. Lv, Z. Sun, Investigation of photocatalytic activities over ZnO–TiO$_2$-reduced graphene oxide composites synthesized via microwave-assisted reaction, *J. Colloid Interface Sci.*, 394 (2013) 441–444.

[56] B.J. Li, H.Q. Cao, ZnO@graphene composite with enhanced performance for the removal of dye from water, *J. Mater. Chem.*, 21 (2011) 3346–3349.

[57] G. Zhu, L.K. Pan, T. Xu, Q.F. Zhao, Z. Sun, Cascade structure of TiO$_2$/ZnO/CdS film for quantum dot sensitized solar cells, *J. Alloys Compd.*, 509 (2011) 7814–7818.

[58] N.L. Yang, J. Zhai, D. Wang, Y.S. Chen, L. Jiang, Two-dimensional graphene bridges enhanced photoinduced charge transport in dye-sensitized solar cells, *ACS Nano*, 4 (2010) 887–894.

[59] J. Wu, B.B. Liu, Z.X. Ren, M.Y. Ni, C. Li, Y.Y. Gong, W. Qin, Y.L. Huang, C.Q. Sun, X.J. Liu, CuS/RGO hybrid photocatalyst for full solar spectrum photoreduction from UV/Vis to near-infrared light, *J. Colloid Interface Sci.*, 517 (2018) 80–85.

[60] Z.X. Ren, L. Li, B.B. Liu, X.J. Liu, Z. Li, X. Lei, C. Li, Y.Y. Gong, L.Y. Niu, L.K. Pan, Cr(VI) reduction in presence of ZnS/RGO photocatalyst under full solar spectrum radiation from UV/vis to near-infrared light, *Catal. Today*, 315 (2018) 46–51.

[61] Y.H. Zhang, N. Zhang, Z.R. Tang, Y.J. Xu, Graphene transforms wide band gap ZnS to visible light photocatalyst. The new role of graphene as a macromolecular photosensitizer, *ACS Nano*, 6 (2012) 9777–9789.

[62] F. Yang, J. Graciani, J. Evans, P. Liu, J. Hrbek, J.F. Sanz, J.A. Rodriguez, CO Oxidation on inverse CeO$_x$/Cu(111) catalysts: High catalytic activity and ceria-promoted dissociation of O$_2$, *J. Am. Chem. Soc.*, 133 (2011) 3444–3451.

[63] C.C. Yeh, P.R. Wu, D.H. Chen, Fabrication and near infrared photothermally-enhanced catalytic activity of Cu nanoparticles/reduced graphene oxide nanocomposite, *Mater. Lett.*, 136 (2014) 274–277.

[64] Z.X. Gan, X.L. Wu, M. Meng, X.B. Zhu, L. Yang, P.K. Chu, Photothermal contribution to enhanced photocatalytic performance of graphene-based nanocomposites, *ACS Nano*, 8 (2014) 9304–9310.

[65] G.M. Neelgund, A. Oki, Photothermal effect: An important aspect for the enhancement of photocatalytic activity under illumination by NIR radiation, *Mater. Chem. Front.*, 2 (2018) 64–75.

[66] T. Inoue, A. Fujishima, S. Konishi, K. Honda, Photoelectrocatalytic reduction of carbon dioxide in aqueous suspensions of semiconductor powders, *Nature*, 277 (1979) 637–638.

[67] L. Chen, M.L. Zhang, J.L. Yang, Y.X. Li, Y. Sivalingam, Q.J. Shi, M.Z. Xie, W.H. Han, Synthesis of $BiVO_4$ quantum dots/reduced graphene oxide composites for $CO_2$ reduction, *Mater. Sci. Semicond. Process.*, 102 (2019) 104578.

[68] N. Shehzad, M. Tahir, K. Johari, T. Murugesan, M. Hussain, Improved interfacial bonding of graphene-$TiO_2$ with enhanced photocatalytic reduction of $CO_2$ into solar fuel, *J. Electrochem. Sci. Eng.*, 6 (2018) 6947–6957.

[69] Z.C. Fu, R.C. Xu, J.T. Moore, F. Liang, X.C. Nie, C. Mi, J. Mo, Y. Xu, Q.Q. Xu, Z. Yang, Z.S. Lin, W.F. Fu, Highly efficient photocatalytic system constructed from CoP/Carbon nanotubes or graphene for visible-light-driven $CO_2$ reduction, *Chem. Eur. J.*, 24 (2018) 4273–4278.

[70] X. Li, J.G. Yu, M. Jaroniec, X.B. Chen, Cocatalysts for selective photoreduction of $CO_2$ into solar fuels, *Chem. Rev.*, 119 (2019) 3962–4179.

[71] W.G. Tu, Y. Zhou, Q. Liu, Z.P. Tian, J. Gao, X.Y. Chen, H.T. Zhang, J.G. Liu, Z.G. Zou, Robust hollow spheres consisting of alternating titania nanosheets and graphene nanosheets with high photocatalytic activity for $CO_2$ Conversion into renewable fuels, *Adv. Funct. Mater.*, 22 (2012) 1215–1221.

[72] J.G. Yu, J. Jin, B. Cheng, M. Jaroniec, A noble metal-free reduced graphene oxide-CdS nanorod composite for the enhanced visible-light photocatalytic reduction of $CO_2$ to solar fuel, *J. Mater. Chem. A*, 2 (2014) 3407–3416.

[73] F. Li, L. Zhang, J.C. Tong, Y.L. Liu, S.G. Xu, Y. Cao, S.K. Cao, Photocatalytic $CO_2$ conversion to methanol by $Cu_2O$/graphene/TNA heterostructure catalyst in a visible-light-driven dual-chamber reactor, *Nano Energy*, 27 (2016) 320–329.

[74] D. Mateo, A.M. Asiri, J. Albero, H. García, The mechanism of photocatalytic $CO_2$ reduction by graphene-supported $Cu_2O$ probed by sacrificial electron donors, *Photochem. Photobio. Sci.*, 17 (2018) 829–834.

[75] L. Hurtado, R. Natividad, H. García, Photocatalytic activity of $Cu_2O$ supported on multi layers graphene for $CO_2$ reduction by water under batch and continuous flow, *Catal. Commun.*, 84 (2016) 30–35.

[76] S.H. Liu, J.S. Lu, Y.C. Chen, Sustainable recovery of $CO_2$ by using visible-light-responsive crystal cuprous oxide/reduced graphene oxide, *Sustainability*, 10 (2018) 4145.

[77] X.Q. An, K.F. Li, J.W. Tang, $Cu_2O$/Reduced graphene oxide composites for the photocatalytic conversion of $CO_2$, *ChemSusChem*, 7 (2014) 1086–1093.

[78] S.H. Liu, J.S. Lu, Y.C. Pu, H.C. Fan, Enhanced photoreduction of $CO_2$ into methanol by facet-dependent $Cu_2O$/reduce graphene oxide, *J. $CO_2$ Util.*, 33 (2019) 171–178.

[79] X.B. Li, J. Xiong, X.M. Gao, J.T. Huang, Z.J. Feng, Z. Chen, Y.F. Zhu, Recent advances in 3D g-$C_3N_4$ composite photocatalysts for photocatalytic water splitting, degradation of pollutants and $CO_2$ reduction, *J. Alloys Compd.*, 802 (2019) 196–209.

[80] D.F. Xu, B. Cheng, W.K. Wang, C.J. Jiang, J.G. Yu, $Ag_2CrO_4$/g-$C_3N_4$/ graphene oxide ternary nanocomposite Z-scheme photocatalyst with enhanced $CO_2$ reduction activity, *Appl. Catal. B: Environ.*, 231 (2018) 368–380.

[81] R.Y. Zhang, Z.A. Huang, C.J. Li, Y.S. Zuo, Y. Zhou, Monolithic g-$C_3N_4$/reduced graphene oxide aerogel with *in situ* embedding of Pd nanoparticles for hydrogenation of $CO_2$ to $CH_4$, *Appl. Surf. Sci.*, 475 (2019) 953–960.

[82] W.J. Ong, L.L. Tan, S.P. Chai, S.T. Yong, A.R. Mohamed, Surface charge modification via protonation of graphitic carbon nitride (g-$C_3N_4$) for electrostatic self-assembly construction of 2D/2D reduced graphene oxide (rGO)/g-$C_3N_4$ nanostructures toward enhanced photocatalytic reduction of carbon dioxide to methane, *Nano Energy*, 13 (2015) 757–770.

[83] C. Han, Y.P. Lei, B. Wang, Y.D. Wang, *In situ*-fabricated 2D/2D heterojunctions of ultrathin SiC/reduced graphene oxide nanosheets for efficient $CO_2$ photoreduction with high $CH_4$ selectivity, *ChemSusChem*, 11 (2018) 4237–4245.

[84] Y.T. Liang, B.K. Vijayan, K.A. Gray, M.C. Hersam, Minimizing graphene defects enhances titania nanocomposite-based photocatalytic reduction of $CO_2$ for improved solar fuel production, *Nano Lett.*, 11 (2011) 2865–2870.

[85] A. Ali, M.R.U.D. Biswas, W.C. Oh, Novel and simple process for the photocatalytic reduction of $CO_2$ with ternary $Bi_2O_3$–graphene–ZnO nanocomposite, *J. Mater. Sci. Mater. Electron.*, 29 (2018) 10222–10233.

[86] X.W. Wang, Q.C. Li, C.X. Zhou, Z.Q. Cao, R.B. Zhang, ZnO rod/reduced graphene oxide sensitized by $\alpha$-$Fe_2O_3$ nanoparticles for effective visible-light photoreduction of $CO_2$, *J. Colloid Interface Sci.*, 554 (2019) 335–343.

[87] J.C. Meng, Q. Chen, J.Q. Lu, H. Liu, Z-Scheme Photocatalytic $CO_2$ reduction on a heterostructure of oxygen-defective ZnO/reduced graphene oxide/UiO-66-$NH_2$ under visible light, *ACS Appl. Mater. Interfaces*, 11 (2019) 550–562.

[88] Y.F. Yu, S.Y. Huang, Y.P. Li, S.N. Steinmann, W.T. Yang, L.Y. Cao, Layer-dependent electrocatalysis of $MoS_2$ for hydrogen evolution, *Nano Lett.*, 14 (2014) 553–558.

[89] W. Qin, T.Q. Chen, L.K. Pan, L.Y. Niu, B.W. Hu, D.S. Li, J.L. Li, Z. Sun, $MoS_2$-reduced graphene oxide composites via microwave assisted synthesis for sodium ion battery anode with improved capacity and cycling performance, *Electrochim. Acta*, 153 (2015) 55–61.

[90] Y.J. Chen, G.H. Tian, Y.H. Shi, Y.T. Xiao, H.G. Fu, Hierarchical $MoS_2/Bi_2MoO_6$ composites with synergistic effect for enhanced visible photocatalytic activity, *Appl. Catal. B: Environ.*, 164 (2015) 40–47.

[91] Y.H. Tan, K. Yu, T. Yang, Q.F. Zhang, W.T. Cong, H.H. Yin, Z.L. Zhang, Y.W. Chen, Z.Q. Zhu, The combinations of hollow $MoS_2$ micro@nanospheres: One-step synthesis, excellent photocatalytic and humidity sensing properties, *J. Mater. Chem. C*, 2 (2014) 5422–5430.

[92] W. Wei, Y. Dai, C.w. Niu, B.B. Huang, Controlling the electronic structures and properties of in-plane transition-metal dichalcogenides quantum wells, *Sci. Rep.*, 5 (2015) 17578.

[93] Q.H. Wang, K. Kalantar-Zadeh, A. Kis, J.N. Coleman, M.S. Strano, Electronics and optoelectronics of two-dimensional transition metal dichalcogenides, *Nat. Nanotechnol.*, 7 (2012) 699–712.

[94] M. Chhowalla, H.S. Shin, G. Eda, L.J. Li, K.P. Loh, H. Zhang, The chemistry of two-dimensional layered transition metal dichalcogenide nanosheets, *Nat. Chem.*, 5 (2013) 263–275.

[95] J. Mann, Q. Ma, P.M. Odenthal, M. Isarraraz, D. Le, E. Preciado, D. Barroso, K. Yamaguchi, G.V.S. Palacio, A. Nguyen, 2-dimensional transition metal dichalcogenides with tunable direct band gaps: $MoS_{2(1-x)}Se_{2x}$ monolayers, *Adv. Mater.*, 26 (2014) 1399–1404.

[96] D. James, T. Zubkov, Photocatalytic properties of free and oxide-supported $MoS_2$ and $WS_2$ nanoparticles synthesized without surfactants, *J. Photochem. Photobiol. A: Chem.*, 262 (2013) 45–51.

[97] Y.L. Min, G.Q. He, Q.J. Xu, Y.C. Chen, Dual-functional $MoS_2$ sheet-modified CdS branch-like heterostructures with enhanced photostability and photocatalytic activity, *J. Mater. Chem. A*, 2 (2014) 2578–2584.

[98] J.L. Li, X.J. Liu, L.K. Pan, W. Qin, T.Q. Chen, Z. Sun, $MoS_2$-reduced graphene oxide composites synthesized via a microwave-assisted method for visible-light photocatalytic degradation of methylene blue, *RSC Adv.*, 4 (2014) 9647–9651.

[99] H. Jung, K.M. Cho, K.H. Kim, H.W. Yoo, A. Al-Saggaf, I. Gereige, H.T. Jung, Highly efficient and stable $CO_2$ reduction photocatalyst with a hierarchical structure of mesoporous $TiO_2$ on 3D graphene with few-layered $MoS_2$, *ACS Sustain. Chem. Eng.*, 6 (2018) 5718–5724.

[100] M.R.U.D. Biswas, A. Ali, K.Y. Cho, W.C. Oh, Novel synthesis of $WSe_2$-graphene-$TiO_2$ ternary nanocomposite via ultrasonic technics for high photocatalytic reduction of $CO_2$ into $CH_3OH$, *Ultrason. Sonochem.*, 42 (2018) 738–746.

[101] Y.L. Zhao, Y.C. Wei, X.X. Wu, H.L. Zheng, Z. Zhao, J. Liu, J.M. Li, Graphene-wrapped $Pt/TiO_2$ photocatalysts with enhanced photogenerated charges separation and reactant adsorption for high selective photoreduction of $CO_2$ to $CH_4$, *Appl. Catal. B: Environ.*, 226 (2018) 360–372.

[102] Z.Z. Zhu, Y. Han, C.P. Chen, Z.X. Ding, J.L. Long, Y.D. Hou, Reduced Graphene oxide-cadmium sulfide nanorods decorated with silver nanoparticles for efficient photocatalytic reduction carbon dioxide under visible light, *ChemCatChem*, 10 (2018) 1627–1634.

[103] N. Karachi, M. Hosseini, Z. Parsaee, R. Razavi, Novel high performance reduced graphene oxide based nanocatalyst decorated with $Rh_2O_3$/Rh-NPs for $CO_2$ photoreduction, *J. Photochem. Photobiol. A: Chem.*, 364 (2018) 344–354.

[104] D. Mateo, J. Albero, H. García, Graphene supported NiO/Ni nanoparticles as efficient photocatalyst for gas phase $CO_2$ reduction with hydrogen, *Appl. Catal. B: Environ.*, 224 (2018) 563–571.

[105] Q. Li, B.D. Guo, J.G. Yu, J.G. Ran, B.H. Zhang, H.J. Yan, J.R. Gong, Highly efficient visible-light-driven photocatalytic hydrogen production of CdS-cluster-decorated graphene nanosheets, *J. Am. Chem. Soc.*, 133 (2011) 10878–10884.

[106] N. Zhang, Y. Zhang, X. Pan, X. Fu, S. Liu, Y.J. Xu, Assembly of CdS nanoparticles on the two-dimensional graphene scaffold as visible-light-driven photocatalyst for selective organic transformation under ambient conditions, *J. Phys. Chem. C*, 115 (2011) 23501–23511.

[107] N. Zhang, M.Q. Yang, Z.R. Tang, Y.J. Xu, CdS-graphene nanocomposites as visible light photocatalyst for redox reactions in water: A green route for selective transformation and environmental remediation, *J. Catal.*, 303 (2013) 60–69.

[108] N. Zhang, Y.H. Zhang, X.Y. Pan, M.Q. Yang, Y.J. Xu, Constructing ternary CdS-graphene-$TiO_2$ hybrids on the flatland of graphene oxide with enhanced visible-light photoactivity for selective transformation, *J. Phys. Chem. C*, 116 (2012) 18023–18031.

[109] N. Zhang, M.Q. Yang, Z.R. Tang, Y.J. Xu, Toward improving the graphene-semiconductor composite photoactivity via the addition of metal ions as generic interfacial mediator, *ACS Nano*, 8 (2014) 623–633.

[110] X.Z. Li, B. Weng, N. Zhang, Y.J. Xu, In situ synthesis of hierarchical $In_2S_3$-graphene nanocomposite photocatalyst for selective oxidation, *RSC Adv.*, 4 (2014) 64484–64493.

[111] M. Yang, B. Weng, Y. Xu, Improving the visible light photoactivity of $In_2S_3$-graphene nanocomposite via a simple surface charge modification approach, *Langmuir*, 29 (2013) 10549–10558.

[112] M.Q. Yang, X.Y. Pan, N. Zhang, Y.J. Xu, A facile one-step way to anchor noble metal (Au, Ag, Pd) nanoparticles on a reduced graphene oxide mat with catalytic activity for selective reduction of nitroaromatic compounds, *CrystEngComm*, 15 (2013) 6819–6828.

[113] W.J. Zhang, D.Z. Li, Z.X. Chen, M. Sun, W.J. Li, Q. Lin, X.Z. Fu, Microwave hydrothermal synthesis of $AgInS_2$ with visible light photocatalytic activity, *Mater. Res. Bull.*, 46 (2011) 975–982.

[114] I. Sopyan, M. Watanabe, S. Murasawa, K. Hashimoto, A. Fujishima, An efficient $TiO_2$ thin-film photocatalyst: Photocatalytic properties in gas-phase acetaldehyde degradation, *J. Photochem. Photobiol. A: Chem.*, 98 (1996) 79–86.

[115] T.X. Wu, G.M. Liu, J.C. Zhao, H. Hidaka, N. Serpone, Photoassisted degradation of dye pollutants. V. Self-photosensitized oxidative transformation

of rhodamine B under visible light irradiation in aqueous $TiO_2$ dispersions, *J. Phys. Chem. B*, 102 (1998) 5845–5851.

[116] J.G. Yu, G.P. Dai, B.B. Huang, Fabrication and characterization of visible-light-driven plasmonic photocatalyst $Ag/AgCl/TiO_2$ nanotube arrays, *J. Phys. Chem. C*, 113 (2009) 16394–16401.

[117] M. Pelaez, N.T. Nolan, S.C. Pillai, M.K. Seery, P. Falaras, A.G. Kontos, P.S.M. Dunlop, J.W.J. Hamilton, J.A. Byrne, K. O'Shea, A review on the visible light active titanium dioxide photocatalysts for environmental applications, *Appl. Catal. B: Environ.*, 125 (2012) 331–349.

[118] S. Ahmed, M.G. Rasul, R. Brown, M.A. Hashib, Influence of parameters on the heterogeneous photocatalytic degradation of pesticides and phenolic contaminants in wastewater: A short review, *J. Environ. Manage.*, 92 (2011) 311–330.

[119] M.E. Simonsen, H. Jensen, Z. Li, E.G. Sogaard, Surface properties and photocatalytic activity of nanocrystalline titania films, *J. Photochem. Photobiol. A: Chem.*, 200 (2008) 192–200.

[120] H.T. Hu, X.B. Wang, F.M. Liu, J.C. Wang, C.H. Xu, Rapid microwave-assisted synthesis of graphene nanosheets-zinc sulfide nanocomposites: Optical and photocatalytic properties, *Synthetic Met.*, 161 (2011) 404–410.

[121] T. Lv, L.K. Pan, X.J. Liu, T. Lu, G. Zhu, Z. Sun, Enhanced photocatalytic degradation of methylene blue by ZnO-reduced graphene oxide composite synthesized via microwave-assisted reaction, *J. Alloys Compd.*, 509 (2011) 10086–10091.

[122] T. Lv, L.K. Pan, X.J. Liu, T. Lu, G. Zhu, Z. Sun, C.Q. Sun, One-step synthesis of $CdS-TiO_2$-chemically reduced graphene oxide composites via microwave-assisted reaction for visible-light photocatalytic degradation of methyl orange, *Catal. Sci. Technol.*, 2 (2012) 754–758.

[123] I.V. Lightcap, T.H. Kosel, P.V. Kamat, Anchoring semiconductor and metal nanoparticles on a two-dimensional catalyst mat. Storing and shuttling electrons with reduced graphene oxide, *Nano Lett.*, 10 (2010) 577–583.

[124] Y. Lin, Z.G. Geng, H.B. Cai, L. Ma, J. Chen, J. Zeng, N. Pan, X.P. Wang, Ternary graphene–$TiO_2$–$Fe_3O_4$ nanocomposite as a recollectable photocatalyst with enhanced durability, *Eur. J. Inorg. Chem.*, 2012 (2012) 4439–4444.

[125] Q.J. Xiang, J.G. Yu, M. Jaroniec, Graphene-based semiconductor photocatalysts, *Chem. Soc. Rev.*, 41 (2012) 782–796.

[126] J.T. Zhang, Z.G. Xiong, X.S. Zhao, Graphene-metal-oxide composites for the degradation of dyes under visible light irradiation, *J. Mater. Chem.*, 21 (2011) 3634–3640.

[127] W.Q. Fan, Q.H. Zhang, Y. Wang, Semiconductor-based nanocomposites for photocatalytic $H_2$ production and $CO_2$ conversion, *Phys. Chem. Chem. Phys.*, 15 (2013) 2632–2649.

[128] N. Zhang, Y.H. Zhang, Y.J. Xu, Recent progress on graphene-based photocatalysts: Current status and future perspectives, *Nanoscale*, 4 (2012) 5792–5813.

[129] N. Zhang, Y.H. Zhang, X.Y. Pan, X.Z. Fu, S.Q. Liu, Y.J. Xu, Assembly of CdS nanoparticles on the two-dimensional graphene scaffold as visible-light-driven photocatalyst for selective organic transformation under ambient conditions, *J. Phys. Chem. C*, 115 (2011) 23501–23511.

[130] Y.J. Xu, Y. Zhuang, X.Z. Fu, New insight for enhanced photocatalytic activity of $TiO_2$ by doping carbon nanotubes: A case study on degradation of benzene and methyl orange, *J. Phys. Chem. C*, 114 (2010) 2669–2676.

[131] Y.H. Zhang, N. Zhang, Z.R. Tang, Y.J. Xu, Improving the photocatalytic performance of graphene-$TiO_2$ nanocomposites via a combined strategy of decreasing defects of graphene and increasing interfacial contact, *Phys. Chem. Chem. Phys.*, 14 (2012) 9167–9175.

[132] Z. Chen, N. Zhang, Y.J. Xu, Synthesis of graphene-ZnO nanorod nanocomposites with improved photoactivity and anti-photocorrosion, *CrystEngComm*, 15 (2013) 3022–3030.

[133] S.G. Pan, X.H. Liu, CdS-Graphene nanocomposite: Synthesis, adsorption kinetics and high photocatalytic performance under visible light irradiation, *New J. Chem.*, 36 (2012) 1781–1787.

[134] X.G. Chen, Y.Q. He, Q. Zhang, L.J. Li, D.G. Hu, T. Yin, Fabrication of sandwich-structured ZnO/reduced graphite oxide composite and its photocatalytic properties, *J. Mater. Sci.*, 45 (2010) 953–960.

[135] H. Zhang, X.J. Lv, Y.M. Li, Y. Wang, J.H. Li, P25-graphene composite as a high performance photocatalyst, *ACS Nano*, 4 (2010) 380–386.

[136] N. Zhang, Y.H. Zhang, M.Q. Yang, Z.R. Tang, Y.J. Xu, A critical and benchmark comparison on graphene-, carbon nanotube-, and fullerene-semiconductor nanocomposites as visible light photocatalysts for selective oxidation, *J. Catal.*, 299 (2013) 210–221.

[137] M.Q. Yang, N. Zhang, Y.J. Xu, Synthesis of fullerene-, carbon nanotube-, and graphene-$TiO_2$ nanocomposite photocatalysts for selective oxidation: A comparative study, *ACS Appl. Mater. Interfaces*, 5 (2013) 1156–1164.

[138] M.Q. Yang, Y.J. Xu, Basic principles for observing the photosensitizer role of graphene in the graphene-semiconductor composite photocatalyst from a case study on graphene-ZnO, *J. Phys. Chem. C*, 117 (2013) 21724–21734.

# PART II
# Functional Nanomaterials for Solar Cells

Chapter 4

# Development of Antimony Selenide Thin-Film Solar Cells

Zhiqiang Li

*National-Local Joint Engineering Laboratory
of New Energy Photoelectric Devices,
College of Physics Science and Technology,
Hebei University, Baoding 071002, China
Hebei Key Laboratory of Optic-Electronic
Information and Materials,
College of Physics Science and Technology,
Hebei University, Baoding 071002, China*

## 1. Introduction

Antimony selenide, a simple binary inorganic compound semiconductor with a chemical formula $V_2$–$VI_3$ [1], has garnered much attention in recent times due to its superior material properties: having a single and fixed phase [2], suitable bandgap, high absorption coefficient at a visible region [3], low toxicity and earth-abundant constituents [4]. Moreover, $Sb_2Se_3$ was reported to have a 1D nanoribbon grain structure comprising covalently bonded $(Sb_4Se_6)_n$ ribbons held together by Van der Waals forces [5, 6]. Theoretical calculation and experimental characterizations confirmed that $Sb_2Se_3$ has benign grain boundaries (GBs) if properly aligned along the ribbon direction. In recent years, through steady improvement in the growth of these aligned $Sb_2Se_3$ films, the power conversion efficiency of $Sb_2Se_3$ has increased to 9.2% [5, 7], which is approaching the efficiency of other inorganic semiconductor compound thin-film solar cells such as copper zinc tin sulfide (CZTS) [8, 9]. However, the conversion efficiency is still far behind copper indium gallium selenide (CIGS, 22.9%) [10, 11] and cadmium telluride (CdTe, 22.1%) [12, 13]. This chapter will review the progress of $Sb_2Se_3$ thin-film solar cells and discuss the issues of $Sb_2Se_3$ thin-film material features and solar cell device physics.

## 2. Device fabrication techniques

### 2.1. *Non-vacuum deposition methods*

Since the minority carrier (electron) diffusion length was beyond several hundreds of nanometers, the $Sb_2Se_3$ compound was appropriate for application of planar heterojunction thin-film solar cells. The first report of $Sb_2Se_3$-based thin-film solar cells was from Prof. Nair's group as early as 2006 [14]. The $Sb_2Se_3$ thin films were obtained via a multistep process. At first, 300 nm of Se thin films were deposited by a chemical deposition method from a solution of sodium selenosulfate. The selenium precursor thin films became crystalline and photoconductive after annealing for 15 min at 150–200°C. Then, about 200 nm of Sb thin film were thermally evaporated onto the Se thin films to obtain the Sb–Se films. Finally, a thermal treatment process was carried out in nitrogen ($N_2$), which led to the formation of the $Sb_2Se_3$ compound after the reaction of Se and Sb. $J–V$ characteristics of the film in the dark and under illumination illustrated that the electrical conductivity of the films in the dark is $10^{-4}$ $(\Omega \text{ cm})^{-1}$ and the photosensitivity was about 4 under an illumination intensity of $1 \text{ kW m}^{-2}$. In 2009, they applied a chemical bath deposition (CBD) method to obtain an $Sb_2Se_3$-based absorber layer and fabricated the cell in the structure of $TCO/CdS/Sb_2Se_3$: $Sb_2O_3/PbS$. The cell had a $V_{oc}$ of 520 mV and $J_{sc}$ of 4.2 mA cm$^{-2}$ and conversion efficiency of 0.66% and the scalability of the cells was demonstrated in four series-connected cells of area $1 \text{ cm}^2$ each, showing a $J_{sc}$ of 1.5 mA and $V_{oc}$ of 1.9 V under sunlight [15].

An encouraging device efficiency of 2.26% was obtained in 2014 by Tang's group [2]. They fabricated a device-quality $Sb_2Se_3$ thin film from a hydrazine-based solution process. The fabrication process is

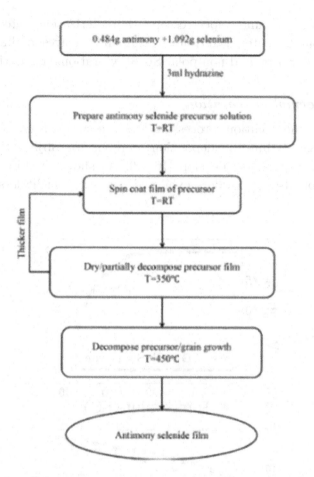

**Fig. 1.** The schematic flow diagram for the detailed procedure of $Sb_2Se_3$ film deposition with non-vacuum deposition methods.

*Source*: Reproduced with permission from Ref. [2] Copyright 2018, Advanced Energy Materials.

shown in Fig. 1. The first step for the preparation of $Sb_2Se_3$ solution was mixing Sb and Se powders in hydrazine under continuous magnetic stirring inside a $N_2$-filled glove box, finally yielding a clear blood-red solution. The $Sb_2Se_3$ thin films were prepared by spin coating of the $Sb_2Se_3$ solution and a post-annealing process. X-ray diffraction (XRD) measurement displayed that major peaks were indexed to the diffraction planes and no secondary phase or impurity was detected. Raman spectroscopy characterization detected two Raman bands centered at 189 and 252 cm$^{-1}$, corresponding to the heteropolar Sb–Se and non-polar Sb–Sb vibrations, respectively.

## 2.2. *Thermal evaporation*

Although the solution processing was possibly of low cost, the vacuum-based thin-film deposition methods for Sb$_2$Se$_3$ thin films were also required to develop [16–18]. As shown in Fig. 2, TGA analysis of Sb$_2$Se$_3$ powders exhibited that the weight loss started

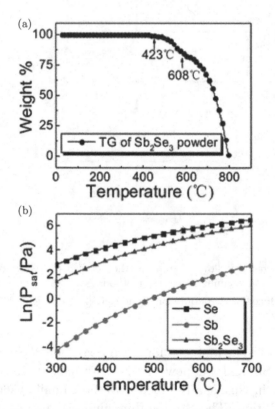

**Fig. 2.** (a) Thermogravimetric analysis (TGA) of Sb$_2$Se$_3$ powder (b) temperature-dependent saturated vapor pressure of Sb, Se and Sb$_2$Se$_3$ in the temperature range from 300°C to 700°C.

*Source*: Reproduced with permission from Ref. [18]. Copyright 2018, ACS Appl Mater Interfaces.

at $\sim$423°C, and sharply increased after 608°C [18]. It indicated that rapid evaporation would occur when $Sb_2Se_3$ was heated up above 423°C. The relative weight-loss temperature suggested that the thermal evaporation of $Sb_2Se_3$ powders might be a proper method for fabrication of $Sb_2Se_3$ thin films. Liu *et al.* obtained rod-like $Sb_2Se_3$ films at a substrate temperature of 290°C by thermal evaporation of $Sb_2Se_3$ powders at an average rate of $\sim$10 Å/s. The XRD pattern of the film showed no preferred orientation. Further investigation of the evaporation of $Sb_2Se_3$ showed that $Sb_2Se_3$ could decompose during thermal evaporation. In the temperature range of 677–822 K, partial $Sb_2Se_3$ decomposed into $Sb_4$ and $Se_2$. By studying the temperature-dependent vapor pressure of $Sb_2Se_3$, element Sb and element Se, some points were concluded. First, $Sb_2Se_3$ had a very high vapor pressure at elevated temperature, 22.5 Pa at 400°C and $3.48 \times 10^3$ Pa at 600°C, confirming its easy evaporation as stated before. Second, for all of the temperature ranges investigated, Se has a larger vapor pressure than $Sb_2Se_3$, and a far larger vapor pressure than Sb. Thus, the slight decomposition of $Sb_2Se_3$ during thermal evaporation would lead to the formation of Se-deficient $Sb_2Se_{3-x}$ product as well as Se vacancy ($V_{Se}$) defect, which was a deep recombination center in the $Sb_2Se_3$ absorber [7].

Two solutions were developed to address this issue [19–21]. Leng *et al.* carried out an additional selenization step in the thermal evaporation process [21]. The solar cell efficiencies of the devices with and without selenization treatment are summarized in Table 1, where 18 devices of each type were analyzed for statistics. For the device without selenization, the average $V_{oc}$ was 0.28 V, $J_{sc}$ was 21.3 mA cm$^{-2}$ and FF was 40.1%, corresponding to a power conversion efficiency of 2.4%. For the $Sb_2Se_3$ solar cells selenized at different substrate temperatures (170°C, 200°C and 230°C), all devices showed substantially increased $V_{oc}$ and FF as compared with the unselenized devices. Selenization at 200°C was found to be the optimal condition, producing an average 3.4% efficiency. The average efficiency of 230°C selenized samples decreased to 1.9% due to high $J_{sc}$ loss.

Admittance spectroscopy (AS) and capacitance–voltage ($C$–$V$) measurements exhibited that the depletion width of the controlled

**Table 1.** Statistics of device performance for the $Sb_2Se_3$ solar cells with selenization at different substrate temperatures or without selenization.

| Treatment | $V_{oc}$ (V) | $J_{sc}$ (mA cm$^{-2}$) | FF (%) | Efficiency (%) |
|---|---|---|---|---|
| Control sample | $0.28 \pm 0.01$ | $21.3 \pm 0.6$ | $40.1 \pm 3.1$ | $2.4 \pm 0.2$ |
| Selenization at 170°C | $0.30 \pm 0.01$ | $21.5 \pm 2.5$ | $42.5 \pm 1.5$ | $2.7 \pm 0.4$ |
| Selenization at 200°C | $0.32 \pm 0.01$ | $23.7 \pm 1.3$ | $44.7 \pm 2.8$ | $3.4 \pm 0.3$ |
| Selenization at 230°C | $0.34 \pm 0.01$ | $12.6 \pm 2.6$ | $43.7 \pm 1.7$ | $1.9 \pm 0.4$ |

*Source*: Reproduced with permission from Ref. [21]. Copyright 2018, Applied Physics Letters.

$Sb_2Se_3$ (C-$Sb_2Se_3$) device was estimated to be 287 nm, while the selenized $Sb_2Se_3$ (S-$Sb_2Se_3$) showed a higher one of 343 nm. Moreover, $C$–$V$ profiling indicated that the doping density in the C-$Sb_2Se_3$ device was substantially unchanged ($1.3 \times 10^{16}$–$2.4 \times 10^{16}$ cm$^{-3}$) in the depletion width, while doping density in the S-$Sb_2Se_3$ device was ($1.7 \times 10^{16}$–$2.6 \times 10^{16}$ cm$^{-3}$). The slightly enhanced doping density increased the built-in potential and boosted device $V_{oc}$. For the $Sb_2Se_3$ film made by thermal evaporation, $Sb_2Se_3$ decomposed during evaporation and the decomposition product Se had much higher vapor pressure than Sb and $Sb_2Se_3$ and produced a Se-poor $Sb_2Se_3$ films. Se vacancies were n-type donors, which not only reduce the effective p-type doping density of $Sb_2Se_3$ film but also act as recombination centers annihilating photogenerated carriers. An additional selenization step provided excess Se to compensate the $V_{Se}$, and hence increases doping density and attenuates recombination loss, as evidenced by the enhanced photoresponse in the increased photocurrent. Selenization also improved CdS–$Sb_2Se_3$ junction quality and produced a device with better $V_{oc}$ and FF.

Another solution to the Se deficiency for thermally evaporated $Sb_2Se_3$ was the online Se compensation process, which supplied a Se-rich environment by introducing an additional Se source during the evaporation of $Sb_2Se_3$ [19, 20]. The theoretical simulation predicted that different kinds of point defects exist in the $Sb_2Se_3$ lattice under Se-rich and Se-poor conditions. As shown in Fig. 3, five intrinsic defects, including the Se vacancy $V_{Se}$, Sb vacancy $V_{Sb}$, $Se_{Sb}$ antisite (Se replacing Sb), $Sb_{Se}$ antisite and Se interstitial $Se_i$,

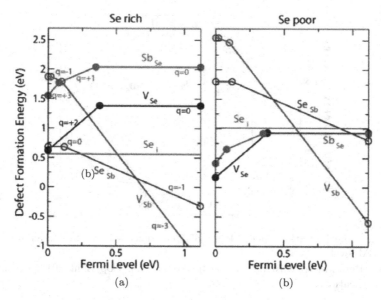

**Fig. 3.** The theoretical simulation predicted that different kinds of point defects exist in the $Sb_2Se_3$ lattice under Se-rich and Se-poor conditions.

*Source*: Reproduced with permission from Ref. [19]. Copyright 2018, Research and Applications.

had formation energies lower than 2 eV. Under the Se-rich condition, $Se_{Sb}$ and $Se_i$ had formation energies as low as 0.6 eV, when they are neutral. $Se_{Sb}$ antisite was a shallow acceptor defect, with the $(-1/0)$ ionization level located as 0.12 eV above the valence band maximum (VBM), as shown in Fig. 4. The high concentration of $Se_{Sb}$ contributes to the p-type conductivity under the Se-rich condition, while the high concentration of $V_{Se}$ and $Sb_{Se}$ does not make the sample show n-type conductivity under the Se-poor condition. Its low formation energy of 0.6 eV corresponds to an equilibrium concentration in the order of $10^{17}$ cm$^{-3}$ at a 400°C substrate temperature and would generate $10^{15}$ cm$^{-3}$ hole carriers at room temperature if the antisite defects were fully ionized [22]. The cation-replace-anion or anion-replace-cation antisite defects had high concentration in $Sb_2Se_3$, which may be due to the special 1D crystal structure of $Sb_2Se_3$. It may also have more tolerance to the intrinsic defects. The anion interstitial defects $Se_i$ also had low formation energy and high

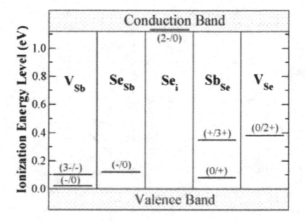

**Fig. 4.** The ionization energy levels of point defects in $Sb_2Se_3$, referenced to the VBM level.
*Source*: Reproduced with permission from Ref. [19]. Copyright 2018, Research and Applications.

concentration in $Sb_2Se_3$ lattice with the void region located between the $Sb_2Se_3$ atomic ribbons. On the contrary, the acceptor defect $Se_{Sb}$ had much lower formation energy and higher concentration compared with the donor defects $V_{Se}$ and $Sb_{Se}$. Thus, $Se_{Sb}$ was the dominant acceptor defect that generated free hole carriers under the Se-rich condition.

Under the Se-poor condition, $Se_{Sb}$ and $Se_i$ were no longer the defects with the lowest formation energy, while the donor defects $V_{Se}$ and $Sb_{Se}$ decreased their formation energies significantly to be slightly lower than 1 eV, as shown in Fig. 3(b) [23]. Despite the low formation energy at the neutral state, they were not easily ionized, with the ionization energy level deep in the bandgap and close to the VBM level, so their contribution to the free carrier concentration could be ignored. As the location of the Fermi level was the result of a competition between the donor and accepter defects, the $Sb_2Se_3$ samples deposited under the Se-poor condition should show intrinsic poor conductivity with low carrier concentration. In short, the density functional calculation results revealed that longer minority carrier lifetime, better p-type conductivity and better conversion efficiency could be achieved under the Se-rich condition.

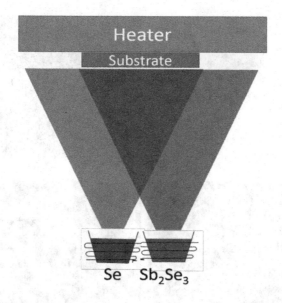

**Fig. 5.** Schematic illustration of the co-evaporation of $Sb_2Se_3$ and Se sources.

Li *et al.* and Liu *et al.* fabricated the $Sb_2Se_3$ thin films with on-line supply of additional Se vapor [19, 20]. As shown in Fig. 5, an additional Se source was introduced and it would supply different Se-rich conditions for deposition of $Sb_2Se_3$ thin films. In comparison with the $Sb_2Se_3$ films without any Se compensation, analysis of XRD patterns suggested that the evaporation of excessive Se enhanced the preferred orientation in the selected orientation directions and a higher Se flux would further strengthen this preference. Besides, Se compensation also reduced the surface roughness and improved the grain size.

### 2.3. *Magnetron sputtering of $Sb_2Se_3$*

Magnetron sputtering, as a widely used vacuum deposition technique, has also been investigated to fabricate the $Sb_2Se_3$ thin films. Liang *et al.* reported the sputtering route to grow uniform, well-crystallized, regular-shape $Sb_2Se_3$ from an $Sb_2Se_3$ (40:60 at.%) ceramic target [24, 25]. The morphology, film composition and crystallization changed significantly at different substrate temperatures. The sample

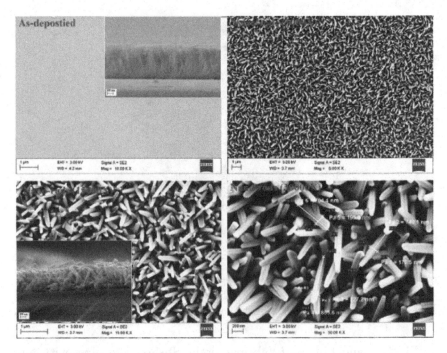

**Fig. 6.** The surface morphology and cross-sectional images of amorphous and Sb₂Se₃ nanorods with magnetron sputtering of Sb₂Se₃.

*Source*: Reproduced with permission from Ref. [24]. Copyright 2018, Solar Energy Materials and Solar Cells.

was amorphous, whereas it was well-crystallized nanorod arrays at a substrate temperature of 375°C, as shown in Fig. 6. For the amorphous sample, the ratio of Sb/Se approached the stoichiometric composition of 0.66, indicating a composition transfer from the Sb₂Se₃ target. On the contrary, for the Sb₂Se₃ thin films deposited at high temperature, an important loss of Se was observed, probably due to partial Se sublimation at a relatively high substrate temperature. Moreover, a large amount of nanorods randomly distributed on the film surface was observed from the low-magnification SEM image. The high-magnification SEM image revealed that the nanorod films consisted of regular shape and vertically arrayed nanorods with a length of about 0.5–1 μm and a width of about 0.12–0.20 μm.

Liang *et al.* ascribed the formation of nanorod array to the thermally induced structural evolution. During the sputtering process, $Ar^+$ generated by ionization was directed at the $Sb_2Se_3$ target to sputter the atoms. The ejected atoms with certain kinetic energy were transported to the substrate, and then these atoms condensed to form the amorphous film. With heated substrate, the deposited atoms with higher energy transferred sufficient momentum, allowing the rearranging of the sputtered atoms, leading to an improvement in quality and crystallinity of the sputtered films. If the substrate was at a proper temperature, the situation would be favorable for the fine grains of $Sb_2Se_3$ to grow with preferential growth in the *c*-axis orientation, leading to the formation of $Sb_2Se_3$ nanorod arrays.

Yuan *et al.* developed the $Sb_2Se_3$ thin-film deposition technique by sputtering of Sb metal precursor and a post-selenization process, where the selenization process was performed in a rapid heating furnace using element Se vapor [16, 26]. This two-step process of $Sb_2Se_3$ thin films reduced the Se vacancy defects in the $Sb_2Se_3$ films. They systemically investigated the effects of annealing temperature on selenization of Sb metal precursor. Figure 7(a) shows the XRD patterns of $Sb_2Se_3$ thin films prepared by the normal thermal process (NTP) and the rapid thermal process (RTP). The diffraction peaks for both the NTP and RTP samples could be corresponding to the $Sb_2Se_3$ orthorhombic phase. NTP-$Sb_2Se_3$ had more diffraction peaks than RTP-$Sb_2Se_3$, indicating that RTP resulted in more diffraction peaks than the NTP process. The peaks corresponding to (020), (120) and (240) for NTP-$Sb_2Se_3$ were much stronger than those for RTP-$Sb_2Se_3$, suggesting that these diffraction peaks were suppressed in the RTP process. As shown in Fig. 7(b), NTP-$Sb_2Se_3$ and RTP-$Sb_2Se_3$ had very different grain shapes. Grains were long, rod-like for the NTP sample, while the grains seemed to be spherical for the RTP sample. Furthermore, the annealing temperatures in the RTP treatment also influenced the morphologies of $Sb_2Se_3$. In the range of 320–440°C, the surface features of grains for all samples were spherical. As the annealing temperature increased, the grains grew larger, as shown in Fig. 8.

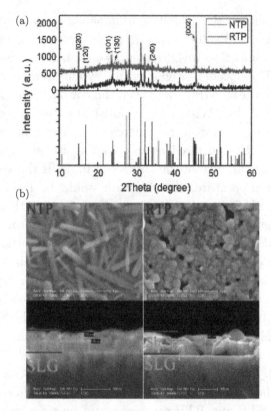

**Fig. 7.** The (a) XRD and (b) SEM of $Sb_2Se_3$ prepared with magnetron sputtering.

*Source*: Reproduced with permission from Ref. [16]. Copyright 2018, Solar Energy.

## 2.4. *Rapid thermal evaporation deposition technique*

Due to the low melting point of 608°C and high saturated vapor pressure (~1200 Pa at 550°C) properties of $Sb_2Se_3$, Zhou *et al.* developed a rapid thermal evaporation (RTE) technique for the deposition of $Sb_2Se_3$ films [5]. The schematic diagram of the RTE equipment is shown in Fig. 9. The $Sb_2Se_3$ powder and FTO–CdS substrate were put in a tube furnace with a high ramp rate. The distance and the temperature distance between the $Sb_2Se_3$ source and the substrate were controlled by the spacer. Both the evaporation source and substrate were heated by the ramp heater.

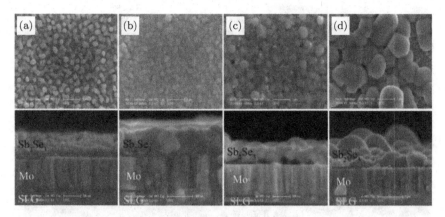

**Fig. 8.** SEM of Sb₂Se₃ films under different anneal temperatures (a) 320°C, (b) 360°C, (c) 400°C and (d) 440°C.

*Source*: Reproduced with permission from Ref. [16]. Copyright 2018, Solar Energy.

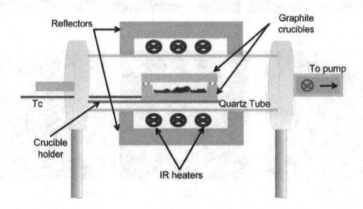

**Fig. 9.** The schematic diagram of RTE equipment.

*Source*: Reproduced with permission from Ref. [5]. Copyright 2018, Nature Photonics.

Once heated, the Sb₂Se₃ powder evaporated and condensed on the substrate to form the Sb₂Se₃ film due to the temperature gradient. The distance between the evaporating source and substrate was kept at a low value of 0.8 cm in the experiment to enable high material usage and a fast deposition rate. The deposition rate could reach as high as $1\,\mu\text{m}\,\text{min}^{-1}$, much greater than that of regular

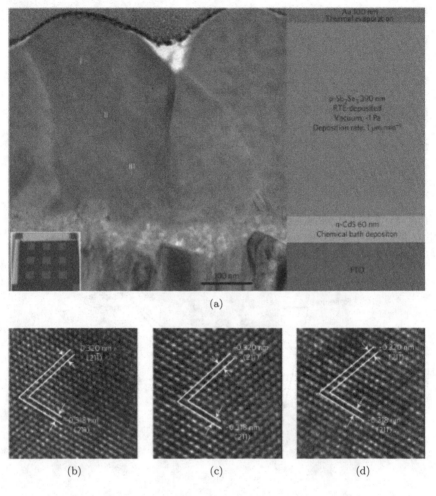

**Fig. 10.** Structure and TEM analysis of Sb₂Se₃ films and devices.
*Source*: Reproduced with permission from Ref. [5]. Copyright 2018, Nature
Photonics.

thermal evaporation (typically $0.01$–$0.1\,\mu\mathrm{m\,min^{-1}}$) or sputtering
(typically $0.01$–$0.05\,\mu\mathrm{m\,min^{-1}}$). An $\sim$390-nm-thick Sb₂Se₃ absorber
layer deposited on the FTO–CdS substrate was obtained by the RTE
process. The cross-sectional TEM image showed that the Sb₂Se₃ film
was compact and composed of large Sb₂Se₃ grains of a size equal to
the film thickness, as shown in Fig. 10(a). They applied HRTEM

to analyze three arbitrarily selected points (I, II and III in Fig. 10) with the corresponding lattice fringes shown in Figs. 10(b)–10(d). The distances between lattice lines were measured as 0.320 nm and 0.318 nm, which corresponded to the separation of the (221) and (211) planes in orthorhombic $Sb_2Se_3$. The crystal planes extended continuously from the top to the bottom of the active region of the devices, which confirmed that this grain was single crystalline. This trend was confirmed with further studies of multiple grains, reinforcing the picture that the $Sb_2Se_3$ films were made up of single-crystalline grains.

Zhou *et al.* further performed composition analysis of $Sb_2Se_3$ thin films deposited onto the substrate at a temperature of 300°C and 350°C. The EDX spectroscopy analysis showed that the measured molar ratio between Se and Sb was exactly the same, 1.515, for the samples deposited at 300°C and 350°C, respectively, indicating that the films were actually slightly Se rich. It also indicated that $Sb_2Se_3$ decomposition occurred during the evaporation because the evaporating source experienced a high temperature. Se with a slightly larger value than the stoichiometric ratio evaporated from the source and condensed into the film, leading to the Se-rich film.

The substrate temperature was found to be a very important factor for controlling the orientational preference of the $Sb_2Se_3$ thin films. For the $Sb_2Se_3$ deposited at 300°C, the (120) peak was much weaker than the (211) peak, whereas in the films deposited at 350°C, the intensities of these two peaks were comparable with each other. The texture coefficient of the peaks was calculated to measure film orientation. As shown in Fig. 11, the TC values of the (hk0) planes were close to zero, and the TC values of (hkl, l ≠ 0) were larger than 1 in the sample deposited at 300°C. It indicated that [hk0] orientated grains were suppressed in this Sb2Se3 films. In contrast, diffraction peaks for the [hk0] orientated grains with poor carrier transport were strengthened in the $Sb_2Se_3$ films deposited at 350°C, suggesting very poor carrier transport capacity. Moreover, the properties of GBs in the RTE-processed $Sb_2Se_3$ films were studied by using Kelvin probe force microscope (KPFM) and electron beam-induced current (EBIC) measurements. Two-dimensional topography spatial maps

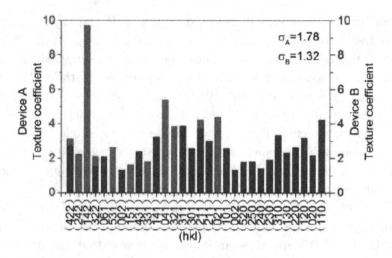

**Fig. 11.**   The TC values of Sb$_2$Se$_3$ films and devices.

*Source*: Reproduced with permission from Ref. [5]. Copyright 2018, Nature Photonics.

and the corresponding surface potentials of Sb$_2$Se$_3$ thin films, as shown in Figs. 12(a) and 12(b), revealed that there was no correlation between GBs and the substantial potential variation in the KPFM image. In an illustrative line scan crossing the GBs in Fig. 12(c), the surface potential difference between two grains was as low as 10 mV, which indicated a lack of significant band bending and surface defects in the Sb$_2$Se$_3$ films.

## 3.  Solar cell configurations and interfaces

As an emerging and attractive light absorber for photovoltaic devices, significant progress of Sb$_2$Se$_3$ solar cells has been made in superstrate and substrate planar thin-film heterojunction configurations during the past few years. By combining the Sb$_2$Se$_3$ absorber layer with another semiconductor material, Sb$_2$Se$_3$-based heterojunctions were constructed. Because Sb$_2$Se$_3$ was p-type, the corresponding n-type semiconductors, such as CdS, ZnO and SnO$_2$, were usually employed as buffer layer in the heterojunction devices.

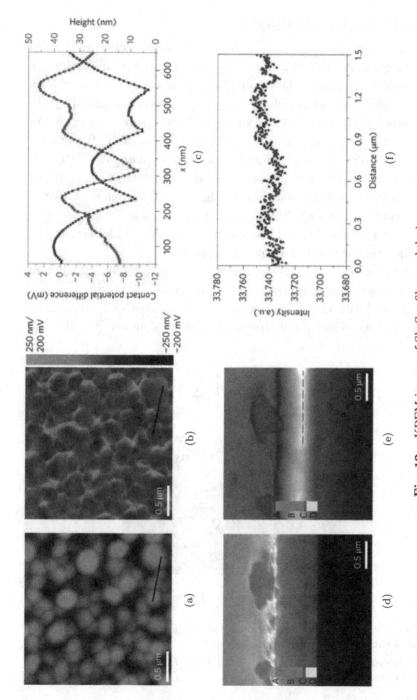

**Fig. 12.** KPFM images of Sb$_2$Se$_3$ films and devices.

*Source:* Reproduced with permission from Ref. [5]. Copyright 2018, Nature Photonics.

### 3.1. *CdS–Sb$_2$Se$_3$ heterojunction*

Based on the CdS buffer layer, there was tremendous progress in Sb$_2$Se$_3$ solar cells. In 2014, Luo *et al.* presented the fabrication of superstrate CdS–Sb$_2$Se$_3$ heterojunction solar cells [7]. The Sb$_2$Se$_3$ absorber film was produced by thermal evaporation and the CdS buffer layer using the CBD method. A CdS film of about 120 nm thickness was deposited on FTO glass by CBD at 65°C for 18 min. Then, the CdS film was annealed on a hot plate at 350°C for 15 min before Sb$_2$Se$_3$ film deposition. In this superstrate CdS–Sb$_2$Se$_3$ heterojunction solar cell (shown in Fig. 13), photogenerated carriers separate at the CdS–Sb$_2$Se$_3$ junction interface. Electrons were injected into CdS and were collected by FTO, while holes flew to the back side and were collected by Au contact.

Wang *et al.* found CdCl$_2$ treatment on the CdS buffer layer before Sb$_2$Se$_3$ deposition had a positive effect on the CdS–Sb$_2$Se$_3$ junction interface and the device performance [27]. The CdCl$_2$ treatment was as follows: CdCl$_2$ anhydrous methanol solution (0.2 M) was

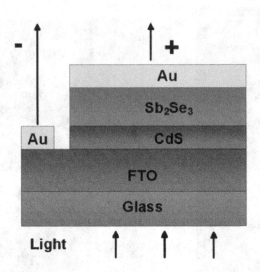

**Fig. 13.** Superstrate CdS–Sb$_2$Se$_3$ photovoltaic device. (a) Schematic demonstration of CdS–Sb$_2$Se$_3$ superstrate solar cell.

*Source*: Reproduced with permission from Ref. [7]. Copyright 2018, Applied Physics Letters.

**Table 2.** Statistics of $Sb_2Se_3$ photovoltaic device performance employing control CdS, $N_2$–CdS, Air–CdS, $CdCl_2$–$N_2$–CdS and $CdCl_2$–Air–CdS.

| Treatment | $V_{oc}$ (V) | $J_{sc}$ (mA/cm$^2$) | FF (%) | PCE (%) |
|---|---|---|---|---|
| Control | 0.37 ± 0.01 | 23.9 ± 1.4 | 45.5 ± 2.4 | 4.07 ± 0.32 |
| $N_2$–CdS | 0.37 ± 0.01 | 24.1 ± 1.5 | 45.8 ± 2.5 | 4.08 ± 0.24 |
| $CdCl_2$–$N_2$–CdS | 0.37 ± 0.01 | 24.3 ± 1.7 | 47.8 ± 2.3 | 4.39 ± 0.27 |
| Air–CdS | 0.38 ± 0.01 | 24.6 ± 1.6 | 50.2 ± 1.5 | 4.69 ± 0.21 |
| $CdCl_2$–Air–CdS | 0.40 ± 0.01 | 25.1 ± 1.2 | 52.8 ± 1.6 | 5.23 ± 0.26 |

*Source*: Reproduced with permission from Ref. [27]. Copyright 2018, Applied Physics Letters.

spin coated onto CdS buffer layer, followed by thermal annealing the CdS layer on a hot plate preheated to 400°C for 5 min in air, and then naturally cooling down the substrates. For simplicity, this type of device was named the $CdCl_2$–Air–CdS device, and devices using CBD-derived CdS without any post treatment were named control devices. Zhou *et al.* also investigated annealing the CdS layer, without $CdCl_2$ treatment, at 400°C on a hot plate for 5 min in air (Air–CdS), or annealing the $CdCl_2$–treated CdS layer at 400°C on a hot plate for 5 min in a $N_2$-filled glove box ($CdCl_2$–$N_2$–CdS). The device performance employing control, Air–CdS, $CdCl_2$–$N_2$–CdS, $N_2$–CdS and $CdCl_2$–Air–CdS is summarized in Table 2. It showed that all these treatments produced devices with better efficiencies than the control devices. The $CdCl_2$–Air–CdS exhibited the most striking improvement in device performance. Zhou *et al.* mainly focused on the comparison of control and $CdCl_2$–Air–CdS devices. The statistics of device performance were from 120 photovoltaic devices. For control devices, the $V_{oc}$, $J_{sc}$, FF and conversion efficiency were $0.37 \pm 0.01$ V, $23.9 \pm 1.4$ mA/cm$^2$, $45.5 \pm 2.4\%$ and $4.07 \pm 0.32\%$, respectively. For the $CdCl_2$–Air–CdS devices, the corresponding values were $0.40 \pm 0.01$ V, $25.1 \pm 1.2$ mA/cm$^2$, $52.8 \pm 1.6\%$ and $5.23 \pm 0.26\%$. Clearly, devices employing $CdCl_2$–Air–CdS showed significantly better performance that the control device. All device parameters were improved, and the FF enjoyed the largest relative improvement.

The effects of ambient $CdCl_2$ treatment were believed to be at least threefold: (i) thermal annealing increased the grain size and crystallinity of chemical bath deposited CdS, enabling better optical transparency; (ii) trace amount of Cl from $CdCl_2$ resides in the film and passivates surface defects on CdS grains, analogous to the passivation of PbS colloidal quantum dots by halide ligands [19]; (iii) oxygen contents increased due to thermal oxidation and hydroxide/carbonate decomposition, passivated defects in CdS buffer layer and improved junction quality, possibly in a similar way to the beneficial effect observed when oxygen was introduced at the initial stage of thermal evaporation of the $Sb_2Se_3$ layer [22]. In addition, by referring to device performance and photoluminescence investigation, it was believed that the effect of oxygen was dominant in the ambient $CdCl_2$ treatment and required further detailed study.

### 3.2. $ZnO$–$Sb_2Se_3$ heterojunction

The CdS buffer was highly toxic, which may cause significant obstacles for its potential applications. Other binary or ternary metal oxide buffers were also developed [19, 20, 28, 29, 31]. In 2017, Zhou *et al.* developed a non-toxic ZnO buffer layer to substitute the CdS buffer [19]. ZnO buffer was produced using spray pyrolysis of $Zn(NO_3)_2$ aqueous solution, an environmentally benign and scalable process. ZnO had a larger bandgap than CdS, formed proper band alignment with $Sb_2Se_3$ and hence was highly preferable as the buffer layer for $Sb_2Se_3$ photovoltaics.

A carton of the spray pyrolysis system is shown in Fig. 14(a). The ZnO thin films were deposited onto the FTO glass at 300°C, 400°C and 500°C. For ZnO deposited at 300°C, poor film morphology with a great amount of flakes and small particles was observed, suggesting incomplete decomposition of zinc nitrate, consistent with the TGA result that $Zn(NO_3)_2$ decomposition occurred at 344°C. As spray temperature increased to 400°C or 500°C, no flake was observed due to complete precursor decomposition, as shown in Fig. 15. However, the grain size was different in these two samples: for the ZnO film sprayed at 400°C, the average size was ~20 nm with uniform distribution; for the ZnO film deposited at 500°C,

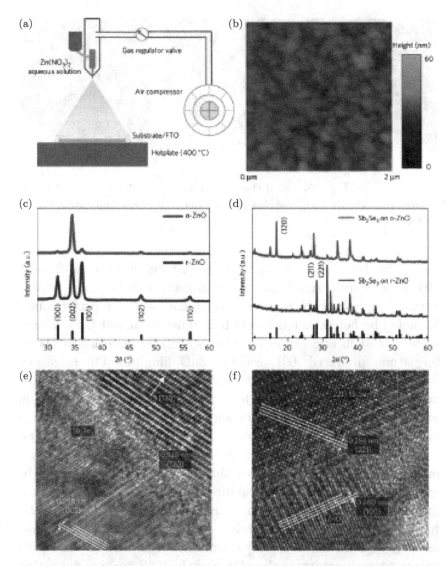

**Fig. 14.** Deposition and characterization of ZnO and Sb$_2$Se$_3$ layer. (a) A schematic of the spray pyrolysis system. (b) Atomic force micrograph of ZnO film prepared. (c) XRD patterns of ZnO. (d) The corresponding Sb$_2$Se$_3$ films produced on top. The standard diffraction patterns for Sb$_2$Se$_3$ (JCPDS: 15-0861) and ZnO (JCPDS: 70-2551) are included for reference. (e,f) HRTEM images of Sb$_2$Se$_3$ (e) and r-ZnO (f). The planes are determined by the interplane distance, and the normal vector for the (hkl) planes in orthorhombic Sb$_2$Se$_3$ is [hkl].

*Source*: Reproduced with permission from Ref. [19]. Copyr ight 2018, Nature Energy.

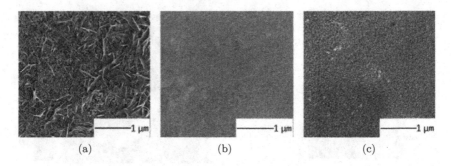

(a)                 (b)                 (c)

**Fig. 15.** SEM characterization of ZnO film.

*Source*: Reproduced with permission from Ref. [19]. Copyright 2018, Nature Energy.

the grain size was larger and widely distributed between 20 and 70 nm, indicating significant grain growth resulting in increased surface roughness, which was not favorable for device performance. It should be noted that the ZnO film annealed at 500°C for 20 min showed random orientation (r-ZnO), while that annealed at 500°C for 60 min produced [001]-oriented ZnO film (o-ZnO), as shown in Fig. 14(b). Following ZnO completion, Zhou *et al.* deposited an $Sb_2Se_3$ absorber layer on top using the RTE technique. Zhou *et al.* noticed a strong orientation correlation between the ZnO substrate and the $Sb_2Se_3$ films on top of ZnO. $Sb_2Se_3$ produced on r-ZnO always demonstrated preferred [221] orientation, while [120] orientation dominated in $Sb_2Se_3$ films on top of o-ZnO (Figs. 14(b) and 14(c)). The calculated texture coefficients, representing the growth of the $Sb_2Se_3$ film on a single-crystalline ZnO substrate, further supported this observation. Furthermore, the $Sb_2Se_3$ films exclusively demonstrated [221] orientation regardless of the r-ZnO substrate temperature employed during the RTE process, suggesting the strong interaction between r-ZnO and the $Sb_2Se_3$ film, controlling the orientation of $Sb_2Se_3$ films. The morphologies of $Sb_2Se_3$ films grown on r-ZnO and o-ZnO were also compared. For the $Sb_2Se_3$ film on r-ZnO, long strips of $Sb_2Se_3$ grains were abundantly observed, suggesting that $(Sb_4Se_6)_n$ ribbons lying on ZnO substrate were [120] oriented. In contrast, for the $Sb_2Se_3$ film on r-ZnO, the films were mostly composed of compact grains with clear edges and polyhedron shapes, indicating that $(Sb_4Se_6)_n$ ribbons stand vertically on the substrate, being [221] oriented. High-resolution transmission electron microscopy (HRTEM) was used to reveal the orientation correlation microscopically. Lattice fringes belonging to (002) planes of ZnO were clearly visible in the o-ZnO–$Sb_2Se_3$ sample, with the corresponding $Sb_2Se_3$ film demonstrating [120] orientation (Fig. 14(e)). In contrast, the [221]-oriented $Sb_2Se_3$ film closely neighboring the ZnO grains exposed with (100) planes was observed in the r-ZnO–$Sb_2Se_3$ sample (Fig. 14(f)).

The top-view SEM image of ZnO films pyrolyzed from $Zn(NO_3)_2$ aqueous solution with substrate temperature set at (a) 300°C, (b) 400°C and (c) 500°C. For ZnO deposited at 300°C, poor film morphology with a great amount of flakes and small particles was observed, suggesting incomplete decomposition of zinc nitrate, consistent with the TGA result that $Zn(NO_3)_2$ decomposition occurred at 344.2°C. As spray temperature increased to 400°C or 500°C, no flake was observed due to complete precursor decomposition. However, the grain size was different in these two samples: for the ZnO film pyrolyzed at 400°C, the average grain size is ~20 nm with uniform distribution; for the ZnO film pyrolyzed at 500°C, the grain size is larger and widely distributed between 20 and 70 nm, indicating significant grain growth resulting in increased surface roughness, which is not favorable for device performance.

Photovoltaic devices on r-ZnO demonstrated significantly better efficiency than those on o-ZnO. Typical device performance measured under $100 \, mW \, cm^{-2}$ illumination revealed a $V_{oc}$ of 357 mV, $J_{sc}$ of $28.1 \, mA \, cm^{-2}$ and FF of 48.0% for the device using r-ZnO, yet the corresponding values for the device using o-ZnO were only 345 mV, $18.3 \, mA \, cm^{-2}$ and 44.6% (Fig. 16(a)). The large efficiency improvement from the o-ZnO device (2.82%) to the o-ZnO device (4.81%) was attributed to the better $Sb_2Se_3$ orientation on r-ZnO. The [221]-oriented $Sb_2Se_3$ film showed better carrier transport along the $(Sb_4Se_6)_n$ ribbons and enjoyed reduced recombination loss at the dangling-bond-free GBs, and thus resulted in devices with higher $J_{sc}$ and FF. This analysis was supported by the EQE and biased EQE

178                                     Z. Li

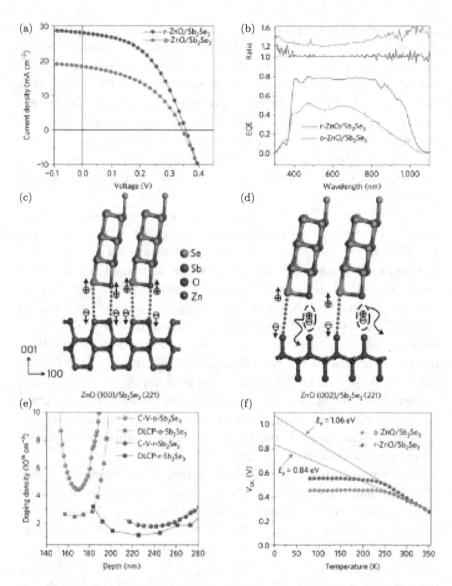

**Fig. 16.** Analysis of orientation-dependent performance. (a) and (b) Current density–voltage ($J$–$V$) characteristics (a) and external quantum efficiency (EQE) spectra (b) of an Sb$_2$Se$_3$ solar cell on r-ZnO and o-ZnO substrate. The inset (top)

results. As shown in Fig. 16(b), the EQE spectra of the r-ZnO/$Sb_2Se_3$ device demonstrated an abrupt increase at longer wavelength. For comparison, the EQE spectrum of the o-ZnO/$Sb_2Se_3$ device showed a much reduced EQE value in the whole responsive spectrum. Note that for the r-ZnO–$Sb_2Se_3$ device, the EQE response between 500 and 560 nm was much better than the previous CdS–$Sb_2Se_3$ heterojunction devices because of the reduced parasitic absorption loss of ZnO as compared with the CdS buffer layer [21, 32]. The biased EQE measurement showed that the EQE ratio, defined as [EQE($-1$ V)/EQE (0 V)], was approximately unity during the whole spectrum for the r-ZnO–$Sb_2Se_3$ device, while the o-ZnO–$Sb_2Se_3$ device was strongly bias dependent, especially in the long-wavelength region, suggesting carrier collection in this device was problematic and required assistance of external bias. The high EQE value and its weak bias dependence revealed that carrier collection was highly efficient for the [221]-oriented $Sb_2Se_3$ film grown on r-ZnO and hence explained the apparently larger $J_{sc}$ value of the r-ZnO–$Sb_2Se_3$ device.

Capacitance–voltage ($C$–$V$) profiling and deep-level capacitance profiling (DLCP) measurements were therefore carried out on these devices to further characterize the orientation-related defects [33].

←——————————————————————————————————————

**Fig. 16.** (*Continued*) in (b) is the ratio EQE ($-1$ V)/EQE (0 V). Applied bias is $-1$ V. (c) and (d) Atomistic model of [221]-oriented Sb2Se3 film on the (100) (c) and (002) (d) plane of ZnO. The dotted light magenta lines represent bond formation leading to successful charge separation at the interface, and the absence of such lines means dangling bonds causing recombination loss. + represents hole, − represents electron, straight arrows represent smooth carrier separation and curvilinear arrows represent recombination loss. (e) $C$–$V$ profiling and DLCP profiling for the r-ZnO–$Sb_2Se_3$ and o-ZnO–$Sb_2Se_3$ devices. (f) Temperature-dependent open circuit voltage ($V_{oc}$–$T$) measurement of the two devices. $V_{oc}$ was measured down to liquid $N_2$ temperature under 1 sun illumination and then extrapolated to $T = 0$ K. The solid lines are the linear fitting, and the $E_a$, the activation energy, can be obtained from the intercepted values with the $y$ axis.

*Source*: Reproduced with permission from Ref. [19]. Copyright 2018, Nature Energy.

In general, the $C$–$V$ measurement includes response from free carriers, defects in the bulk and interfacial defects, while the DLCP measurement is only sensitive to free carriers and bulk defects [34]. Thus, the difference between $N_{C-V}$ (defect density calculated from $C$–$V$ measurement) and $N_{DLCP}$ (defect density calculated from DLCP measurement) reflects defect density at the ZnO–Sb$_2$Se$_3$ interface [7, 34]. As shown in Fig. 16(e), the DLCP curves were approximately overlapped, confirming a similar quality of Sb$_2$Se$_3$ bulk film since they were produced using identical RTE processes. However, the $C$–$V$ curve of the o-ZnO–Sb$_2$Se$_3$ device shifted to significantly higher values than that of the r-ZnO–Sb$_2$Se$_3$ device, indicating more defects present at the o-ZnO–Sb$_2$Se$_3$ interface. Since the depletion width ($W_d$) was mainly located in the Sb$_2$Se$_3$ layer because it had much lower doping density than ZnO, and since the volume-to-surface ratio was Wd, the interfacial defect density could be calculated to be $1.22 \times 10^{11}\,\mathrm{cm}^{-2}$ in the r-ZnO–Sb$_2$Se$_3$ device, which was two times lower than that in the o-ZnO–Sb$_2$Se$_3$ device ($3.77 \times 10^{11}\,\mathrm{cm}^{-2}$). The temperature-dependent current density–voltage ($J$–$V$–$T$) measurements were further performed to obtain the activation energy of the recombination process in these devices. By extrapolating the $V_{oc}$ to absolute zero (Fig. 16(f)), the activation energy of the r-ZnO–Sb$_2$Se$_3$ device (1.06 eV) was equal to the indirect bandgap of Sb$_2$Se$_3$ (1.03 eV), indicating that bulk recombination dominated in this device [33]. In contrast, the activation energy of the o-ZnO–Sb$_2$Se$_3$ device was only 0.84 eV, confirming that interfacial recombination took preponderance in this device.

Adopting r-ZnO as the buffer layer and RTE-produced Sb$_2$Se$_3$ as the absorber layer, and optimization of the back contacts by subjecting the Sb$_2$Se$_3$ film with a carbon disulfide (CS$_2$) rinse prior to Au electrode deposition, a certified device demonstrated a power conversion efficiency of 5.93% with the corresponding $J_{sc}$ of $26.2\,\mathrm{mA\,cm}^{-2}$, $V_{oc}$ of 391 mV and FF of 57.8%. Zhou *et al.* further checked device performance under low light intensities of 100, 50 and $10\,\mathrm{mW\,cm}^{-2}$, and a representative device demonstrated efficiency values of 5.94%, 6.41% and 7.04%, respectively. The $J_{sc}$ depended linearly on the illumination intensities, but the FF improved significantly due to the reduced resistive loss, leading to increased conversion efficiency under weaker irradiations, which was one of the very competitive advantages of thin-film solar cells over crystalline Si in real-world applications where the solar insolation is often much lower than the standard 100-mW cm$^{-2}$ testing conditions.

Device performance experienced no degradation after 3 months of ambient storage in a typical laboratory environment. Further damp heat testing of the same device for 1100 h revealed minor device degradation, with efficiency dropping from the initial 5.83% to 5.74% after the test. For comparison, the efficiency of a typical CdS–Sb$_2$Se$_3$ device dropped from the initial 5.67% to 5.16% after an even sorter test (100 h). To understand the underlying mechanism for the enhanced stability of Sb$_2$Se$_3$ photovoltaics using ZnO buffer from an ion diffusion and interfacial defects perspective, Zhou *et al.* applied a high-angle annular dark-field scanning transmission electron microscope (HAADF-STEM) equipped with energy-dispersive spectroscopy (EDX) to characterize the interface of the r-ZnO–Sb$_2$Se$_3$ device. As shown in Fig. 17(a), the void-free interface suggested good adhesion and bond formation between the [221] Sb$_2$Se$_3$ film and the r-ZnO substrate. A rectangular area in the Z-contrast HAADF cross-sectional image was chosen to analyze the Sb, Se, Zn and O element distribution at the ZnO–Sb$_2$Se$_3$ interface. Zn, Se and Sb showed sharp edges in their spatial distribution maps, indicating negligible interfacial interdiffusion at the junction, at least at a level below the detection limit of the EDX facility. A line scan further supported this claim. The uniformly scattered signal of oxygen in the Sb$_2$Se$_3$ layer could be ascribed to oxygen contamination during the RTE process because it was carried out in a low-vacuum environment. Significant oxygen diffusion could be ruled out because otherwise an oxygen concentration gradient from the ZnO–Sb$_2$Se$_3$ interface to the Sb$_2$Se$_3$ bulk should be observed. In sharp contrast, significant Cd diffusion into the Sb$_2$Se$_3$ absorber layer was observed when similar characterization was applied to the CdS–Sb$_2$Se$_3$ device, as shown in Fig. 18. The diffusion depth was estimated to be around 50 nm based on the line scan. Severe Cd diffusion probably ruins the rectifying junction and serves as the culprit for performance

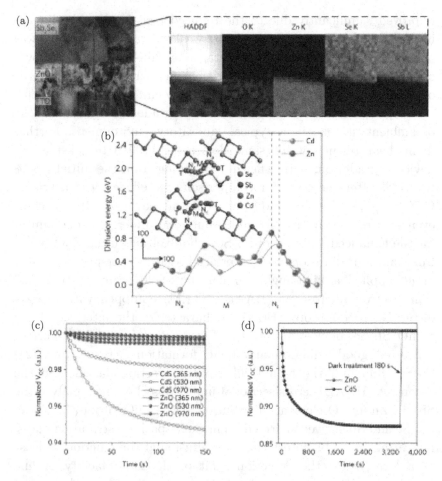

**Fig. 17.** Mechanistic investigation of improved device stability. (a) HAADF-STEM image and energy-dispersive spectroscopy elemental mapping of the ZnO–Sb$_2$Se$_3$ device. (b) The calculated diffusion energy curves of Zn and Cd in Sb$_2$Se$_3$. (c) and (d) Temporal evolution of the normalized $V_{oc}$ of ZnO–Sb$_2$Se$_3$ and CdS–Sb$_2$Se$_3$ devices under monochromatic illumination of different wavelengths (c) and under 1 sun AM1.5 illumination (d). The recovery time for the CdS–Sb$_2$Se$_3$ device.

*Source*: Reproduced with permission from Ref. [19]. Copyright 2018, Nature Energy.

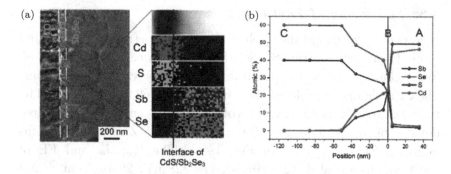

**Fig. 18.** HADDF and EDX analysis at the CdS–Sb$_2$Se$_3$ interface. (a) HAADF-STEM image and EDX elemental mapping of the ZnO–Sb$_2$Se$_3$ device. (b) Right at the interface ($x = 0$ nm, spot B, black line), signals from all the four elements were collected. In contrast with the ZnO case, significant Cd, and S to a less extent, was observed to diffuse into the Sb$_2$Se$_3$ up to a depth of 50 nm.

*Source*: Reproduced with permission from Ref. [19]. Copyright 2018, Nature Energy.

degradation. The first-principles density functional theory (DFT) simulations on the Zn and Cd diffusion in Sb$_2$Se$_3$ showed that both Zn and Cd atoms prefer to diffuse along the gap between $(Sb_4Se_6)_n$ ribbons to lower the strain energy induced by the Zn or Cd atom. Furthermore, in each diffusion path, because there is a relatively large space at the anion tetrahedron site (labeled as T) or near the anion site (labeled as $N_a$), they have a small strain energy basin. However, near the cation site (labeled as $N_c$), the smaller space and the larger strain energy cost make it the diffusion energy barrier. For Zn diffusion, the energy barrier at the $N_c$ site was 0.9 eV, which was 0.1 eV higher than that of Cd diffusion, consistent with the experimental observations. This is because the atom size of Cd was matching more with an Sb atom, and it has smaller strain energy than that of Zn diffusion at the $N_c$ site (Fig. 17(b)). DFT simulation and EDX mapping confirmed that Cd atoms diffuse easier in Sb$_2$Se$_3$, and device stability is substantially enhanced by inserting an Al$_2$O$_3$ layer at the CdS–Sb$_2$Se$_3$ interface, both of which strongly suggest that Cd diffusion was most likely responsible for device degradation. The r-ZnO–Sb$_2$Se$_3$ devices enjoyed negligible interdiffusion and hence had better stability.

### 3.3. *Sputtering of ZnO buffer*

Wen *et al.* investigated the application of a magnetron-sputtered ZnO buffer layer in superstrate $Sb_2Se_3$ thin-film solar cells [28]. Zhou *et al.* found that the annealing, under ambient air or vacuum conditions, played a significant role in the $Sb_2Se_3$ solar cell performance. The representative current density–voltage ($J$–$V$) curves of the three devices using as-deposited ZnO, air-annealed ZnO and vacuum-annealed ZnO are given in Fig. 19(a). The $V_{oc}$, $J_{sc}$ and FF of the best air-annealed ZnO–$Sb_2Se_3$ are $328\,\mathrm{mV}$, $28.0\,\mathrm{mA\,cm^{-2}}$ and 44.5%, respectively, corresponding to a device efficiency of 4.08%. For comparison, the efficiency of as-deposited ZnO–$Sb_2Se_3$ and vacuum-annealed ZnO–$Sb_2Se_3$ was 1.88% and 0.76%, respectively. The statistics of the device efficiency, along with the standard deviation, are summarized in Table 3. The general efficiency trend was air-annealed ZnO > as-deposited ZnO > vacuum-annealed ZnO. The EQE curves of these three samples showed a rapid augmentation for incident photons of 380 nm, which represented the absorption onset of the ZnO layer. It was well worth mentioning that the EQE response prior to 560-nm spectra was better than traditional CdS–$Sb_2Se_3$ heterojunction devices [5] since the parasite absorption loss by the CdS buffer layer was reduced in the ZnO–$Sb_2Se_3$ solar cells due to the larger bandgap of ZnO over CdS. The EQE curve of the device with air ambient-annealed ZnO even reached 80% during 400–800-nm range and then gradually decreased at a longer wavelength. Among these three devices, the ambient-annealed ZnO device had a more superior EQE value in the whole EQE spectrum than the others, indicating the best electron collection efficiency, considering the identical $Sb_2Se_3$ film thickness in the three samples, which should yield the same photon absorption and photocarrier generation.

To characterize the electrical properties of ZnO/$Sb_2Se_3$ devices, DLCP, $C$–$V$ profiling and capacitance–frequency ($C$–$f$) tests were adopted. Air-annealed ZnO–$Sb_2Se_3$ and as-deposited ZnO–$Sb_2Se_3$ devices were chosen to do the tests for their better performances, and vacuum-annealed ZnO–$Sb_2Se_3$ devices were excluded from the

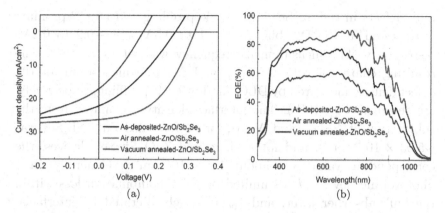

**Fig. 19.** (a) The illuminated current density–voltage ($J$-$V$) characteristic and (b) corresponding external quantum efficiency (EQE) curves with Sputtering of ZnO buffer.

*Source*: Reproduced with permission from Ref. [17]. Copyright 2018, Solar Energy Materials and Solar Cells.

**Table 3.** Statistics of $Sb_2Se_3$ photovoltaic devices using as-deposited ZnO, air-annealed ZnO and vacuum-annealed ZnO buffer layer.

| Devices | $V_{oc}$ (V) | $J_{sc}$ (mA/cm$^2$) | FF (%) | PCE (%) |
|---|---|---|---|---|
| As-deposited ZnO–$Sb_2Se_3$ | $0.24 \pm 0.01$ | $20.7 \pm 1.2$ | $31.6 \pm 2.3$ | $1.57 \pm 0.31$ |
| Air-annealed ZnO–$Sb_2Se_3$ | $0.32 \pm 0.01$ | $26.6 \pm 1.4$ | $43.5 \pm 1.5$ | $3.92 \pm 0.16$ |
| Vacuum-annealed ZnO–$Sb_2Se_3$ | $0.15 \pm 0.01$ | $16.7 \pm 1.5$ | $28.7 \pm 2.7$ | $0.65 \pm 0.11$ |

*Source*: Reproduced with permission from Ref. [17]. Copyright 2018, Solar Energy Materials and Solar Cells.

comparison due to their poor rectification because a high-quality diode is the prerequisite for credible $C$–$V$ and diode parameter analysis. $C$–$f$ was measured from $10^3$ to $10^6$ Hz in the dark. An AC voltage of 30 mV was used, and DC bias was kept at zero during the measurement. As shown in Fig. 19(a), the capacitance of the as-deposited ZnO–$Sb_2Se_3$ device showed larger frequency dependence than the air-annealed ZnO–$Sb_2Se_3$ device, indicating larger trap densities in as-annealed ZnO–$Sb_2Se_3$ devices. The capacitance was mainly influenced by the defects at heterojunction interfaces and bulk defects in the $Sb_2Se_3$ layer. $C$–$V$ profiling and DLCP profiling were tested at the DC bias from $-1$ to $0.5\,\mathrm{V}$ and $-0.2$ to $0.2\,\mathrm{V}$, respectively. The measurement frequency was set as 10 kHz. The resulting charge densities, $N_{DLCP}$ and $N_{C-V}$, which stand for the defect density measured by DLCP and by $C$–$V$ profiling, respectively, are shown in Fig. 19(b). The interface density, calculated from the subtraction of $N_{DLCP}$ and $N_{C-V}$, for air-annealed $ZnO$–$Sb_2Se_3$ was $\sim\!5.03 \times 10^{16}\,\mathrm{cm}^{-3}$, and for as-deposited $ZnO$–$Sb_2Se_3$ devices, the corresponding value was about $1.2 \times 10^{17}\,\mathrm{cm}^{-3}$. Generally, in a photovoltaic device, $J_{sc}$ is limited by the recombination loss within the bulk absorber layer, and $V_{oc}$ is largely dictated by interfacial defects because interfacial defects often pin the Fermi level and reduce the Fermi level splitting under illumination, suppressing achievable $V_{oc}$ [35, 36]. Air annealing reduced the interfacial defect density at the $ZnO$–$Sb_2Se_3$ interface, leading to substantial $V_{oc}$ improvement. It also resulted in a top $Sb_2Se_3$ absorber with less bulk defects and hence reduced recombination loss, thus increasing the $J_{sc}$ to a less significant extent.

## 3.4. $TiO_2$–$Sb_2Se_3$ heterojunction

Chen *et al.* employed a combinatorial strategy to accelerate the optimization of our $TiO_2$–$Sb_2Se_3$ solar cells. For glass–fluorine-doped tin oxide (FTO)–$TiO_2$–$Sb_2Se_3$–Au devices, optical absorption and parasitic resistance in the $TiO_2$ layer were non-ignorable losses in photocurrent and fill factor, making it necessary to minimize the $TiO_2$ thickness without pinholes and cracks because otherwise the $Sb_2Se_3$ absorber will be in direct contact with the FTO and create parallel shunting junctions degrading device performance. The $TiO_2$ layer was deposited by spraying. This spray track, controlled by a CNC $X$–$Y$ scanner, was used to produce a 1D thickness gradient along the $X$-axis (keep the thickness constant along the $Y$-axis) as shown in Fig. 20(a). The numbers beside the solid arrows stand for the spray cycles, and the dashed arrows are the connecting track. Naturally, $TiO_2$ thickness along $X$-direction monotonously increased from left to right as more precursors were sprayed onto the right side. The precursor solution of titanium diisopropoxide

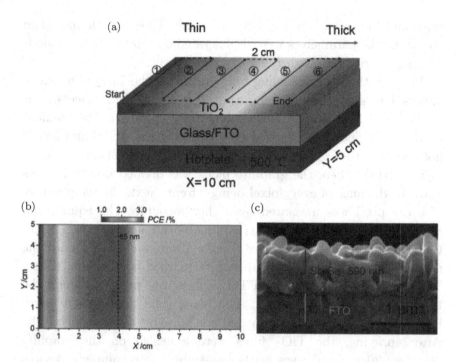

**Fig. 20.** Combinatorial optimization of TiO$_2$ buffer thickness. (a) The schematic diagram of combinatorial spraying pyrolysis process. The numbers beside the solid arrows stand for the spraying cycles, and the dashed arrows are the connecting track (no spraying here). (b) The spatially resolved PCE of assembled device. (c) The cross-sectional SEM morphologies of device with maximum efficiency.

*Source*: Reproduced with permission from Ref. [38]. Copyright 2018, Advanced Energy Materials.

bis(acetylacetonate) in anhydrous ethanol with 7.5 wt.% concentration was sprayed at 450°C with the constant flow rate of 10 mL min$^{-1}$ and the substrate was further annealed at 500°C for 30 min [37]. Zhou *et al.* fabricated 72 devices on the substrate and measured these pixels under simulated AM 1.5 G illumination. The spatially resolved device PCE is shown in Fig. 20(b). The optimal TiO$_2$ thickness was identified as 65 nm based on the cross-sectional scanning electron microscope (SEM) characterization of the corresponding maximal PCE device. When TiO$_2$ was thicker than 65 nm, the $J_{sc}$ and FF dramatically reduced due to more optical absorption of the TiO$_2$

layer and increased series resistance. While $TiO_2$ was thinner than 65 nm, all the parameters were degenerated due to the discontinuous coverage or pinholes with the $TiO_2$ buffer layer.

Further, the post-annealing temperature of the $TiO_2$ film, which always has great influence on the crystallinity and defects, was studied by a high-throughput combinatorial approach. The gradient temperature field was achieved by connecting the 400°C and 550°C hot plates with a graphite block of 1 cm × 6 cm × 11 cm as shown in Fig. 21(a). Then, the graphite block was divided into 264 pixels with the distance of every pixel being 0.5 cm. Next, the temperature of every pixel was measured by a thermocouple. The equilibrium temperature distribution of the graphite block is shown in Fig. 21(b). Cleaned FTO glass of the size 5 cm × 10 cm was placed onto the graphite block. When the temperature reached equilibrium, 65-nm $TiO_2$ film was deposited on FTO using spray pyrolysis along the track as shown in Fig. 21(a). The black solid and dashed arrow lines correspond to the spraying and non-spraying track, respectively. After spraying, the $TiO_2$ films were annealed on the graphite block for 30 min and were further assembled into complete devices as per the description above. The mapping of spatially resolved PCE of these 72 pixels (6 × 12) revealed that the highest PCE of 4.2% was obtained using 450°C annealed $TiO_2$. Combining the temperature distribution (Fig. 21(b)), the representative PCEs at typical temperatures (400°C, 450°C, 500°C and 550°C) are shown in Fig. 21(d) and the corresponding detailed device parameters are shown in Table 4. The XPS measurement showed that $TiO_2$ films annealed at 450°C were closest to stoichiometric ratio compared to those annealed at 400°C, 500°C and 550°C, meaning that the $TiO_2$ films annealed at 450°C had the fewest oxygen vacancies $(V_O)$.

The interfacial defects at the $TiO_2$–$Sb_2Se_3$ heterojunction were characterized through combined $C$–$V$ and DLCP measurements. The $N_{DLCP}$ curve of the device with $TiO_2$ annealed at 450°C was approximately overlapped with its $N_{C-V}$ curve, confirming the lowest interfacial defects. Furthermore, the interfacial defect densities of these devices were proportional to the deviation of oxygen from the stoichiometric ratio, indicating that $V_o$ defects were the culprit that

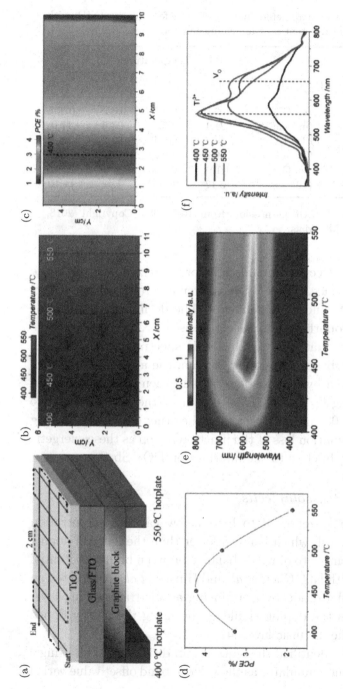

**Fig. 21.** Combinatorial optimization of the post-annealing temperature. (a) The schematic diagram of combinatorial spray pyrolysis of TiO$_2$. The black solid and dashed arrows correspond to the spraying and non-spraying track, respectively. (b) The measured temperature distribution of the graphite block, which was in direct contact with the FTO substrate. (c) The spatially resolved PCE of the assembled device and (d) PCE of TiO$_2$–Sb$_2$Se$_3$ devices using TiO$_2$ buffer annealed at 400°C, 450°C, 500°C and 550°C. (e) The spatially resolved photoluminescence of TiO$_2$ buffer and (f) photoluminescence of TiO$_2$ buffer annealed at 400°C, 450°C, 500°C and 550°C.

*Source:* Reproduced with permission from Ref. [38]. Copyright 2018, Advanced Energy Materials.

**Table 4.** The device parameters and interface defect density (NIF) with TiO$_2$ buffer layer at different post-annealing temperatures.

| Measurement | Parameters | Temperature (°C) | | | |
|---|---|---|---|---|---|
| | | 400 | 450 | 500 | 550 |
| J–V | Voc (V) | 0.341 | 0.361 | 0.358 | 0.355 |
| | Jsc (mA cm$^{-2}$) | 21.1 | 25.4 | 23.5 | 17.0 |
| | FF (%) | 45.6 | 46.1 | 42.3 | 31.4 |
| | PCE (%) | 3.3 | 4.2 | 3.6 | 1.9 |
| XPS | Ti:O | 1:1.90 | 1:1.96 | 1:1.93 | 1:1.76 |
| CV and DLCP | N$_{iF}$ (10$^{16}$ cm$^{-3}$) | 6.17 | 0.64 | 4.62 | 11.10 |

*Source*: Reproduced with permission from Ref. [38]. Copyright 2018, Advanced Energy Materials.

induced interface defects and enhanced carrier recombination at the TiO$_2$–Sb$_2$Se$_3$ interface. The $C$–$f$ measurement reflected the defect response via the frequency dependence. The declining quantities at low frequency proportional to defect density from small to large were TiO$_2$ annealed at 450°C. The EQE spectra of these devices demonstrated an abrupt increase at 380 nm, the absorption onset of TiO$_2$ layer, followed by a short plateau, and then rapidly decreased at longer wavelength. The EQE spectrum of the TiO$_2$-450 device between 400 and 600 nm was larger than the other devices, indicating the lowest recombination loss at the junction region as these energetic photons were largely close to the interface of TiO$_2$–Sb$_2$Se$_3$.

### 3.5. *p–i–n Sb$_2$Se$_3$ solar cells*

Sb$_2$Se$_3$ is commonly reported to have a low carrier concentration ($\sim$10$^{13}$ cm$^{-3}$) [23, 39], which is much lower than the optimal doping density of $\sim$10$^{16}$ cm$^{-3}$, to obtain a balance between built-in potential and depletion width [40]. Chen *et al.* and Hutter *et al.* constructed an n–i–p device configuration by introducing a hole-transporting layer (HTL) to address the doping challenge by using the lightly doped Sb$_2$Se$_3$ layer as the intrinsic layer [41].

Chen *et al.* first identified the requisite conditions for HTL using a solar cell capacitance simulator as the valence band offset value varied from $-0.2$ to $0.1\,\text{eV}$. A PbS colloidal quantum dot (CQD) film was selected as HTL, and the size of CQDs, the ligand treatment and the film thickness were all well optimized. Figure 22(a) presented the $J$–$V$ curves of the best device with PbS CQD film HTL and the control device. Compared with the control devices, $V_{\text{oc}}$, $J_{\text{sc}}$ and FF were relatively enhanced by 7.3%, 7.4% and 10.1%, respectively, after using PbS CQD film as HTL, resulting in a 27% relative improvement in PCE from 5.4% to 6.87%. To check the reproducibility, over 100 separate devices with and without HTL were tested. The histograms of the device performance are shown in Fig. 22(b). The average efficiency of the control devices was 5.08% with a standard deviation of 0.27%, while the devices with optimal HTL demonstrated an average efficiency of 6.62% with a standard deviation of 0.08%. Obviously, the PbS CQD HTL not only significantly increased device efficiency but also decreased the standard deviation of device efficiency. Light-beam-induced current (LBIC) measurements were carried out to check the spatial uniformity of the devices. The photocurrent mappings, as shown in Figs. 22(c) and 22(d), evaluated the spatial non-uniformity and localized performance of the investigated photovoltaic devices. The scanning results made by LBIC revealed that the control device showed a larger photocurrent difference of $136\,\text{pA}$ with a standard deviation of $1.7\,\text{pA}$, while the corresponding values for the device with PbS CQD HTL were 118 and $1.0\,\text{pA}$, respectively. Whereas the net value of these numbers allowed no direct comparison between different samples because the system was not calibrated and the connection could be varied from sample to sample, the ratio of standard deviation to mean value was physically meaningful. The calculated ratios for the devices with and without HTL are $8.47 \times 10^{-3}$ and $1.25 \times 10^{-2}$, respectively. The calculated ratio unambiguously confirmed that device homogeneity was significantly enhanced upon the employment of a PbS CQD HTL. The more uniform back electrode contact leads to the more homogeneous device. The large-area ($1.02\,\text{cm}^2$) devices as shown in Fig. 22(e) achieved the highest efficiency of 6.39%, over 18% higher than the previous record 5.4% ($1.08\,\text{cm}^2$ area) [41]. This result was very encouraging because there is often a substantial efficiency gap

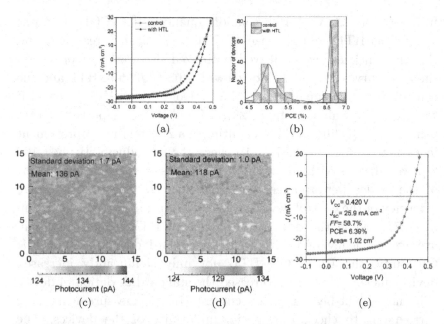

**Fig. 22.** (a) $J$–$V$ characteristics of device with and without PbS CQD HTL. (b) Histogram of device performance measured on 100 separate devices with and without HTL. (c) Without and (d) with PbS CQD HTL, respectively. (e) $J$–$V$ characteristics of a large ($1.02\,\mathrm{cm}^2$) device employing PbS CQD HTL tested under standard AM1.5G illumination.

*Source*: Reproduced with permission from Ref. [41]. Copyright 2018, Advanced Energy Materials.

between small and relatively large-sized devices, particularly for some next-generation photovoltaic devices. This result also meant that the RTE-derived $Sb_2Se_3$ absorber films were homogeneous and the back contact played a critical role in determining device uniformity.

Hutter *et al.* employed poly [N-9′-heptadecanyl-2,7-carbazole-alt-5,5-(4′,7′-di-2-thienyl-2′, 1′, 3′-benzothiadiazole)] (PCDTBT) as the HTL and fabricated a glass |FTO|TiO$_2$|Sb$_2$Se$_3$|PCDTBT| Au solar cell structure. The ionization and electron affinity of PCDTBT were at 5.4 and 3.6 eV, respectively [41]. Photooxidation is known to generate sub-bandgap states close in energy to the valence band of $Sb_2Se_3$ and therefore allows efficient hole extraction [15, 16]. The LUMO of PCDTBT was also high enough to block the transfer of minority carriers (electrons) from the $Sb_2Se_3$ to Au and thereby reduce recombination at the back contact. Various PCDTBT concentrations were investigated, with the average PCE of devices for 2, 4 and 5 mg/mL of PCDTBT in chloroform being 4.26%, 6.06% and 3.55%, respectively, when spin coated at 6,000 rpm. This demonstrated that PCDTBT coverage is a balance between covering the pin holes, which affects the performance negatively in the device, and making the PCDTBT layer too resistive within the device. This demonstrated that the $Sb_2Se_3$ layers used in this work are highly absorbent and very little light gets past the $Sb_2Se_3$ through to the PCDTBT layer; the PCDTBT layer is acting only as a pinhole blocking layer, and not as an absorber layer within the device. Furthermore, uniformity of performance is vastly improved without any loss in peak performance.

## 4. Admittance spectroscopy and deep-level transient spectroscopy analysis

### 4.1. *Admittance spectroscopy of $Sb_2Se_3$*

Hu *et al.* performed admittance measurements on $Sb_2Se_3$-based thin-film solar cells with energy conversion efficiencies from 3.85% to 5.91% [42]. Admittance spectroscopy involved the measurement of the complex admittance $Y(\omega, T) = G(\omega, T) + i\omega C(\omega, T)$ as a function of angular frequency $\omega$ and the temperature $T$, where $G(\omega, T)$ was the conductance term that entailed full spectral information that was related to the Kramers–Kronig relations [43]. Therefore, only the capacitance spectra $C(\omega, T)$ would be discussed. In addition to the capacitance of the space charge region (SCR) for a semiconductor junction, electronically active defects or traps in the SCR of this junction could also contribute to the capacitance spectra at lower frequencies and/or higher temperatures [44]; this was due to trapping and detrapping of defects at a location where the energy level $E_t$ of electron (hole) traps crosses the quasi-Fermi level $E_{Fn}$ ($E_{Fp}$) in the SCR. The results in the appearance of a differential capacitance peak $(dC/d\ln\omega-\omega)$ spectrum at this crossing point at a temperature when the time constant $\tau_0$ associated with the trapping (capture) and detrapping (or emission) processes matches the modulation frequency. The characteristic frequancy ($\omega_0 = 2\pi f_0$) of this differential capacitance peak is related to the emission rate ($1/\tau_0$) of trapped carriers according to the equation [40]

$$\omega_0 = 2/\tau 0 = 2N_{C,V}v_{\text{th}}\sigma_{n,p}\exp\left(-\frac{E_a}{kT}\right) = \xi_0 T^2 \exp\left(-\frac{E_a}{kT}\right),$$

where $N_{C,V} \sim T^{3/2}$ was the effective density of states in the conduction or valence band; $v_{\text{th}} \sim T^{1/2}$ is the thermal velocity of carriers; $E_a$ was the activation energy of the defect, which was the energetic distance to the conduction or valence band; KT was the thermal energy; $\sigma_{n,p}$ was the electron/hole capture cross section; and $\xi_0$ was a temperature-independent emission pre-factor defined as $N_{C,V}v_{\text{th}}\sigma_{n,p}/T^2$. In the case of a p-type semiconductor, hole traps within the SCR having an activation energy of $E_a$ could be charged and discharged at frequencies of $\omega < \omega_0$.

Figure 23 shows the capacitance spectra measured at various temperatures from 180 to 334 K at 4–5-k increments for three typical Sb$_2$Se$_3$ solar cells with efficiencies of 5.91%, 4.93% and 3.85%. The equivalent circuit model for the solar cell is given as the inset of Fig. 23(a). The low-frequency region (20–$10^5$ Hz) of all spectra showed three obvious capacitance steps at different frequencies. This capacitance response was attributed to three defect levels in the bandgap of Sb$_2$Se$_3$ that were named D1, D2 and D3. The capacitance response in the high-frequency region ($10^5$–$10^6$ Hz) was attributed mainly to parasitic effects that will be discussed later. C$_g$ showed the geometric capacitance of the solar cells, which occurred when free carriers could not follow the external excitation, for example, at low enough temperatures or high enough frequencies, and the semiconductor acted as an insulator.

To determine details regarding the possible defects, frequency derivative ($-dC/d\ln f$) vs. frequency was plotted, with results of the highest-efficiency cell sample as an example. For clarity, only parts of the measured curves are displayed. The steps in the $C$–$f$ spectra transform into maxima $f_{\text{max}}$ of the derivative $-dC/d\ln f$ that was used to determine the inflection frequencies for each defect level as a function of temperature. Figure 24(b) shows the Arrhenius

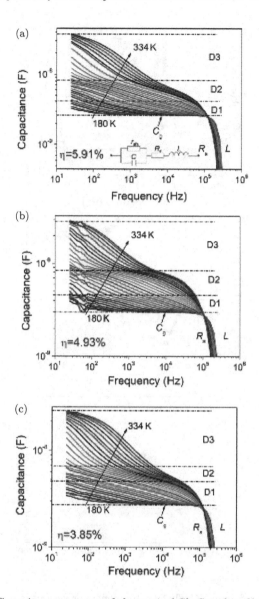

**Fig. 23.** Capacitance spectra of the typical Sb$_2$Se$_3$ thin-film solar cells.

*Source*: Reproduced with permission from Ref. [45]. Copyright 2018, Solar Energy Materials and Solar Cells.

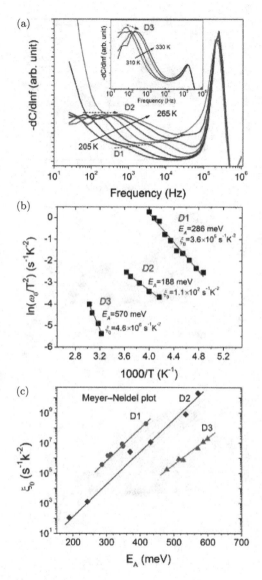

**Fig. 24.** (a) Derivative $-dC/d\ln f$ vs. frequency for various temperatures for the 5.91% cell sample. (b) Arrhenius plots of $\omega_0/T_2$ vs. $1000/T$ for the three defects measured from the 5.91% cell sample. (c) Meyer–Neldel relations for determining the same defect types.

*Source*: Reproduced with permission from Ref. [45]. Copyright 2018, Solar Energy. Materials and Solar Cells.

plots of $\ln(\omega_0/T^2)$ vs. $1000/T$ for the three transitions in Fig. 24(a), where $T$ is the temperature at which the characteristic frequency $(\omega_0)$ of the capacitance peak occurs. The activation energy $(E_A)$ of the transition was obtained from the slope of the least-squares fit to the data and the value of $\xi_0$ from the $y$-axis intercept. The energy positions of the three defects in different samples were found to vary. Therefore, to determine the same-type defects in different samples, Meyer–Neldel relations were plotted for each defect; these describe the relationship between the activation energy $(E_A)$ and the carrier emission pre-factor $(\xi_0)$ as follows [44]:

$$\xi_0 = \xi_{00} \exp\left(\frac{E_A}{E_{\text{char}}}\right), \tag{1}$$

where $\xi_{00}$ and $E_{\text{char}}$ were characteristic constants of the relations and should be identical for the same type of defect. As shown in Fig. 24(c), three types of defects could be clearly distinguished by plotting the Meyer–Neldel relations. This allowed the categorization of defects with different activation energies in different samples.

Table 5 shows the summed parameters of the detected defects for the 5.91% device obtained via admittance spectroscopy. The capture lifetime of holes was calculated and is shown in Table 5. Only the capture lifetime of majority carriers (holes) could be obtained from the admittance measurements. Although solar cells are minority carrier devices, the capture lifetime of majorities could reflect minority carrier lifetime to some extent. When a hole trap was

**Table 5.** Defect parameters of VTD- and RTE-fabricated Sb$_2$Se$_3$ solar cells.

| Defects | $E_T$ (eV) | $\sigma$ (cm$^2$) | $N_T$ (cm$^{-3}$) |
| --- | --- | --- | --- |
| VTD-H1 | $E_V + 0.48 \pm 0.07$ | $1.5 \times 10^{-17}$ | $1.2 \times 10^{15}$ |
| VTD-H2 | $E_V + 0.71 \pm 0.02$ | $4.9 \times 10^{-13}$ | $1.1 \times 10^{14}$ |
| VTD-E1 | $E_C - 0.61 \pm 0.03$ | $4.0 \times 10^{-13}$ | $2.6 \times 10^{14}$ |
| RTE-H1 | $E_V + 0.49 \pm 0.03$ | $2.2 \times 10^{-16}$ | $1.2 \times 10^{14}$ |
| RTE-H2 | $E_V + 0.74 \pm 0.04$ | $7.7 \times 10^{-13}$ | $2.3 \times 10^{15}$ |
| RTE-E1 | $E_C - 0.60 \pm 0.02$ | $1.6 \times 10^{-12}$ | $1.7 \times 10^{15}$ |

*Source*: Reproduced with permission from Ref. [46]. Copyright 2018, Nature Communications.

filled by a hole, it could act as a recombination center by attracting an electron. Once the electron was captured, recombination occurred, and the trap may become empty again and may capture another hole. If the lifetime of the captured hole in the trap was very short, then there may be more opportunities for recombination and therefore the lifetime of electrons in such a material may be shorter and solar cell parameters may suffer. It was found that the capture cross sections of holes for D1 and D3 were much larger than for D2; usually, capture cross sections in the range of $10^{-16}$ to $10^{-17}\,\mathrm{cm^2}$ indicated that defects may act as effective trapping or recombination centers [47]. However, although the capture cross section of D1 was large, the defect density was low, which in conjunction led to capture lifetime of holes in the order of microseconds, suggesting that D1 may not act as an effective trapping center to influence the performance of the solar cells. D2 was assigned as a shallow acceptor level since the capture cross section was so small ($10^{-20}$ to $10^{-21}\,\mathrm{cm^2}$) and, in conjunction with the defect density, led to a very long capture lifetime of holes in the order of milliseconds. D3 with both high defect density and large capture cross section had a very short capture lifetime of holes in the order of nanoseconds, and therefore might act as a very effective trapping or recombination center that limits the efficiency of the $Sb_2Se_3$ solar cell.

## 4.2. DLTS of $Sb_2Se_3$

DLTS is well-accepted powerful tool to investigate defect energy level, type and concentration in thin-film photovoltaics [35, 48–52]. It uses the transient capacitance of the p–n junction at different temperatures as a probe to monitor the changes in charge state of a deep defect center. Traps in the device are filled by carriers by applying a voltage pulse to the device, which changes the capacitance associated with the p–n junction of the device [53].

Wen *et al.* comparatively analyzed the deep-level defects in vapor transport deposition (VTD)-processed and RTE-fabricated $Sb_2Se_3$ films by DLTS [54]. DLTS was employed with minority carrier injection (inj-DLTS) to detect both electron and hole traps in $Sb_2Se_3$ films [49, 50, 52, 55]. As shown in Fig. 25(a), during the measurement,

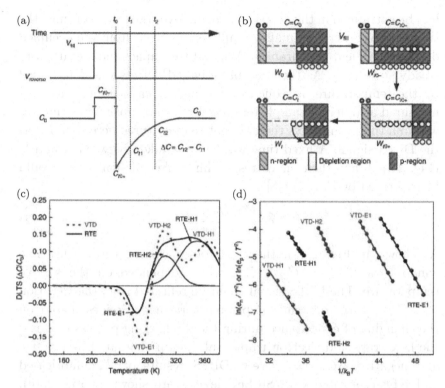

**Fig. 25.** DLTS analysis of VTD- and RTE-fabricated CdS–Sb$_2$Se$_3$ solar cells. (a) Schematic demonstration of the mechanism of DLTS measurement. (b) Variation of depletion width and the process of holes being trapped and emitted during the measurement. (c) DLTS signals of VTD-fabricated and RTE-fabricated devices.

*Source*: Reproduced with permission from Ref. [54]. Copyright 2018, Nature Communications.

a quiescent reverse bias ($V_{\text{reverse}}$) was first applied to the junction to form a depletion width in the Sb$_2$Se$_3$ layer. Then, a forward pulse voltage ($V_{\text{fill}}$) was applied to fill the traps and emitted, before and after the pulse voltage was applied. $C_0$ represents the steady-state junction capacitance at $V_{\text{reverse}}$ bias. As shown in Fig. 25(b), when $V_{\text{fill}}$ was applied and held for a while, $W_d$ narrowed down to $W_{t0-}$ and the hole trap defects were filled. Once $V_{\text{fill}}$ pulse relaxed, $W_d$ broadened to $W_{t0-}$ and capacitance decreased to $C_{t0+}$ instantaneously. $W_{t0+}$ was even larger than $W_0$ because some holes had been trapped in the depletion region. Over the course of time, the trapped holes were gradually completely emitted from the occupied deep level. Then, $W_d$ shrank to $W_0$ and the capacitance returned to steady state ($C_0$). As the state of activated defects was determined by the temperature, a sequence of transient capacitance was thus measured at different sample temperatures, and capacitance changes within a time window vs the different temperatures were sampled as the DLTS signal. A fixed time window was taken between $t_1$ and $t_2$ (Fig. 25(a)), and then the corresponding hole emission rate $e_p$ could be expressed by Eq. (2) [48]

$$e_p = \frac{\ln(t_2/t_1)}{t_2 - t_1}. \tag{2}$$

As shown in Fig. 25(a), the capacitance change within the time window was $\Delta C = C_{t2} - C_{t1}$, which depended on the sample temperature. The DLTS signal could be reflected from the variation of $\Delta C/C_0$ with different sample temperatures. Besides, based on the changing of capacitance during the discharging process of traps, the hole traps and electron traps could be differentiated by positive and negative $\Delta C$, respectively. DLTS results of VTD-fabricated and RTE-fabricated CdS/Sb$_2$Se$_3$ devices are shown in Fig. 25(c). One negative and two positive peaks were found in both devices, indicating one electron trap (E1) and two hole trap defects (H1 and H2) in Sb$_2$Se$_3$ films. The activation energy and capture cross section of the trap could be obtained from the Arrhenius plot based on Eqs. (3) and (4) [48, 50]:

$$\ln\left(\frac{e_p}{T^2}\right) = \ln\left(\sigma_p \frac{16\pi k_B^2 m_p^*}{h^3}\right) - \frac{E_T - E_V}{k_B T}, \tag{3}$$

$$\ln\left(\frac{e_n}{T^2}\right) = \ln\left(\sigma_n \frac{16\pi k_B^2 m_n^*}{h^3}\right) - \frac{E_C - E_T}{k_B T}, \tag{4}$$

where $e_n$ and $e_p$ represented the electron and hole emission rate, respectively, which was obtained by Eq. (2); $\sigma_p$ and $\sigma_n$ were capture cross section of hole and electron traps, respectively, T was the temperature and K$_B$T was the thermal energy. In addition, $m_p$ and $m_n$, respectively, represented effective mass of hole and electron, and $E_T$, $E_C$ and $E_V$ were the energy level of defect, conduction and valence bands, respectively. Fig. 25(d) shows the Arrhenius plot obtained from Fig. 25(c), by the varied $e$ (hole or electron emission rate) corresponding to the DLTS peak positions in temperature according to Eqs. (3) and (4). The activation energy ($E_T$–$E_V$ or $E_C$–$E_T$) of the trap could be calculated from the slope of $\ln(e_p/T_2)$ or $\ln(e_n/T_2)$ vs. $1/k_B T$ plot, and the capture cross-section could be extracted from the $y$-intercept. The trap concentration ($N_T$) can be obtained from Eq. (5): [48, 51]

$$N_T = \frac{2\Delta C_{\text{MAX}}}{C_0} N_A. \tag{5}$$

Here, $N_A$ is the net acceptor concentration in the $Sb_2Se_3$ film, which can be obtained from $C$–$V$ profiling. The properties of deep defects in the $Sb_2Se_3$ film were experimentally uncovered. By comparing the activation energies of these trap defects, they were found to be similar in both devices, indicating the same origin for these defects. To investigate whether Cd diffusion affected the DLTS results, Zhou *et al.* measured the variation of depletion width with the applied bias. The depletion width of VTD- and RTE-fabricated devices decreased from 430 and 324 nm to 240 and 255 nm, respectively, with the bias pulse changing from −0.5 to 0.4 V. However, the Cd diffusion depths in VTD- and RTE-fabricated $Sb_2Se_3$ layers were about 200 and 100 nm from SIMS results, respectively. Thus, the Cd ions located outside the detected depletion region had no effect on the DLTS result. The defects only originated from the intrinsic $Sb_2Se_3$. In the VTD and RTE facilities, Se vapor was always in excess as compared to the vapor pressure of Sb and $Sb_2Se_3$, so the $Sb_2Se_3$ films were actually slightly Se rich [5, 19]. Previous *ab initio* calculation of the intrinsic defects in $Sb_2Se_3$ had demonstrated that the dominant acceptor defects are antimony vacancy ($V_{Sb}$) and selenium antisite ($Se_{Sb}$) defects under the Se-rich condition. Therefore, Zhou *et al.* tentatively attributed H1 and H2 defects to $V_{Sb}$ and $Se_{Sb}$ defects, respectively. As reported by Tumelero *et al.* [23], the antisite defects dominated the distribution of defects in trichalcogenides due to the similar sizes of the constituent atoms.

The previous simulation also showed that $Se_{Sb}$ and antimony antisite ($Sb_{Se}$) are acceptor and donor defects, respectively. Consequently, the E1 defect was most likely associated with the formation of SbSe antisite defects. Interestingly, E1 and H2 always had similar densities in both VTD- and RTE-fabricated samples, which indicated that antisite defect pairs formed in $Sb_2Se_3$ films, presumably forming the [$Sb_{Se}$ + $Se_{Sb}$] complex. Note that using the VTD process reduced the density of these antisite defect pairs by more than an order of magnitude. The conduction band, valence band and Fermi levels were obtained from ultraviolet photoelectron spectroscopy (UPS) and Tauc plots by transmission spectra of $Sb_2Se_3$ films. Clearly, the energy levels of H2 and E1 were above the Fermi level ($E_F$) in both samples, and H1 was under EF. In the CdS–$Sb_2Se_3$ device, the photongenerated electrons were driven into the CdS layer, prompting the electron quasi-Fermi level ($E_{Fn}$) to move upward in CdS. Meanwhile the photogenerated holes resulted in the hole quasi-Fermi level ($E_{Fp}$) shifting downward in the $Sb_2Se_3$ layer. The shift of quasi-Fermi levels was positively correlated with the non-equilibrium carrier concentration. The difference between the two quasi-Fermi levels determined the $V_{oc}$ of the solar cell [35]. The position of $E_{Fp}$ was always dependent on illumination intensity. The intersections of $E_F$ and defect levels can be used as boundaries to differentiate whether the defects are charged or not during the shifting of the quasi-Fermi level [56]. Clearly, the H1 state was under the Fermi level and submersed in electrons, so H1 defects stayed inert. In contrast, H2 and E1 defect states were mostly above $E_F$ and they were active in trapping holes and electrons, respectively. These trapped photogenerated carriers would most probably contribute to recombination loss. Furthermore, the energy levels of H2 and E1 were located near the mid-gap, which significantly increased the recombination possibility of photogenerated carriers [57]. Therefore, the H2 and E1 were the dominant defects that influenced the shift of quasi-Fermi levels and trap-assisted recombination, and then the $V_{oc}$ and $J_{sc}$ of the solar cells. Moreover, due to the higher defect density of H2 and E1 than carrier concentration (about $10^{13}$ cm$^{-3}$)

in the $Sb_2Se_3$ layer [3], the $E_{Fp}$ would be more likely to be pinned near E1 and H2 levels. Obviously, both H2 and E1 had lower defect density in the VTD-fabricated sample, which could lead to relatively larger $E_{Fp}$ downshifting and suppressed trap-assisted recombination, explaining the better $V_{oc}$ and $J_{sc}$ in VTD-fabricated $Sb_2Se_3$ solar cells.

# References

[1] J. Petzelt, J. Grigas, Far infrared dielectric dispersion in $Sb_2S_3$, $Bi_2S_3$ and $Sb_2Se_3$ single crystals, *Ferroelectrics*, 5 (1973) 59–68.

[2] Y. Zhou, M.Y. Leng, Z. Xia, J. Zhong, H.B. Song, X.S. Liu, B. Yang, J.P. Zhang, J. Chen, K.H. Zhou, J.B. Han, Y.B. Cheng, J. Tang, Solution-processed antimony selenide heterojunction solar cells, *Adv. Energy Mater.*, 4 (2014) 1079–1083.

[3] C. Chen, D.C. Bobela, Y. Yang, S.C. Lu, K. Zeng, C. Ge, B. Yang, L. Gao, Y. Zhao, M.C. Beard, J. Tang, Characterization of basic physical properties of $Sb_2Se_3$ and its relevance for photovoltaics, *Front. Optoelectron.*, 10 (2017) 18–30.

[4] G. Ghosh, The sb-se (antimony-selenium) system, *J. Phase Equilib.*, 14 (1993) 753–763.

[5] Y. Zhou, L. Wang, S.Y. Chen, S.K. Qin, X.S. Liu, J. Chen, D.J. Xue, M. Luo, Y.Z. Cao, Y.B. Cheng, E.H. Sargent, J. Tang, Thin-film $Sb_2Se_3$ photovoltaics with oriented one-dimensional ribbons and benign grain boundaries, *Nat. Photon.*, 9 (2015) 409–415.

[6] H.B. Song, T.Y. Li, J. Zhang, Y. Zhou, J.J. Luo, C. Chen, B. Yang, C. Ge, Y.Q. Wu, J. Tang, Highly anisotropic $Sb_2Se_3$ nanosheets: Gentle Exfoliation from the bulk precursors possessing 1D crystal structure, *Adv. Mater.*, 29 (2017) 1700441.

[7] M. Luo, M.Y. Leng, X.S. Liu, J. Chen, C. Chen, S.K. Qin, J. Tang, Thermal evaporation and characterization of superstrate $CdS/Sb_2Se_3$ solar cells, *Appl. Phys. Lett.*, 104 (2014) 173904.

[8] C. Yan, J. Huang, K. Sun, S. Johnston, Y. Zhang, H. Sun, A. Pu, M. He, F. Liu, K. Eder, L. Yang, J.M. Cairney, N.J. Ekins-Daukes, Z. Hameiri, J.A. Stride, S. Chen, M.A. Green, X. Hao, $Cu_2ZnSnS_4$ solar cells with over 10% power conversion efficiency enabled by heterojunction heat treatment, *Nat. Energy*, 3 (2018) 764–772.

[9] W. Wang, M.T. Winkler, O. Gunawan, T. Gokmen, T.K. Todorov, Y. Zhu, D.B. Mitzi, Device characteristics of CZTSSe thin-film solar cells with 12.6% efficiency, *Adv. Energy Mater.*, 4 (2014) 403–410.

[10] R. Kamada, T. Yagioka, S. Adachi, A. Handa, K.F. Tai, T. Kato, H. Sugimoto, New world record $Cu(In, Ga)(Se, S)_2$ thin film solar cell efficiency beyond 22%, (2016) 1287–1291.

[11] S. Frontier, Solar frontier achieves world record thin-film solar cell efficiency of 22.9% (2017). Available at: http://www.solar-frontier.com/eng/news/2017/1220_press.html, accessed: June 2018.

[12] http://www.pv-tech.org/news/first-solar-pushes-cdte-cell-efficiency-to-record-22.1, accessed: February 23, 2016.

[13] M.A. Green, Y. Hishikawa, E.D. Dunlop, D.H. Levi, J. Hohl-Ebinger, A.W.Y. Ho-Baillie, Solar cell efficiency tables (version 51), *Prog. Photovoltaics Res. Appl.*, 26 (2018) 3–12.

[14] K. Bindu, M.T.S. Nair, P.K. Nair, Chemically deposited Se thin films and their use as a planar source of selenium for the formation of metal selenide layers, *J. Electrochem. Soc.*, 153 (2006) C526–C534.

[15] M. Sarah, M.T.S. Nair, P.K. Nair, Antimony selenide absorber thin films in all-chemically deposited solar cells, *J. Electrochem. Soc.*, 156 (2009) H327–H332.

[16] C.C. Yuan, L.J. Zhang, W.F. Liu, C.F. Zhu, Rapid thermal process to fabricate $Sb_2Se_3$ thin film for solar cell application, *Solar Energy*, 137 (2016) 256–260.

[17] Z.Q. Li, X. Chen, H.B. Zhu, J.W. Chen, Y.T. Guo, C. Zhang, W. Zhang, X.N. Niu, Y.H. Mai, $Sb_2Se_3$ thin film solar cells in substrate configuration and the back contact selenization, *Sol. Energy Mater. Sol. Cells*, 161 (2017) 190–196.

[18] X. Liu, J. Chen, M. Luo, M. Leng, Z. Xia, Y. Zhou, S. Qin, D.J. Xue, L. Lv, H. Huang, D. Niu, J. Tang, Thermal evaporation and characterization of $Sb_2Se_3$ thin film for substrate $Sb_2Se_3$/CdS solar cells, *ACS Appl. Mater. Interfaces*, 6 (2014) 10687–10695.

[19] X.S. Liu, X. Xiao, Y. Yang, D.J. Xue, D.B. Li, C. Chen, S.C. Lu, L. Gao, Y.S. He, M.C. Beard, G. Wang, S.Y. Chen, J. Tang, Enhanced $Sb_2Se_3$ solar cell performance through theory-guided defect control, *Prog. Photovoltaics Res. Appl.*, 25 (2017) 861–870.

[20] L. Zhiqiang, Z. Hongbing, G. Yuting, N. Xiaona, C. Xu, Z. Chong, Z. Wen, L. Xiaoyang, Z. Dong, C. Jingwei, M. Yaohua, Efficiency enhancement of $Sb_2Se_3$ thin-film solar cells by the co-evaporation of Se and $Sb_2Se_3$, *Appl. Phys. Exp.*, 9 (2016) 052302.

[21] M.Y. Leng, M. Luo, C. Chen, S.K. Qin, J. Chen, J. Zhong, J. Tang, Selenization of $Sb_2Se_3$ absorber layer: An efficient step to improve device performance of $CdS/Sb_2Se_3$ solar cells, *Appl. Phys. Lett.*, 105 (2014) 083905.

[22] S. Chen, A. Walsh, X.G. Gong, S.H. Wei, Classification of lattice defects in the kesterite $Cu_2ZnSnS_4$ and $Cu_2ZnSnSe_4$ earth-abundant solar cell absorbers, *Adv. Mater.*, 25 (2013) 1522–1539.

[23] M.A. Tumelero, R. Faccio, A.A. Pasa, Unraveling the Native conduction of trichalcogenides and its ideal band alignment for new photovoltaic interfaces, *J. Phys. Chem. C*, 120 (2016) 1390–1399.

[24] G.X. Liang, X.H. Zhang, H.L. Ma, J.G. Hu, B. Fan, Z.K. Luo, Z.H. Zheng, J.T. Luo, P. Fan, Facile preparation and enhanced photoelectrical performance of $Sb_2Se_3$ nano-rods by magnetron sputtering deposition, *Sol. Energy Mater. Sol. Cells*, 160 (2017) 257–262.

[25] G.X. Liang, Z.H. Zheng, P. Fan, J.T. Luo, J.G. Hu, X.H. Zhang, H.L. Ma, B. Fan, Z.K. Luo, D.P. Zhang, Thermally induced structural evolution and performance of $Sb_2Se_3$ films and nanorods prepared by an easy sputtering method, *Sol. Energy Mater. Sol. Cells*, 174 (2018) 263–270.

[26] C.C. Yuan, X. Jin, G.S. Jiang, W.F. Liu, C.F. Zhu, $Sb_2Se_3$ solar cells prepared with selenized dc-sputtered metallic precursors, *J. Mater. Sci. Mater. Electron.*, 27 (2016) 8906–8910.

[27] L. Wang, M. Luo, S.K. Qin, X.S. Liu, J. Chen, B. Yang, M.Y. Leng, D.J. Xue, Y. Zhou, L. Gao, H.S. Song, J. Tang, Ambient $CdCl_2$ treatment on CdS buffer layer for improved performance of $Sb_2Se_3$ thin film photovoltaics, *Appl. Phys. Lett.*, 107 (2015) 143902.

[28] X.X. Wen, Y.S. He, C. Chen, X.S. Liu, L. wang, B. Yang, M.Y. Leng, H.B. Song, K. Zeng, D.B. Li, K.H. Li, L. Gao, J. Tang, Magnetron sputtered ZnO buffer layer for $Sb_2Se_3$ thin film solar cells, *Sol. Energy Mater. Sol. Cells*, 172 (2017) 74–81.

[29] T. Minemoto, T. Negami, S. Nishiwaki, H. Takakura, Y. Hamakawa, Preparation of $Zn_{1-x}Mg_xO$ films by radio frequency magnetron sputtering, *Thin Solid Films*, 372 (2000) 173–176.

[30] L.S. Cheng, Z. Yang, C. Chao, Z. Ying, L.D. Bing, L.K. Hua, C.W. Hao, W.X. Xing, W. Chong, K. Rokas, L. Nathan, T. Jiang, $Sb_2Se_3$ Thin-film photovoltaics using aqueous solution sprayed $SnO_2$ as the buffer layer, *Adv. Electron. Mater.*, 4 (2018) 1700329.

[31] C. Chen, Y. Zhao, S.C. Lu, K.H. Li, Y. Li, B. Yang, W. H. Chen, L. Wang, D.B. Li, H. Deng, F. Yi, J. Tang, Accelerated optimization of $TiO_2/Sb_2Se_3$ Thin film solar cells by high-throughput combinatorial approach, *Adv. Energy Mater.*, 7 (2017) 1700866.

[32] X.S. Liu, C. Chen, L. Wang, J. Zhong, M. Luo, J. Chen, D.J. Xue, D.B. Li, Y. Zhou, J. Tang, Improving the performance of $Sb_2Se_3$ thin film solar cells over 4% by controlled addition of oxygen during film deposition, *Prog. Photovoltaics Res. Appl.*, 23 (2015) 2892.

[33] J.T. Heath, J.D. Cohen, W.N. Shafarman, Bulk and metastable defects in $CuIn_{1-x}Ga_xSe_2$ thin films using drive-level capacitance profiling, *J. Appl. Phys.*, 95 (2004) 1000–1010.

[34] H.S. Duan, W.B. Yang, B. Bob, C.J. Hsu, B. Lei, Y. Yang, The role of sulfur in solution-processed $Cu_2ZnSn(S, Se)_4$ and its effect on defect properties, *Adv. Funct. Mater.*, 23 (2013) 1466–1471.

[35] S. Heo, G. Seo, Y.H. Lee, D. Lee, M. Seol, J. Lee, J.B. Park, K. Kim, D.J. Yun, Y.S. Kim, J.K. Shin, T.K. Ahn, M.K. Nazeeruddin, Deep level trapped defect analysis in $CH_3NH_3PbI_3$ perovskite solar cells by deep level transient spectroscopy, *Energy Environ. Sci.*, 10 (2017) 1128–1133.

[36] T. Song, A. Kanevce, J.R. Sites, Emitter/absorber interface of CdTe solar cells, *J. Appl. Phys.*, 119 (2016) 233104.

[37] F. Giordano, A. Abate, J.P. Correa Baena, M. Saliba, T. Matsui, S.H. Im, S.M. Zakeeruddin, M.K. Nazeeruddin, A. Hagfeldt, M. Graetzel, Enhanced electronic properties in mesoporous $TiO_2$ via lithium doping for high-efficiency perovskite solar cells, *Nat. Commun.*, 7 (2016) 10379.

[38]  A. Chirila, P. Reinhard, F. Pianezzi, P. Bloesch, A.R. Uhl, C. Fella, L. Kranz, D. Keller, C. Gretener, H. Hagendorfer, D. Jaeger, R. Erni, S. Nishiwaki, S. Buecheler, A.N. Tiwari, Potassium-induced surface modification of $Cu(In, Ga)Se_2$ thin films for high-efficiency solar cells, *Nat. Mater.*, 12 (2013) 1107–1111.

[39]  K. Zeng, D.J. Xue, J. Tang, Antimony selenide thin-film solar cells, *Semicond. Sci. Technol.*, 31 (2016) 063001.

[40]  A. Jasenek, U. Rau, V. Nadenau, H.W. Schock, Electronic properties of $CuGaSe_2$-based heterojunction solar cells. Part II. Defect spectroscopy, *J. Appl. Phys.*, 87 (2000) 594–602.

[41]  C. Chen, L. Wang, L. Gao, D. Nam, D.B. Li, K.H. Li, Y. Zhao, C. Ge, H. Cheong, H. Liu, H.S. Song, J. Tang, 6.5% Certified efficiency $Sb_2Se_3$ solar cells using PbS colloidal quantum dot film as hole-transporting layer, *ACS Energy Lett.*, 2 (2017) 2125–2132.

[42]  X.B. Hu, J.H. Tao, G. Weng, J.C. Jiang, S.Q. Chen, Z.Q. Zhu, J.H. Chu, Investigation of electrically-active defects in $Sb_2Se_3$ thin-film solar cells with up to 5.91% efficiency via admittance spectroscopy, *Sol. Energy Mater. Sol. Cells*, 186 (2018) 324–329.

[43]  R. de L. Kronig, On the theory of dispersion of X-rays, *J. Opt. Soc. Am.*, 12 (1926) 547–557.

[44]  R. Herberholz, T. Walter, C. Müller, T. Friedlmeier, H.W. Schock, M. Saad, M.C. Lux-Steiner, V. Alberts, Meyer-Neldel behavior of deep level parameters in heterojunctions to $Cu(In, Ga)(S, Se)_2$, *Appl. Phys. Lett.*, 69 (1996) 2888–2890.

[45]  X. Hu, J. Tao, G. Weng, J. Jiang, S. Chen, Z. Zhu, J. Chu, Investigation of electrically-active defects in $Sb_2Se_3$ thin-film solar cells with up to 5.91% efficiency via admittance spectroscopy, *Solar Energy Mater. Solar Cells*, 186 (2018) 324–329.

[46]  X. Wen, C. Chen, S. Lu, K. Li, R. Kondrotas, Y. Zhao, W. Chen, L. Gao, C. Wang, J. Zhang, G. Niu, J. Tang, Vapor transport deposition of antimony selenide thin film solar cells with 7.6% efficiency, *Nat. Commun.*, 9 (2018) 2179.

[47]  K. Taretto, U. Rau, J.H. Werner, Numerical simulation of grain boundary effects in $Cu(In, Ga)Se_2$ thin-film solar cells, *Thin Solid Films*, 480 (2005) 8–12.

[48]  D.V. Lang, Deep-level transient spectroscopy: A new method to characterize traps in semiconductors, *J. Appl. Phys.*, 45 (1974) 3023–3032.

[49]  F.D. Auret, M. Nel, Detection of minority-carrier defects by deep level transient spectroscopy using Schottky barrier diodes, *J. Appl. Phys.*, 61 (1987) 2546–2549.

[50]  N. Fourches, A quantitative treatment for deep level transient spectroscopy under minority-carrier injection, *J. Appl. Phys.*, 70 (1991) 209–214.

[51]  L.L. Kerr, S.S. Li, S.W. Johnston, T.J. Anderson, O.D. Crisalle, W.K. Kim, J. Abushama, R.N. Noufi, Investigation of defect properties in $Cu(In, Ga)Se_2$ solar cells by deep-level transient spectroscopy, *Solid State Electron.*, 48 (2004) 1579–1586.

[52] R.M. Fleming, C.H. Seager, D.V. Lang, J.M. Campbell, Injection deep level transient spectroscopy: An improved method for measuring capture rates of hot carriers in semiconductors, *J. Appl. Phys.*, 118 (2015) 015703.

[53] T.P. Nguyen, Defects in organic electronic devices, *Phys. Status Solidi A*, 205 (2008) 162–166.

[54] X. Wen, C. Chen, S. Lu, K. Li, R. Kondrotas, Y. Zhao, W. Chen, L. Gao, C. Wang, J. Zhang, G. Niu, J. Tang, Vapor transport deposition of antimony selenide thin film solar cells with 7.6% efficiency, *Nat. Commun.*, 9 (2018) 2179.

[55] N. Dharmarasu, M. Yamaguchi, J.C. Bourgoin, T. Takamoto, T. Ohshima, H. Itoh, M. Imaizumi, S. Matsuda, Majority- and minority-carrier deep level traps in proton-irradiated n+/p-InGaP space solar cells, *Appl. Phys. Lett.*, 81 (2002) 64–66.

[56] P.P. Boix, G. Garcia-Belmonte, U. Muñecas, M. Neophytou, C. Waldauf, R. Pacios, Determination of gap defect states in organic bulk heterojunction solar cells from capacitance measurements, *Appl. Phys. Lett.*, 95 (2009) 233302.

[57] C.M. Tejas S. Sherkar, Lid, J. on Gil-Escrig, M. Avila, H.J.B. Sessolo, and L. Jan Anton Koster, Recombination in Perovskite solar cells: significance of grain boundaries, interface traps and defect ions, *ACS Energy Lett.*, 2 (2017) 1214–1222.

Chapter 5

# Microwave-Assisted Chemical Bath Deposition Method for Quantum Dot-Sensitized Solar Cells

Ke Ning and Guang Zhu*

*Key Laboratory of Spin Electron and Nanomaterials
of Anhui Higher Education Institutes,
Suzhou University, Suzhou 234000, P. R. China*
*guangzhu@ahszu.edu.cn

## 1. Introduction

Sensitized solar cells (SSCs) have attracted considerable attention and represent a key class of cell architecture that has emerged as a promising candidate for the development of next-generation solar cells due to their acceptable power conversion efficiency and low production cost [1–15]. As sensitizers for SSCs, inorganic semiconductor quantum dots (QDs), such as PbS, CdS, CdSe and InAs, have been suggested along with organometallic or organic dyes because QD sensitizers have advantages of high extinction coefficients, spectral tunability by particle size and good stability, which are known to increase the overall power conversion efficiency of solar cells [16–19], and their working mechanism is depicted in Fig. 1 involving key processes (1)–(7). In brief, electrons, excited from the valance band (VB) to the conduction band (CB) of QDs by absorbing light (path (1)), are rapidly injected into the CB of $TiO_2$ particles (path (2)) and then transported to F-doped $SnO_2$ (FTO) (path (3)). The oxidized QDs are regenerated by accepting

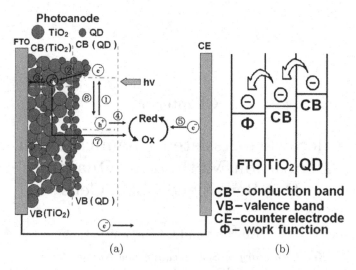

**Fig. 1.** Schematic diagram of (a) working mechanism and (b) energy band structure of QDSSCs.

the electrons from the counter electrode (CE) (path (5)) via a redox pair (path (4)). However, the charge recombination (paths (6) and (7)) will happen in the meantime and deteriorate the cell performance. Therefore, the improvement of the performance of QDs and their interconnectivity with the $TiO_2$ substrate will facilitate the electron transfer from QDs to $TiO_2$ and decrease the back transport electrons from $TiO_2$ to electrolyte, which is very important for the cell efficiency. Many studies have also been devoted to exploring different fabrication techniques to attach QDs onto $TiO_2$. Therefore, in quantum dot-sensitized solar cells (QDSSCs), the performance of QDs and their interconnectivity with the $TiO_2$ substrate are essential for device performance. Many studies have been devoted to exploring different fabrication techniques to attach QDs onto $TiO_2$. So far, the QDs have been successfully sensitized on the surface of $TiO_2$ by a self-assembled monolayer (SAM) through the linking of bifunctional molecules or direct attachment [16, 20], chemical bath deposition (CBD) [19, 21, 22], electrochemical deposition [23, 24] or photodeposition [25] techniques. Recently, Gao *et al.* [26] and

Lee *et al.* [27] studied a close-space sublimation technique in which CdS powder was heated at 500°C in a furnace with Ar or Ar/H$_2$ gas flow to form CdS nanoparticles on the surface of TiO$_2$ for QDSSCs. Lee *et al.* [28] reported that QDSSCs based on CdS-coated TiO$_2$ electrodes using the spray pyrolysis deposition (SPD) method showed a power conversion efficiency of 1.84% in I$^-$/I$^{3-}$ electrolyte and 0.87% in polysulfide electrolyte.

As an inexpensive, quick, clean, versatile technique, microwave irradiation induces interaction of the dipole moment of polar molecules or molecular ionic aggregates with alternating electronic and magnetic fields, causing molecular-level heating which leads to homogeneous and quick thermal reactions [29, 30]. If such heating is performed in closed vessels under high pressure, more energy input at the same temperature and enormous accelerations in reaction time can be achieved, which means that a reaction in the conventional solvothermal method that takes several hours can be completed over the course of minutes or a reaction that would not have proceeded previously will now proceed, typically with higher yields [31]. However, the microwave technique has been seldom studied until now for its abilities to coat different QDs onto TiO$_2$ electrode for QDSSCs although such a method has been used successfully to fabricate QDs [32–34], nanowires [35] and nanotubes [36].

Herein, we have synthesized different QDSSCs (such as CdS, CdSe, PbS and CdS/CdSe) using a simple, rapid and effective microwave-assisted chemical bath deposition (MACBD) technique [37–39]. Compared with the conventional CBD technique, the MACBD technique can rapidly synthesize QDs and precisely control their sizes with a narrow distribution, improve the wettability of the TiO$_2$ surface and form good contact between QDs and the TiO$_2$ layer due to rapidly elevated temperature during microwave irradiation. Furthermore, this technique can offer an easy control over all experimental parameters without the requirement of a repetitive immersing operation, organic linker or high-temperature heating. The results indicated that the as-prepared cells have high conversion efficiency under one sun illumination (AM 1.5 G, 100 mW cm$^{-2}$).

## 2. Experimental section

The $TiO_2$ electrode was prepared by screen printing of $TiO_2$ paste on the F-doped $SnO_2$ (FTO) (resistivity: 14 $\Omega/\square$, Nippon Sheet Glass, Japan) glass [40]. The electrode configuration involved a transparent layer of nanocrystalline $TiO_2$ (P25, Degussa) with a mean size of 25 nm and a scattering layer microcrystalline $TiO_2$ (Dalian HeptaChroma SolarTech Co., Ltd.) with a mean size of 200 nm. Such a double-layer structure favors contact between the substrate and electrode and enhances the light scattering ability, which can improve the performance of the cells [56]. The as-prepared electrodes were sintered at 500°C for 30 min.

The as-prepared $TiO_2$ electrodes were sensitized by different QDs by the MACBD process [37–39], as shown in Fig. 2. For QD deposition, the electrodes were immersed into a sealed vessel with a precursor aqueous solution, and then the vessel was put into an automated focused microwave system (Explorer-48, CEM Co.) and treated at 150°C with the microwave irradiation power of 100 W for 30 min. In experiments, the effect of synthesis conditions on the device performance has been carefully studied and their selection is based on our previous works [37–39]. In this work, in order to compare the performances of QDSSCs using different methods, QDs were also deposited on the $TiO_2$ film by the SCBD method.

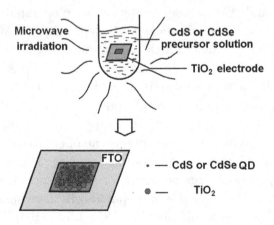

Fig. 2. Schematic diagram of MACBD process.

The morphology and structure of the as-prepared electrodes were characterized using a Hitachi 4800 field emission scanning electron microscope (FESEM) and a JEOL-2010 high-resolution transmission electron microscope (HRTEM), respectively. The UV–Vis absorption spectra of electrodes were detected using a UV–Vis spectrophotometer (Hitachi U-3900).

The different QDSSCs were sealed in a sandwich structure with a 25-$\mu$m spacer (Surlyn) using thin Au-sputtered FTO glass as the counter electrode. Water/methanol (3:7 by volume) solution was used as a co-solvent of the polysulfide electrolyte [57]. Electrolyte solution consists of 0.5 M $Na_2S$, 2 M S and 0.2 M KCl. The active area of the cell is 0.2 cm$^2$. $J$–$V$ measurement was performed with a Keithley model 2440 Source Meter and a Newport solar simulator system (equipped with a 1-kW xenon arc lamp, Oriel) under one sun illumination (AM 1.5 G, 100 mW cm$^{-2}$). Incident photon to current conversion efficiency (IPCE) was measured as a function of wavelength from 300 to 800 nm using an Oriel 300W xenon arc lamp and a lock-in amplifier M 70104 (Oriel) under monochromator illumination. The electrochemical impedance spectroscopy (EIS) measurements were carried out in dark conditions at a forward bias of 0–0.5 V, applying a 10-mV AC sinusoidal signal over the constant applied bias with the frequency ranging between 100 kHz and 0.1 Hz (Autolab, PGSTAT 302N and FRA2 module).

## 3. Results and discussion

### 3.1. *CdS-dot-sensitized solar cells*

Figures 3(a) and 3(b) show the FESEM images (top view) of NC-$TiO_2$ underlayer and MC-$TiO_2$ overlayer films, respectively. The $TiO_2$ electrode is constructed by a random agglomeration of $TiO_2$ particles. The porous structure of $TiO_2$ favors easy penetration of electrolyte, as well as Cd and S precursors, during deposition. The cross-sectional FESEM image in Fig. 3(c) displays clearly the two-layer $TiO_2$ electrode consisting of a 10-$\mu$m thick compact P25 NC-$TiO_2$ transparent layer and a 3-$\mu$m thick loose MC-$TiO_2$ scattering layer. Such a double-layer structure can favor the contact between

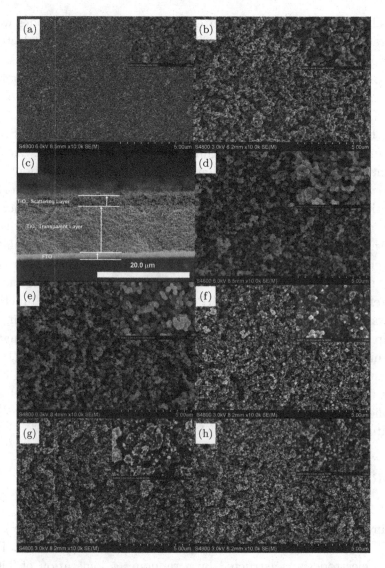

**Fig. 3.** Surface morphologies of (a) NC-TiO$_2$ underlayer, (b) MC-TiO$_2$ overlayer, (d) electrode 1, (e) electrode 2, (f) electrode 3, (g) electrode 4 and (h) electrode 5 measured by FESEM; insets are corresponding magnified FESEM images. (c) Cross-sectional FESEM image of as-prepared TiO$_2$ electrode. (TiO$_2$/CdS films fabricated using 0.01, 0.015, 0.025, 0.05 and 0.1 M Cd(NO$_3$)$_2$ and CH$_4$N$_2$S solutions are named electrodes 1, 2, 3, 4 and 5, respectively.)

the substrate and electrode and enhance the light scattering ability, which can improve the performance of the cells [41]. Figures 3(d)–3(h) show the FESEM images (top view) of the electrodes 1–5, respectively. It can be observed from Figs. 3(d) and 3(e) that a small amount of CdS has been distributed on the surface of the $TiO_2$ film as compared with the bare $TiO_2$ film (Fig. 3(b)) due to a low concentration of precursor solution. A more compact surface with a porous structure is observed from Figs. 3(f) and 3(g), indicating that with the increase of the precursor concentration, the amount of CdS deposited on the $TiO_2$ film increases gradually. However, excessive deposition of CdS blocks most pores in the electrode, as seen from Fig. 3(h), which is not beneficial to access the electrolyte deep in the surface porous structure. The composition of electrode 4 was identified by energy-dispersive X-ray spectroscopy (EDS) linked to FESEM, as shown in Fig. 4. Quantitative analysis of the EDS spectrum gives a Cd:S atomic ratio of about 1, indicating that high-grade CdS particles are formed. Electrodes 1, 2, 3 and 5 have similar EDS results.

Figure 5(a) shows a high-magnification HRTEM image of CdS QDs prepared using a 0.05-M precursor aqueous solution of $Cd(NO_3)_2$ and $CH_4N_2S$ with microwave irradiation power of 100 W

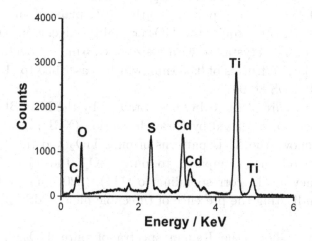

**Fig. 4.** EDS spectrum of electrode 4.

**Fig. 5.** (a) High-magnification HRTEM image of CdS prepared by MACBD; (b) low-magnification and (c) high-magnification HRTEM images of electrode 4, inset is the corresponding SAED pattern.

at 150°C for 30 min. It is clearly found that the as-prepared CdS QDs have a fine crystallite and a narrow size distribution of $\sim$5 nm. Figure 5(b) shows a low-magnification HRTEM image of electrode 4. It is observed that the aggregation is composed of small NC-$TiO_2$ and bigger MC-$TiO_2$ particles. CdS is difficult to be observed in the low-magnification HRTEM image. The corresponding selected area electron diffraction (SAED) pattern in an upper right inset in Fig. 5(b) indicates that the $TiO_2$/CdS film is a polycrystalline structure. Figure 5(c) shows a high-magnification HRTEM image of electrode 4. The larger crystallites are identified as MC-$TiO_2$ (left) and NC-$TiO_2$ (right). The lattice spacing measured for the crystalline plane is 0.352 nm, corresponding to the (101) plane of anatase $TiO_2$ (JCPDS 21-1272). Around the $TiO_2$ crystallite edge, a fine crystallite is observed. The crystallite with a size of ca 5 nm connecting to the $TiO_2$ has lattice fringes of 0.335 nm, which is ascribed to (111) plane of CdS (JCPDS 80-0019).

The structure of the CdS QDs obtained by the MACBD method was further characterized by X-ray diffraction (XRD) measurement. Figure 6 shows the XRD patterns of pure $TiO_2$ and the $TiO_2$/CdS film (electrode 4). Compared to pure $TiO_2$, the $TiO_2$/CdS film exhibits new peaks corresponding to (111), (220) and (311) planes of CdS, indicating the presence of the cubic phase CdS (JCPDS 80-0019).

Figure 7 shows the Raman spectra of pure $TiO_2$, $TiO_2$/CdS (MACBD, electrode 4), $TiO_2$/CdS(CBD) and $TiO_2$/CdS(SPD)

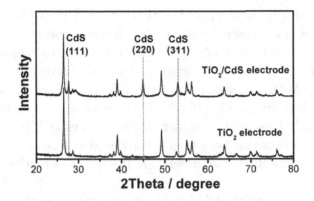

**Fig. 6.** XRD patterns of pure $TiO_2$ electrode and $TiO_2/CdS$ electrode 4.

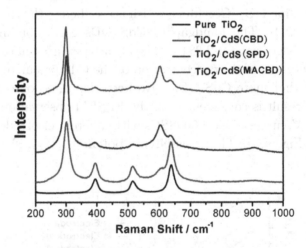

**Fig. 7.** Raman spectra of pure $TiO_2$, $TiO_2/CdS(MACBD)$, $TiO_2/CdS(CBD)$ and $TiO_2/CdS(SPD)$ electrodes.

films. The spectrum of $TiO_2$ is dominated by 395, 516 and 637 cm$^{-1}$ bands, characteristic of the $TiO_2$ anatase phase. Compared with a pure $TiO_2$ film, a longitudinal optical (LO) mode at around 300 cm$^{-1}$, together with its overtones at 600 and 900 cm$^{-1}$ for CdS, is observed in the spectra of all $TiO_2/CdS$ films, showing a combination of two semiconducting characteristic bands. It should be noticed that the LO peaks of $TiO_2/CdS(CBD)$ and $TiO_2/CdS(SPD)$ exhibit a small redshift as compared with that of TiO$_2$/CdS(MACBD), which reflects a certain degree of phonon confinement and indicates that the particle size of CdS QDs via MACBD deposition is somewhat larger [42]. Furthermore, the LO bandwidths of both TiO$_2$/CdS(MACBD) and TiO$_2$/CdS(SPD) are almost the same but narrower than that of TiO$_2$/CdS(CBD), indicating that as compared with the CBD method, a higher degree of crystallinity and fewer structural defects can be attained via the MABCD method, which does not require the high-temperature treatment as used in the SPD method.

Figure 8 displays the UV–Vis absorption spectra of the pure TiO$_2$ film and TiO$_2$/CdS electrodes 1–5. Compared with the absorption spectra of the pure TiO$_2$ film, there is an obvious absorption peak near 500 nm for TiO$_2$/CdS films, which is ascribed to the contribution from CdS QDs. The bandgap of CdS QDs corresponding to the absorption edge is about 2.38 eV. The absorbance gradually increases with the increase of the concentration of the CdS precursor solution, indicating that more CdS QDs have been deposited onto the TiO$_2$ film. This result is consistent with the FESEM observation.

The $J$–$V$ curves of CdS QDSSCs with different electrodes 1–5 are shown in Fig. 9(a). The open-circuit potential ($V_{oc}$), short-circuit

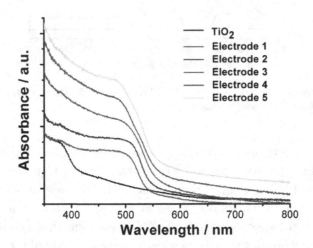

**Fig. 8.** UV–Vis absorption spectra of pure TiO$_2$ and TiO$_2$/CdS electrodes 1–5.

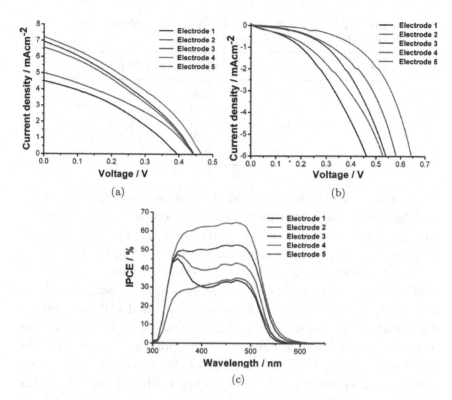

**Fig. 9.** *J–V* curves of CdS QDSSCs with electrodes 1–5 (a) under one sun illumination (AM 1.5 G, $100\,\text{mW cm}^{-2}$) and (b) in dark conditions; (c) IPCE curves of CdS QDSSCs with electrodes 1–5.

current density $(J_{\text{sc}})$, fill factor (FF) and conversion efficiency $(\eta)$ of all cells are listed in Table 1. The $J_{\text{sc}}$ and $V_{\text{oc}}$ of the cell with electrode 1 are $4.49\,\text{mA cm}^{-2}$ and $0.40\,\text{V}$, respectively, resulting in a very low value of conversion efficiency (0.63%). When more CdS QDs are deposited onto the $TiO_2$ film, $J_{\text{sc}}$, $V_{\text{oc}}$ and $\eta$ increase obviously and reach maximum values of $7.20\,\text{mA cm}^{-2}$, $0.46\,\text{V}$ and 1.18%, respectively, at a precursor concentration of 0.05 M, and then decrease with a further increase of CdS amount. At the beginning of the deposition, the MACBD process is supposed to increase the coverage ratio of CdS on the $TiO_2$ surface by replenishing the uncovered area, and the thickness of the CdS layer increases with the increase of the concentration of the CdS precursor solution. Such

**Table 1.** Photovoltaic parameters of CdS QDSSCs with electrodes 1–5.

| Electrode | $J_{sc}$ (mA cm$^{-2}$) | $V_{oc}$ (V) | FF (%) | $\eta$ (%) |
|---|---|---|---|---|
| 1 | 4.49 | 0.40 | 35.5 | 0.63 |
| 2 | 6.56 | 0.44 | 33.8 | 0.98 |
| 3 | 6.93 | 0.44 | 33.8 | 1.04 |
| 4 | 7.20 | 0.46 | 35.1 | 1.18 |
| 5 | 4.98 | 0.44 | 35.8 | 0.79 |

an increment of CdS loading leads to more excited electrons under the illumination of light, which is advantageous to the photocurrent. An optimized thickness of the CdS layer is obtained at some concentration of precursor solution, which results in the highest $J_{sc}$ and $\eta$ by providing a good interfacial structure between TiO$_2$ and CdS films and reducing the recombination of the injected electrons from TiO$_2$ to the electrolyte due to a well-covered CdS layer on the TiO$_2$ surface. However, as the thickness of the CdS layer further increases, it will be more difficult to transport an electron from the CdS layer into the TiO$_2$ film because of the increase in the average distance and electron injection time from CdS to TiO$_2$ [42]. In the meantime, the CdS/electrolyte contacting area will decrease with the increase of the CdS amount because more pores are probably blocked by the additional loading of CdS, leading to unfavorable electron transportation at the TiO$_2$/CdS/electrolyte interface [43, 44]. This inference was confirmed by the $J$–$V$ curves of the cells with different electrodes 1–5 in dark conditions, as shown in Fig. 9(b). The applied voltage required to drive the electrons across the photoelectrodes is highest for the cell with electrode 4, which is due to a well-covered CdS layer on the TiO$_2$ surface. This result indicates that the cell with electrode 4 has a superior interfacial structure to inhibit the interfacial recombination of the injected electrons from TiO$_2$ to the electrolyte [45, 46], which is also responsible for its higher conversion efficiency. The incident photon to current conversion efficiency (IPCE) curves of CdS QDSSCs with electrodes 1–5, as shown in Fig. 9(c), exhibit a similar trend as $J$–$V$ curves. A maximum IPCE value of 65% at 470 nm is obtained for the cell with electrode 4.

The charge transfer and recombination behavior in the CdS QDSSCs were further studied by analyzing the EIS spectra at various applied voltages in dark conditions. Figure 8(a) shows the typical Nyquist plots of the cells with electrodes 1–5 obtained at an applied voltage of 0.4 V. The EIS spectra are characterized by the presence of two semicircles in a Nyquist plot [47, 48]. The high-frequency semicircle is related to the charge transfer resistance ($R_{ct}$) at the interfaces of the electrolyte/counter electrode and the low-frequency one is due to the contribution from the chemical capacitance of nanostructured $TiO_2$ ($C_\mu$) and the charge recombination resistance ($R_{rec}$) between $TiO_2$ and the polysulfide electrolyte [47, 48]. The corresponding equivalent circuit is shown in the inset of Fig. 10(a) [49, 50]. Figure 10(b) shows the $R_{rec}$ of the cells with electrodes 1–5 at various applied potentials (0–0.6 V) obtained from the EIS fitting. It can be observed that $R_{rec}$ decreases with the increase of applied potential and the cell with electrode 4 shows the highest recombination resistance (i.e., lowest recombination) compared to other cells, which explains the highest $J_{sc}$ measured for the cell with electrode 4. It should be pointed out that the lower $R_{rec}$ of the cell with electrode 5 as compared with the cell with electrode 4 should be ascribed to the loss of effective electrochemical active area of the electrode because excessive CdS hinders the access of electrolyte deep

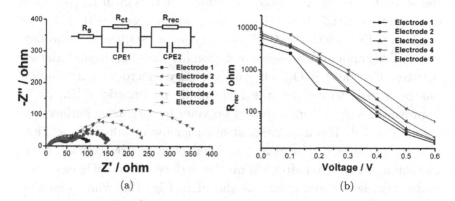

**Fig. 10.** (a) Nyquist plots of the cells with electrodes 1–5 at an applied voltage of 0.4 V. Inset displays the corresponding equivalent circuit and (b) recombination resistance ($R_{rec}$), as a function of applied voltage.

**Table 2.** Comparison of photovoltaic performances of CdS QDSSCs fabricated using CBD, SPD and MACBD methods.

| Method | $J_{sc}$ (mA cm$^{-2}$) | $V_{oc}$ (V) | FF (%) | $\eta$ (%) |
|--------|-------------------------|--------------|--------|------------|
| CBD    | 5.45                    | 0.49         | 39     | 1.06       |
| SPD    | 4.00                    | 0.50         | 37     | 0.76       |
| MACBD  | 7.20                    | 0.46         | 35     | 1.18       |

in the surface porous structure. The EIS result is consistent with the dark current measurement in Fig. 10(b).

The photovoltaic performances of QDSSCs based on $TiO_2/$ CdS(MACBD), $TiO_2$/CdS(CBD) and $TiO_2$/CdS(SPD) electrodes under one sun illumination (AM 1.5 G, 100 mW cm$^{-2}$) are compared and the results are summarized in Table 2. It can be observed that higher $J_{sc}$ and $\eta$ values are achieved via the MACBD method as compared with the CBD and SPD methods. The higher $J_{sc}$ and $\eta$ for the cell using the MACBD method should be ascribed to the following reasons: (i) Good contact between CdS QDs and the $TiO_2$ layer is formed due to rapidly elevated temperature during microwave irradiation. CdS deposition on the surface of $TiO_2$ via the MACBD method is a three-step process (as shown in Fig. 11). In the first step, positively charged $Cd^{2+}$ ions form a complex with the $TiO_2$ film. In the second step, thermal decomposition of $CH_4N_2S$ in the presence of microwave irradiation releases $S^{2-}$, which reacts with $Cd^{2+}$ to produce CdS nuclei. The CdS nuclei grow, crystallize and stabilize on the $TiO_2$ under microwave irradiation [51], leading to good contact between CdS and $TiO_2$. On the contrary, microwave irradiation has been used to prepare the hydrophilic nanoparticles [52] or to improve the surface wettability of polymer by increasing surface free energy [53, 54]. The improvement of surface wettability of the $TiO_2$ film by microwave treatment has been observed by contact angle measurement of water droplets on the surface of the $TiO_2$ electrode under microwave treatment (as shown in Fig. 12), which can also favor CdS deposition on the $TiO_2$ film [55] and form good contact between them. Such a superior interface between $TiO_2$ and CdS via the MACBD method can inhibit the interfacial recombination of the

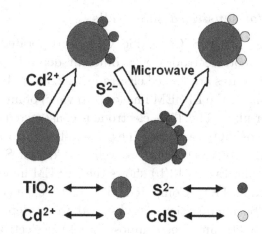

**Fig. 11.** Schematic diagram of the process of CdS deposition on TiO$_2$ film under microwave irradiation.

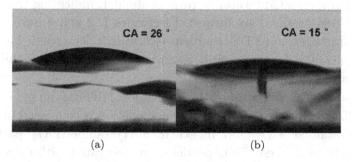

**Fig. 12.** Photographic images of water droplets on the surface of TiO$_2$ electrode (a) before and (b) after microwave treatment.

injected electrons from TiO$_2$ to the electrolyte, which is responsible for its higher $J_{sc}$ and $\eta$. (ii) Microwave irradiation can heat up the aqueous solution homogeneously and fast due to the penetration characteristic of microwaves and the high utilization factor of microwave energy [56, 57]. Therefore, the nucleation and growth of CdS QDs can be finished in an extremely short period of time, which is extraordinarily beneficial for reducing the concentration of surface defects of QDs [58, 59]. The carrier recombination at surface defects of QDs is correspondingly suppressed and thus the cell performance is increased [60].

## 3.2. *CdSe-dot-sensitized solar cells*

In the process of QDSSCs, CdSe QDs incorporated with $TiO_2$ ($TiO_2$/CdSe) photoelectrodes with different deposition times of 5, 10, 20 and 30 minutes are named electrodes 1, 2, 3 and 4, respectively. Figure 13(a) shows the FESEM images (top view) of microcrystalline $TiO_2$ overlayer films. The $TiO_2$ electrode is constructed by a random agglomeration of $TiO_2$ particles. The porous structure of $TiO_2$ favors easy penetration of electrolyte, as well as Cd and Se precursors, during deposition. Figure 13(b) shows the FESEM image (top view) of electrode CdSe/$TiO_2$. Compared with the bare $TiO_2$ film, a great volume of CdSe nanoparticles is observed to be deposited on the surface of the $TiO_2$ film. The composition of the electrode was identified by energy-dispersive X-ray spectroscopy (EDS) measurement, as shown in the inset of Fig. 13(c). Quantitative analysis of the EDS spectrum gives a Cd:Se atomic ratio of about 1, indicating that high-grade CdSe particles are formed. Electrodes 1, 2 and 4 have similar FESEM images and EDS results as electrode 3. The result indicates that CdSe QDs are successfully assembled on the surface of the $TiO_2$ film via the MACBD method.

Figure 14(a) shows a low-magnification HRTEM image of electrode 3. It is clearly found that the aggregation is composed of small nanocrystalline $TiO_2$ and bigger microcrystalline $TiO_2$ particles. The surfaces of these $TiO_2$ particles are decorated with CdSe QDs

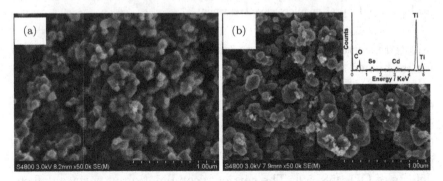

**Fig. 13.** The top view of (a) microcrystalline $TiO_2$ overlayer and (b) $TiO_2$/CdSe electrode 3 by FESEM. Inset is corresponding EDS spectrum.

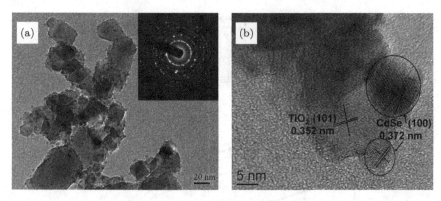

**Fig. 14.** (a) Low-magnification and (b) high-magnification HRTEM images of electrode 3. Inset in (a) is the corresponding SAED pattern.

(black dots). The corresponding selected area electron diffraction (SAED) pattern in an upper right inset in Fig. 14(a) indicates that the $TiO_2$/CdSe film is a polycrystalline structure. Figure 14(b) shows a high-magnification HRTEM image of electrode 3. The lattice spacing measured for the crystalline plane is 0.352 nm, corresponding to the (101) plane of anatase $TiO_2$ (JCPDS 21-1272). Around the $TiO_2$ crystallite edge, fine crystallites are observed. The crystallites connecting to the $TiO_2$ have lattice fringes of 0.372 nm, which is ascribed to (100) plane of CdSe (JCPDS 77-2307). The HRTEM image also indicates that CdSe with size of about 5–10 nm is presented in a crystalline structure.

Figure 15 shows the UV–Vis absorption spectra of the pure $TiO_2$ film and $TiO_2$/CdSe electrodes 1–4. Compared with the absorption spectra of the pure $TiO_2$ film, there is an obvious absorption peak near 700 nm for $TiO_2$/CdSe films, which is ascribed to the contribution from CdSe QDs. The bandgap of CdSe QDs corresponding to the absorption edge is about 1.77 eV, which is slightly higher than that of bulk CdSe due to the quantum size effect [61]. Therefore, the quantum size effect does not play a significant role in the performance of the cells. The absorbance gradually increases with the increase of the deposition time, indicating that more CdSe QDs have been deposited onto the $TiO_2$ film. This is also confirmed by the color

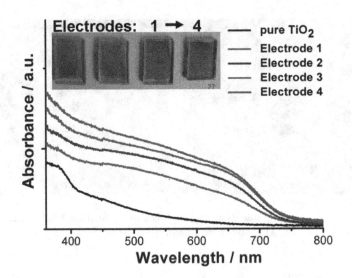

**Fig. 15.** UV–Vis absorption spectra of $TiO_2$ and $TiO_2/CdSe$ electrodes 1–4. Inset shows the photographs of $TiO_2/CdSe$ electrodes 1–4.

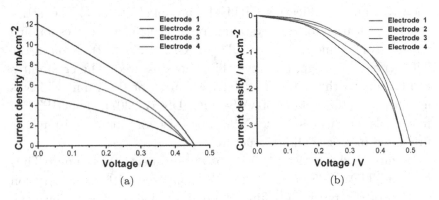

**Fig. 16.** $J–V$ curves of CdSe QDSSCs with electrodes 1–4 (a) under one sun illumination of (AM 1.5 G, $100\,mW\,cm^{-2}$) and (b) in dark conditions.

change, observed through the naked eye, from light brown to dark brown for electrode 1 to electrode 4 in the inset of Fig. 15.

The $J–V$ curves of CdSe QDSSCs with different electrodes 1–4 are shown in Fig. 16(a). The $V_{oc}$, $J_{sc}$, FF and $\eta$ of all cells are listed in Table 3. The $J_{sc}$ and $V_{oc}$ of the cell with electrode 1 are $4.77\,mA\,cm^{-2}$ and $0.44\,V$, respectively, resulting in a very low value of $\eta$ (0.70%).

**Table 3.** Photovoltaic parameters of CdSe QDSSCs with electrodes 1–4.

| Electrode | $J_{sc}$ (mA cm$^{-2}$) | $V_{oc}$ (V) | FF (%) | $\eta$ (%) |
|:---:|:---:|:---:|:---:|:---:|
| 1 | 4.77 | 0.44 | 32.8 | 0.70 |
| 2 | 9.58 | 0.44 | 32.2 | 1.36 |
| 3 | 12.1 | 0.45 | 32.1 | 1.75 |
| 4 | 7.42 | 0.44 | 36.2 | 1.18 |

When the deposition time is prolonged, $J_{sc}$, $V_{oc}$ and $\eta$ increase obviously and reach maximum values of 12.1 mA cm$^{-2}$, 0.45 V and 1.75%, respectively, at a deposition time of 20 min, and then decrease with a further increase of the deposition time. At the beginning of the deposition, the MACBD process is supposed to increase the coverage ratio of CdSe on the TiO$_2$ surface by replenishing the uncovered area and the thickness of CdSe layer increases with the increase of the deposition time. Such an increment of CdSe loading leads to more excited electrons under the illumination of light, which is advantageous to the photocurrent. An optimized thickness of the CdSe layer is obtained at a deposition time of 20 min, which results in the highest $J_{sc}$ and $\eta$ by providing a good interfacial structure between TiO$_2$ and CdSe films and reducing the recombination of the injected electrons from TiO$_2$ to the electrolyte due to a well-covered CdSe on the TiO$_2$ surface. However, as the thickness of the CdSe layer further increases, it will be more difficult to transport an electron from the CdSe outer layer into the TiO$_2$ film. In the meantime, the CdSe/electrolyte contacting area will decrease with the increase of the CdSe amount because more pores are probably blocked by the additional loading of CdSe, leading to unfavorable electron transportation at the TiO$_2$/CdSe/electrolyte interface. This inference was confirmed by the *J–V* curves of the cells with different electrodes 1–4 in dark conditions, as shown in Fig. 16(b). The applied voltage required to drive the electrons across the photoelectrodes is highest for the cell with electrode 3, which is due to a well-covered CdSe on the TiO$_2$ surface. This result indicates that the cell with electrode 3 has a superior interfacial structure to inhibit the

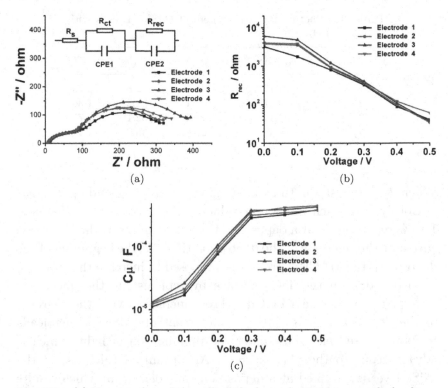

**Fig. 17.** (a) Nyquist plots of the cells with electrodes 1–4 at an applied voltage of 0.3 V. Inset displays the corresponding equivalent circuit. (b) Recombination resistance ($R_{rec}$) and (c) Chemical capacitance ($C_\mu$) as a function of applied voltage.

interfacial recombination of the injected electrons from $TiO_2$ to the electrolyte, which is also responsible for its higher energy conversion efficiency.

The charge transfer and recombination behavior in the CdSe QDSSCs were further studied by analyzing the electrochemical impedance spectroscopy (EIS) at various applied voltages in dark conditions. Figure 17(a) shows the typical Nyquist plots of the cells with electrodes 1–4 obtained at an applied voltage of 0.3 V. The EIS spectra are characterized by the presence of two semicircles in a Nyquist plot. The high-frequency semicircle is related to the charge transfer resistance ($R_{ct}$) at the interfaces of the electrolyte/counter electrode and the low-frequency one is due to the contribution from the chemical capacitance of nanostructured TiO$_2$ ($C_\mu$) and the charge recombination resistance ($R_{rec}$) between TiO$_2$ and the polysulfide electrolyte. The corresponding equivalent circuit is shown in the inset of Fig. 17(a). Figures 17(b) and 17(c) show the $R_{rec}$ and $C_\mu$ of the cells with electrodes 1–4 at various applied potentials (0–0.5 V) obtained from EIS fitting. It can be observed that $R_{rec}$ decreases with an increase of applied potential and the cell with electrode 3 shows the highest recombination resistance (i.e., lowest recombination) compared to other cells, which explains the highest $J_{sc}$ measured for the cell with electrode 3. The result of $R_{rec}$ is consistent with the dark current measurement in Fig. 17(b). In addition, Fig. 17(c) indicates a downward displacement of the TiO$_2$ conduction band (CB), which increases with the increase of the deposition time. The increasing amount of CdSe with the increase of deposition time prevents the direct contact of the polysulfide electrolyte with TiO$_2$, which reduces the TiO$_2$ CB position.

Table 4 compares the photovoltaic performances of QDSSCs with the TiO$_2$ electrode, polysulfide electrolyte and CdSe sensitizer fabricated by the MACBD and other methods under one sun illumination (AM 1.5 G, 100 mW cm$^{-2}$). It can be observed that a higher $J_{sc}$ value of 12.1 mA cm$^{-2}$ is achieved via the MACBD method as compared with SAM, CBD, ED, SPD and a combination of SAM and CBD methods.

**Table 4.** Comparison of photovoltaic performances of CdSe QDSSCs fabricated by different methods.

| Method | $J_{sc}$ (mA cm$^{-2}$) | $V_{oc}$ (V) | FF (%) | $\eta$ (%) |
|---|---|---|---|---|
| SAM [18] | 5.24 | 0.539 | 46 | 0.94 |
| CBD [21] | 8.47 | 0.493 | 36 | 1.5 |
| SAM and CBD [21] | 8.43 | 0.47 | 46 | 1.8 |
| ED [22] | 3.0 | 0.52 | 27 | 0.4 |
| SPD [17] | 10.7 | 0.45 | 35.8 | 1.7 |
| MACBD | 12.1 | 0.45 | 32.1 | 1.75 |

### 3.3. *CdS/CdSe quantum dot-co-sensitized solar cells*

Figure 18(a) shows the FESEM images (top view) of pure $TiO_2$ films. The $TiO_2$ electrode is constructed by a random agglomeration of $TiO_2$ particles. The porous structure of $TiO_2$ favors easy penetration of electrolyte, as well as Cd, S and Se precursors, during deposition. Figure 18(b) shows the FESEM image (top view) of $TiO_2/CdS$ electrode obtained by the MACBD method. Compared with a pure $TiO_2$ electrode, a volume of CdS nanoparticles is observed to be deposited on the surface of the $TiO_2$ film. The composition of the electrode was identified by energy-dispersive X-ray spectroscopy (EDS) measurement, as shown in the inset of Fig. 18(b). Quantitative analysis of the EDS spectrum gives a Cd:S atomic ratio of about 1, indicating that high-grade CdS particles are formed. A more compact surface layer for the $TiO_2/CdS/CdSe$ electrode is clearly observed

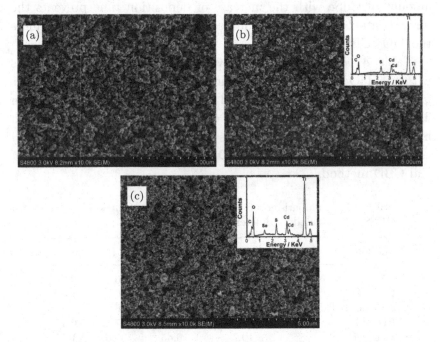

**Fig. 18.** Surface morphologies of (a) pure $TiO_2$ electrode, (b) $TiO_2/CdS$ electrode and (c) $TiO_2/CdS/CdSe$ electrode by FESEM measurement. Insets in (b) and (c) are corresponding EDS spectra.

in Fig. 18(c) as compared with Fig. 18(b), which indicates that an amount of CdSe QDs is assembled on the surface of the $TiO_2/CdS$ electrode by the MACBD method. The Cd, Se and S peaks are clearly observed in the EDS spectrum of the electrode, as shown in inset of Fig. 18(c). Quantitative analysis of the EDS spectrum reveals that the atomic ratio of Cd versus S plus Se is nearly 1, indicating that the deposited CdS and CdSe QDs are likely to be stoichiometric. The result confirms that CdS and CdSe QDs are successfully assembled on the surface of the $TiO_2$ film via the MACBD process.

Figure 19(a) shows a low-magnification HRTEM image of the $TiO_2/CdS/CdSe$ electrode. It is clearly found that the surface of the $TiO_2$ particles is decorated with QDs (black dots). The corresponding selected area electron diffraction (SAED) pattern in the inset of Fig. 19(a) indicates that the electrode is a polycrystalline structure. Figure 19(b) shows a high-magnification HRTEM image of the $TiO_2/CdS/CdSe$ electrode. The lattice spacing measured for the crystalline plane is 0.352 nm, corresponding to the (101) plane of $TiO_2$ (JCPDS 21-1272). Around the $TiO_2$ crystallite, fine crystallites of various orientations and lattice spacing are observed. By carefully measuring and comparing the lattice parameters with the data in JCPD, the crystallites connecting to $TiO_2$ have lattice fringes of 0.335 nm, which is ascribed to the (111) plane of CdS (JCPDS 80-0019), and outer-layer crystallites with lattice spacing of 0.372 nm, next to CdS layer, correspond to the (100) plane of CdSe (JCPDS 77-2307). These results further prove that CdS and CdSe nanocrystals

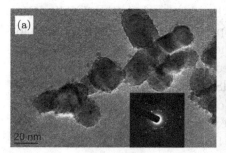

**Fig. 19.** (a) Low-magnification and (b) high-magnification HRTEM images of $TiO_2/CdS/CdSe$ electrode. Inset in (a) is the corresponding SAED pattern.

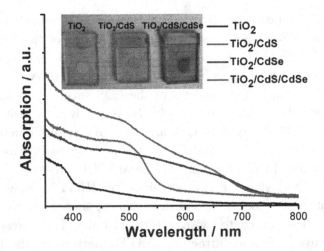

**Fig. 20.** UV–Vis absorption spectra of TiO$_2$, TiO$_2$/CdS, TiO$_2$/CdSe and TiO$_2$/CdS/CdSe electrodes. Inset shows the photographs of TiO$_2$, TiO$_2$/CdS and TiO$_2$/CdS/CdSe electrodes.

have been successfully deposited on the surface of the porous TiO$_2$ to finally form a TiO$_2$/CdS/CdSe cascade structure.

Figure 20 shows the UV–Vis absorption spectra of pure TiO$_2$, TiO$_2$/CdS, TiO$_2$/CdSe and TiO$_2$/CdS/CdSe electrodes. The inset of Fig. 20 shows the photographic images for pure TiO$_2$, TiO$_2$/CdS and TiO$_2$/CdS/CdSe electrodes. After CdSe QDs are deposited, the color of the electrode changes from yellow to orange. Compared with the absorption spectra of the pure TiO$_2$ film, there are obvious absorption peaks near 500 and 650 nm for TiO$_2$/CdS and TiO$_2$/CdSe electrodes, respectively, and the bandgaps of CdS and CdSe QDs are about 2.30 and 1.80 eV, respectively, based on the Tauc plot calculation [62, 63], which are higher than the values of bulk CdS and CdSe due to the quantum size effect. In the meantime, it can be seen that the optical range of TiO$_2$/CdS/CdSe is broader than TiO$_2$/CdS and TiO$_2$/CdSe. Therefore, the co-sensitization effect of CdS and CdSe QDs can be observed by the extension of the absorption range and the increase of absorbance [64].

The IPCE curves of QDSSCs with TiO$_2$/CdS, TiO$_2$/CdSe and TiO$_2$/CdS/CdSe electrodes, as shown in Fig. 21, exhibit a similar

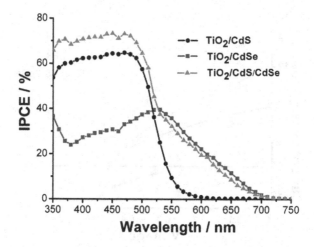

**Fig. 21.** IPCE curves of QDSSCs with $TiO_2/CdS$, $TiO_2/CdSe$ and $TiO_2/CdS/CdSe$ electrodes.

trend as UV–Vis spectra. CdS-sensitized and CdSe-sensitized cells demonstrate IPCE peak values of 65% and 40%, respectively. However, the IPCE peak value reaches 73% for the CdS/CdSe-co-sensitized cell, which is ascribed to the joint contributions from CdS and CdSe QDs in the light harvest.

The *J–V* curves of QDSSCs with $TiO_2/CdS$, $TiO_2/CdSe$ and $TiO_2/CdS/CdSe$ electrodes under one sun illumination (AM 1.5 G, $100\,mW\,cm^{-2}$) and in dark conditions are shown in Fig. 22. The $J_{sc}$, $V_{oc}$, FF and $\eta$ of CdS-sensitized and CdSe-sensitized cells are $7.20\,mA\,cm^{-2}$, 0.46 V, 0.35, 1.18% and $12.1\,mA\,cm^{-2}$, 0.45 V, 0.32, 1.75%, respectively. The higher efficiency of CdSe-sensitized QDSSCs is attributed to their broader light absorption range than that of CdS, and the lower FF of CdSe-sensitized QDSSCs should be due to low driving force of the electron injection. When both QDs are combined sequentially in a co-sensitized structure, synergistic improvement in $J_{sc}$ ($16.1\,mA\,cm^{-2}$) and $V_{oc}$ (0.56 V) is observed with a slight change in FF (0.34), leading to a high conversion efficiency of 3.06%. The higher $V_{oc}$ may be due to efficient QD coverage on the $TiO_2$ film and reduction of recombination loss at interfaces [63]. In addition to the enhanced visible light absorption, the large increase

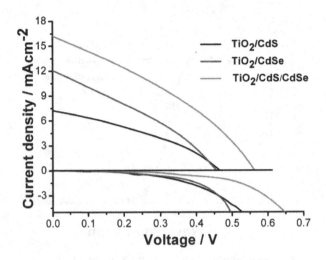

**Fig. 22.** $J$–$V$ curves of QDSSCs with $TiO_2$/CdS, $TiO_2$/CdSe and $TiO_2$/CdS/CdSe electrodes under one sun illumination (AM 1.5 G, $100\,mW\,cm^{-2}$) and in dark conditions.

in $J_{sc}$ and $\eta$ for CdS/CdSe-co-sensitized QDSSCs is ascribed to the following reasons [65, 66]: (1) The cascade energy-level structure (Fig. 23) in the order of $TiO_2 <$ CdS $<$ CdSe is formed via the combination of CdS and CdSe, as reported by others [67]. The favorable alignment of the Fermi levels at $TiO_2$/CdS/CdSe interfaces facilitates the electron injection and hole recovery for both inner CdS and outer CdSe layers. (2) The CdS underlayer promotes the growth of the CdSe layer due to less mismatched constants and more similar chemistries between them [68], and more QD loading might enhance the photocurrent density. The latter effect is confirmed by the absorption spectra presented in Fig. 20. The smaller dark current for CdS/CdSe QDSSCs observed from the $J$–$V$ curves of the cells in dark conditions indicates that CdS/CdSe QDSSCs have a superior interfacial structure to inhibit the interfacial recombination of the injected electrons from $TiO_2$ to the electrolyte, which is also responsible for its higher conversion efficiency.

The stability test of CdS/CdSe QDSSCs was carried out at different time intervals and the results are shown in Fig. 24. It is found that about 20% degradation of $\eta$ is observed after 100 h mainly due to the

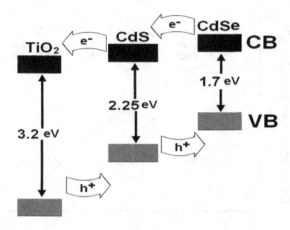

**Fig. 23.** Energy band structure of TiO$_2$, CdS and CdSe in bulk. CB-conduction band; VB-valence band.

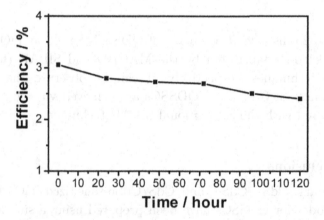

**Fig. 24.** Variation on the conversion efficiency of QDSSCs with TiO$_2$/CdS/ CdSe electrode as a function of time.

photoanodic corrosion [69]. Several studies have introduced a wide bandgap semiconductor (typically ZnS) passivation layer onto the surface of the QD sensitizer to suppress the photoanodic corrosion and improve the cell efficiency and stability [70, 71]. Therefore, further improvement in the cell performance can be expected by introducing and optimizing the passivation layer in future work.

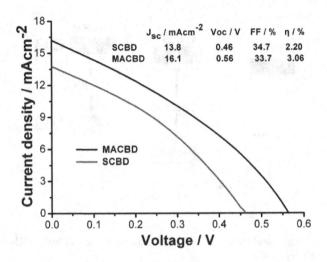

**Fig. 25.**  *J–V* curves of QDSSCs with TiO$_2$/CdS/CdSe electrodes by MACBD and SIAR techniques under one sun illumination (AM 1.5 G, 100 mW cm$^{-2}$).

Figure 25 displays *J–V* curves of QDSSCs based on TiO$_2$/CdS/CdSe electrodes fabricated by the MACBD and SILAR (for comparison) techniques, respectively. It can be observed that higher performances of CdS/CdSe QDSSCs are achieved via the MACBD as compared with the conventional SILAR technique.

## 4.  Conclusions

In summary, CdS-, CdSe- and CdS/CdSe-sensitized TiO$_2$ films as photoanodes for QDSSCs have been prepared using a simple, rapid and effective MACBD technique. This technique can synthesize CdS, CdSe and CdS/CdSe rapidly and suppress their surface defects as well as form good contact between QDs and the TiO$_2$ film. The highest short-circuit current density of 16.1 mA cm$^{-2}$ and a conversion efficiency of 3.06% under one sun illumination have been achieved using the MACBD method, which is comparable to the values achieved using the conventional SILAR method. However, MACBD avoids a repetitive immersing operation required in SILAR.

The present synthetic strategy should be a promising fabrication technique for highly efficient QDSSCs.

## References

[1] Y.L. Lee, Y.S. Lo, Highly efficient quantum-dot-sensitized solar cell based on Co-sensitization of CdS/CdSe, *Adv. Funct. Mater.*, 19 (2009) 604–609.

[2] P. Sudhagar, J.H. Jung, S. Park, R. Sathyamoorthy, H. Ahn, Y.S. Kang, Self-assembled CdS quantum dots-sensitized $TiO_2$ nanospheroidal solar cells: Structural and charge transport analysis, *Electrochim. Acta*, 55 (2009) 113–117.

[3] O. Niitsoo, S.K. Sarkar, C. Pejoux, S. Ruhle, D. Cahen, G. Hodes, Chemical bath deposited CdS/CdSe-sensitized porous $TiO_2$ solar cells, *J. Photochem. Photobiol. A*, 181 (2006) 306–313.

[4] W. Lee, J. Lee, S. Lee, W. Yi, S.H. Han, B.W. Cho, Enhanced charge collection and reduced recombination of quantum-dots sensitized solar cells in the presence of single-walled carbon nanotubes, *Appl. Phys. Lett.*, 92 (2008) 153510.

[5] P. Sudhagar, J.H. Jung, S. Park, Y.G. Lee, R. Sathyamoorthy, Y.S. Kang, H. Ahn, The performance of coupled (CdS:CdSe) quantum dot-sensitized $TiO_2$ nanofibrous solar cells, *Electrochem. Commun.*, 11 (2009) 2220–2224.

[6] J.Y. Hwang, S.A. Lee, Y.H. Lee, S.I. Seok, Improved photovoltaic response of nanocrystalline CdS-sensitized solar cells through interface control, *ACS Appl. Mater. Inter.*, 2 (2010) 1343–1348.

[7] M. Shalom, S. Ruhle, I. Hod, S. Yahav, A. Zaban, Energy level alignment in CdS quantum dot sensitized solar cells using molecular dipoles, *J. Am. Chem. Soc.*, 131 (2009) 9876–9877.

[8] E. Martinez-Ferrero, I.M. Sero, J. Albero, S. Gimenez, J. Bisquert, E. Palomares, Charge transfer kinetics in CdSe-quantum dot sensitized solar cells, *Phys. Chem. Chem. Phys.*, 12 (2010) 2819–2821.

[9] J. Chen, D.W. Zhao, J.L. Song, X.W. Sun, W.Q. Deng, X.W. Liu, W. Lei, Directly assembled CdSe quantum dots on $TiO_2$ in aqueous solution by adjusting pH value for quantum dot sensitized solar cells, *Electrochem. Commun.*, 11 (2009) 2265–2267.

[10] S.Q. Fan, D. Kim, J.J. Kim, D.W. Jung, S.O. Kang, J. Ko, Highly efficient CdSe quantum-dot-sensitized $TiO_2$ photoelectrodes for solar cell applications, *Electrochem. Commun.*, 11 (2009) 1337–1339.

[11] E.M. Barea, M. Shalom, S. Gimenez, I. Hod, I. Mora-Sero, A. Zaban, J. Bisquert, Design of injection and recombination in quantum dot sensitized solar cells, *J. Am. Chem. Soc.*, 132 (2010) 6834–6839.

[12] H.J. Lee, J.H. Yum, H.C. Leventis, S.M. Zakeeruddin, S.A. Haque, P. Chen, S.I. Seok, M. Grätzel, M.K. Nazeeruddin, CdSe quantum dot-sensitized solar cells exceeding efficiency 1% at full-sun intensity, *J. Phys. Chem. C*, 112 (2008) 11600–11608.

[13] L.J. Diguna, Q. Shen, J. Kobayashi, T. Toyoda, High efficiency of CdSe quantum-dot-sensitized $TiO_2$ inverse opal solar cells, *Appl. Phys. Lett.*, 91 (2007) 023116.

[14] G. Zhu, T. Xu, T. Lv, L.K. Pan, Q.F. Zhao, Z. Sun, Graphene-incorporated nanocrystalline $TiO_2$ films for CdS quantum dot-sensitized solar cells, *J. Electroanal. Chem.*, 650 (2011) 248–251.

[15] G. Zhu, Z.J. Cheng, T. Lv, L.K. Pan, Q.F. Zhao, Z. Sun, Zn-doped nanocrystalline $TiO_2$ films for CdS quantum dot sensitized solar cells, *Nanoscale*, 2 (2010) 1229–1232.

[16] I. Robel, V. Subramanian, M. Kuno, P.V. Kamat, Quantum dot solar cells. harvesting light energy with CdSe nanocrystals molecularly linked to mesoscopic $TiO_2$ films, *J. Am. Chem. Soc.*, 128 (2006) 2385–2393.

[17] P. Yu, K. Zhu, A.G. Norman, S. Ferrere, A.J. Frank, A.J. Nozik, Nanocrystalline $TiO_2$ solar cells sensitized with InAs quantum dots, *J. Phys. Chem. B*, 11 (2006) 25451–25454.

[18] G. Zhu, F.F. Su, T. Lv, L.K. Pan, Z. Sun, Au nanoparticles as interfacial layer for CdS quantum dot-sensitized solar cells, *Nanoscale Res. Lett.*, 5 (2010) 1749–1754.

[19] C.H. Chang, Y.L. Lee, Charge control analysis of transistor laser operation, *Appl. Phys. Lett.*, 91 (2007) 053501.

[20] I. Mora-Seró, S. Giménez, T. Moehl, F. Fabregat-Santiago, T. Lana-Villareal, R. Gomez, J. Bisquert, Factors determining the photovoltaic performance of a CdSe quantum dot sensitized solar cell: the role of the linker molecule and of the counter electrode, *Nanotechnology*, 19 (2008) 424007.

[21] W.T. Sun, Y. Yu, H.Y. Pan, X.F. Gao, Q. Chen, L.M. Peng, CdS quantum dots sensitized $TiO_2$ nanotube-array photoelectrodes, *J. Am. Chem. Soc.*, 130 (2009) 1124–1125.

[22] W. Lee, S.K. Min, V. Dhas, S.B. Ogale, S.H. Han, Chemical bath deposition of CdS quantum dots on vertically aligned ZnO nanorods for quantum dots-sensitized solar cells, *Electrochem. Commun.*, 11 (2009) 103–106.

[23] S. Banerjee, S.K. Mohapatra, P.P. Das, M. Misra, Synthesis of coupled semiconductor by filling 1D $TiO_2$ nanotubes with CdS, *Chem. Mater.*, 20 (2008) 6784–6791.

[24] J.A. Seabold, K. Shankar, R.H.T. Wilke, M. Paulose, O.K. Varghese, C.A. Grimes, K.S. Choi, Photoelectrochemical properties of heterojunction $CdTe/TiO_2$ electrodes constructed using highly ordered $TiO_2$ nanotube arrays, *Chem. Mater.*, 20 (2008) 5266–5273.

[25] Y. Jin-nouchi, S. Naya, H. Tada, Quantum-dot-sensitized solar cell using a photoanode prepared by in situ photodeposition of CdS on nanocrystalline $TiO_2$ films, *J. Phys. Chem. C*, 114 (2010) 16837–16842.

[26] X.F. Gao, W.T. Sun, Z.D. Hu, G. Ai, Y.L. Zhang, S. Feng, F. Li, L.M. Peng, An efficient method to form heterojunction $CdS/TiO_2$ photoelectrodes using highly ordered $TiO_2$ nanotube array films, *J. Phys. Chem. C*, 113 (2009) 20481–20485.

[27] J.C. Lee, T.G. Kim, W. Lee, S.H. Han, Y.M. Sung, Growth of CdS nanorod-coated TiO$_2$ nanowires on conductive glass for photovoltaic applications, *Cryst. Growth Des.*, 9 (2009) 4519–4523.

[28] Y.H. Lee, S.H. Im, J.H. Rhee, J.H. Lee, S.I. Seok, Performance enhancement through post-treatments of CdS-sensitized solar cells fabricated by spray pyrolysis deposition, *ACS Appl. Mater. Interfaces*, 2 (2010) 1648–1652.

[29] J.A. Gerbec, D. Magana, A. Washington, G.F. Strouse, Microwave-enhanced reaction rates for nanoparticle synthesis, *J. Am. Chem. Soc.*, 127 (2005) 15791–15800.

[30] A.V. Murugan, T. Muraliganth, A. Manthiram, Comparison of microwave assisted solvothermal and hydrothermal syntheses of LiFePO$_4$/C nanocomposite cathodes for lithium ion batteries, *J. Phys. Chem. C*, 112 (2008) 46–54.

[31] A.B. Panda, G. Glaspell, M.S. El-Shall, Microwave synthesis of highly aligned ultra narrow semiconductor rods and wires, *J. Am. Chem. Soc.*, 128 (2006) 2790–2791.

[32] Y. Wada, H. Kuramoto, J. Anand, T. Kitamura, T. Sakata, H. Mori, S. Yanagida, Microwave-assisted size control of CdS nanocrystallites, *J. Mater. Chem.*, 11 (2001) 1936–1940.

[33] S. Karan, B. Mallik, Tunable visible-light emission from CdS nanocrystallites prepared under microwave irradiation, *J. Phys. Chem. C*, 111 (2007) 16734–16741.

[34] E. Caponetti, D.C. Martino, M. Leone, L. Pedone, M.L. Saladino, V. Vetri, Microwave-assisted synthesis of anhydrous CdS nanoparticles in a water-oil microemulsion, *J. Colloid Interf. Sci.*, 304 (2006) 413–418.

[35] J. He, X.N. Zhao, J.J. Zhu, J. Wang, Preparation of CdS nanowires by the decomposition of the complex in the presence of microwave irradiation, *J. Cryst. Growth*, 240 (2002) 389–394.

[36] M.W. Shao, F. Xu, Y.Y. Peng, J. Wu, Q. Li, S.Y. Zhang, Y.T. Qian, Microwave-templated synthesis of CdS nanotubes in aqueous solution at room temperature, *New J. Chem.*, 26 (2002) 1440–1442.

[37] G. Zhu, L.K. Pan, T. Xu, Z. Sun, One-step synthesis of CdS sensitized TiO$_2$ photoanodes for quantum dot-sensitized solar cells by microwave assisted chemical bath deposition method, *ACS Appl. Mater. Interfaces*, 3 (2011) 1472–1478.

[38] G. Zhu, L.K. Pan, T. Xu, Q.F. Zhao, B. Lu, Z. Sun, Microwave assisted CdSe quantum dot deposition on TiO$_2$ films for dye-sensitized solar cells, *Nanoscale*, 3 (2011) 2188–2193.

[39] G. Zhu, L.K. Pan, T. Xu, Z. Sun, CdS/CdSe-cosensitized TiO$_2$ photoanode for quantum-dot-sensitized solar cells by a microwave-assisted chemical bath deposition method, *ACS Appl. Mater. Interfaces*, 3 (2011) 3146–3151.

[40] Y. Jin-nouchi, S. Naya, H. Tada, Quantum-dot-sensitized solar cell using a photoanode prepared by *in situ* photodeposition of CdS on nanocrystalline TiO$_2$ films, *J. Phys. Chem. C*, 114 (2010) 16837–16842.

[41] V.G. Pedro, X.Q. Xu, I. Mora-Sero, J. Bisquert, From flat to nanostructured photovoltaics: balance between thickness of the absorber and charge screening in sensitized solar cells, *ACS Nano*, 4 (2010) 5783–5790.

[42] N. Guijarro, T. Lana-Villarreal, Q. Shen, T. Toyoda, R. Gomez, Sensitization of titanium dioxide photoanodes with cadmium selenide quantum dots prepared by SILAR: photoelectrochemical and carrier dynamics studies, *J. Phys. Chem. C*, 114 (2010) 21928–21937.

[43] S.C. Lin, Y.L. Lee, C.H. Chang, Y.J. Shen, Y.M. Yang, Quantum-dot-sensitized solar cells: assembly of CdS-quantum-dots coupling techniques of self-assembled monolayer and chemical bath deposition, *Appl. Phys. Lett.*, 90 (2007) 143517.

[44] Q.X. Zhang, Y.D. Zhang, S.Q. Huang, X.M. Huang, Y.H. Luo, Q.B. Meng, D.M. Li, Application of carbon counter electrode on CdS quantum dot-sensitized solar cells (QDSSCs), *Electrochem. Commun.*, 12 (2010) 327–330.

[45] J. Krüger, R. Plass, M. Grätzel, Improvement of the photovoltaic performance of solid-state dye-sensitized device by silver complexation of the sensitizer cis–bis (4,4′-dicarboxy-2,2′bipyridine)-bis(isothiocyanato) ruthenium (II), *Appl. Phys. Lett.*, 81 (2002) 367–369.

[46] J. Krüger, R. Plass, L. Cevey, M. Piccirelli, M. Grätzel, High efficiency solid-state photovoltaic device due to inhibition of interface charge recombination, *Appl. Phys. Lett.*, 79 (2001) 2085–2087.

[47] I. Mora-Sero, S. Gimenez, F. Fabregat-Santiago, R. Gomez, Q. Shen, T. Toyoda, J. Bisquert, Recombination in quantum dot sensitized solar cells, *Acc. Chem. Res.*, 42 (2009) 1848–1857.

[48] E.M. Barea, M. Shalom, S. Gimenez, I. Hod, I. Mora-Sero, A. Zaban, J. Bisquert, Design of injection and recombination in quantum dot sensitized solar cells, *J. Am. Chem. Soc.*, 132 (2010) 6834–6839.

[49] N. Yang, J. Zhai, D. Wang, Y. Chen, L. Jiang, Two-dimensional graphene bridges enhanced photoinduced charge transport in dye-sensitized solar cells, *ACS Nano*, 4 (2010) 887–894.

[50] Q. Wang, J.E. Moser, M. Grätzel, Electrochemical impedance spectroscopic analysis of dye-sensitized solar cells, *J. Phys. Chem. B*, 109 (2005) 14945–14953.

[51] S. Kundu, H. Lee, H. Liang, Synthesis and application of DNA-CdS nanowires within a minute using microwave irradiation, *Inorg. Chem.*, 48 (2009) 121–127.

[52] W. Tu, H. Liu, Continuous synthesis of colloidal metal nanoclusters by microwave irradiation, *Chem. Mater.*, 12 (2000) 564–567.

[53] S.M. Mirabedini, H. Rahimi, S. Hamedifar, S.M. Mohseni, Microwave irradiation of polypropylene surface: a study on wettability and adhesion, *Int. J. Adhes. Adhes.*, 24 (2004) 163–170.

[54] C.Q. Sun, Y. Sun, Y. Ni, X. Zhan, J. Pan, X.H. Wang, J. Zhou, L.T. Li, W. Zheng, S. Yu, L.K. Pan, Z. Sun, Coulomb repulsion at the nanometer-sized contact: a force driving superhydrophobicity, superfluidity, superlubricity, and supersolidity, *J. Phys. Chem. C*, 113 (2009) 20009–20019.

[55] I. Gonzalez-Valls, M. Lira-Cantu, Dye sensitized solar cells based on vertically-aligned ZnO nanorods: effect of UV light on power conversion efficiency and lifetime, *Energy Environ. Sci.*, 3 (2010) 789–795.

[56] S. Komarneni, D. Li, B. Newalkar, H. Katsuki, A.S. Bhalla, Microwave-polyol process for Pt and Ag nanoparticles, *Langmuir*, 18 (2002) 5959–5962.

[57] D. Chen, G. Shen, K. Tang, S. Lei, H. Zheng, Y.J. Qian, Microwave-assisted polyol synthesis of nanoscale $SnS_x$ ($x = 1, 2$) flakes, *Cryst. Growth*, 260 (2004) 469–474.

[58] H. Zhang, L. Wang, H. Xiong, L. Hu, B. Yang, W. Li, Hydrothermal synthesis for high-quality CdTe nanocrystals, *Adv. Mater.*, 15 (2003) 1712–1715.

[59] L. Li, H. Qian, J. Ren, Rapid synthesis of highly luminescent CdTe nanocrystals in the aqueous phase by microwave irradiation with controllable temperature, *Chem. Commun.*, 34 (2005) 528–530.

[60] M. Shanmugam, M.F. Baroughi, D. Galipeau, Effect of atomic layer deposited ultra thin $HfO_2$ and $Al_2O_3$ interfacial layers on the performance of dye sensitized solar cells, *Thin Solid Films*, 518 (2010) 2678–2682.

[61] Z.F. Liu, Y.J. Li, Z.G. Zhao, Y. Cui, K. Hara, M. Miyauchi, Solar cell blends: high-resolution spectroscopic mapping of the chemical contrast from nanometer domains in $P_3HT:PCBM$ organic blend films for solar-cell applications, *J. Mater. Chem.*, 20 (2010) 492–499.

[62] C.Q. Sun, Size dependence of nanostructures: Impact of bond order deficiency, *Prog. Solid State Chem.*, 35 (2007) 1–159.

[63] J. Hiie, T. Dedova, V. Valdna, K. Muska, Comparative study of nano-structured CdS thin films prepared by CBD and spray pyrolysis: Annealing effect, *Thin Solid Films*, 511 (2006) 443–447.

[64] M. Li, Y. Liu, H. Wang, H. Shen, W.X. Zhao, H. Huang, C.L. Liang, CdS/CdSe cosensitized oriented single-crystalline $TiO_2$ nanowire array for solar cell application, *J. Appl. Phys.*, 108 (2010) 094304.

[65] J. Chen, J. Wu, W. Lei, J.L. Song, W.Q. Deng, X.W. Sun, Co-sensitized quantum dot solar cell based on ZnO nanowire, *Appl. Surf. Sci.*, 256 (2010) 7438–7441.

[66] Z. Yang, C.Y. Chen, C.W. Liu, H.T. Chang, Electrocatalytic sulfur electrodes for CdS/CdSe quantum dot-sensitized solar cells, *Chem. Commun.*, 46 (2010) 5485–5487.

[67] G.M. Wang, X.Y. Yang, F. Qian, Y. Li, J.Z. Zhang, Double-sided CdS and CdSe quantum dot Co-sensitized ZnO nanowire arrays for photoelectrochemical hydrogen generation, *Nano Lett.*, 10 (2010) 1088–1092.

[68] H.J. Lee, J. Bang, J. Park, S.J. Kim, S.M. Park, Multilayered semiconductor (CdS/CdSe/ZnS)-sensitized $TiO_2$ mesoporous solar cells: All prepared by successive ionic layer adsorption and reaction processes, *Chem. Mater.*, 22 (2010) 5636–5643.

[69] V. Chakrapani, D. Baker, P.V. Kamat, Understanding the role of the sulfide redox couple ($S^{2-}/S_n^{2-}$) in quantum dot-sensitized solar cells, *J. Am. Chem. Soc.*, 133 (2011) 9607–9615.

[70] M.A. Hossain, J.R. Jennings, Z.Y. Koh, Q. Wang, Carrier generation and collection in CdS/CdSe-sensitized $SnO_2$ solar cells exhibiting unprecedented photocurrent densities, *ACS Nano*, 5 (2011) 3172–3181.

[71] M.A. Hossain, G.W. Yang, M. Parameswaran, J.R. Jennings, Z.Y. Koh, Q. Wang, Mesoporous $SnO_2$ spheres synthesized by electrochemical anodization and their application in CdSe-sensitized solar cells, *J. Phys. Chem. C*, 114 (2010) 21787–21884.

Chapter 6

# Recent Advances in the Deposition Technique of Quantum Dots on Photoanodes for Quantum Dot-Sensitized Solar Cells

Xinjuan Liu*, Taiqiang Chen*, Hengchao Sun[†],
Yan Guo[†] and Likun Pan[‡,§]

*Institute of Optoelectronic Materials and Devices,
College of Optical and Electronic Technology,
China Jiliang University, Hangzhou 310018, China
[†]Beijing Smart-Chip Microelectronics Technology Co., Ltd.,
Beijing 100192, China
[‡]Shanghai Key Laboratory of Magnetic Resonance,
School of Physics and Electronic Science,
East China Normal University, Shanghai 200062, China
[§]lkpan@phy.ecnu.edu.cn

## 1. Introduction

Although solar energy could provide enough power to satisfy worldwide energy demand, the fabrication of efficient solar cells that are competitive with fossil fuels remains a challenge [1–5]. Sensitized solar cells have attracted considerable attention and represent a key class of cell architecture that has emerged as a promising candidate for the development of next-generation solar cells due to their acceptable power conversion efficiency and low production cost [6–13]. Pursuing high efficiency is always a core task for solar cells, and one of the current key issues is to search for suitable panchromatic sensitizers that enhance the light harvest in a visible light region [14–16]. In recent years, inorganic semiconductor quantum dots (QDs), such as CdS [17, 18], CdSe [19], CdTe [20], PbS [21], PbSe [22], $Bi_2S_3$ [23], InP [24], InAs [25], $In_2S_3$ [26], $Ag_2S$ [27], $Ag_2Se$ [28], $Sb_2S_3$ [29], CdZnSSe [30], $Cu_{2-x}S$ [31], Si [32], $SnSe_2$ [33], HgTe [34], $(CH_3NH_3)PbI_3$ [35], $CuInS_2$ [36] and some combinations of them [37], have been suggested along with organometallic or organic dyes to photosensitize semiconductor photoanodes, typically $TiO_2$ or ZnO, due to their high extinction coefficient, spectral tunability by particle size, good stability and resistance toward oxygen and water over their organic counterparts. Furthermore, the large intrinsic dipole moment of QDs leads to rapid charge separation [38] and QDs open up new way to make use of hot electrons or generate multiple electron–hole pairs with one single proton through the impact ionization effect [39, 40]. The theoretical efficiency of quantum dye-sensitized solar cells (QDSSCs) can reach 44%, considerably higher than that of dye-sensitized solar cells (DSSCs) [41, 42]. However, the best energy conversion efficiencies reported so far are still quite modest, revealing both the poor understanding of the fundamental processes controlling the efficiency and the remarkable difficulties found in the preparation of nanoscaled hybrid assemblies.

Unfortunately, a QDSSC, which promises a high theoretical efficiency of up to 44% for its special multielectron generation character, still presents lower energy conversion efficiency that is far below the theoretical value. QDSSCs have been evolving rapidly over the last 2 years, resulting in significant improvement of light to electric power conversion efficiencies, up to ∼6% [43–46]. A typical QDSSC consists of three components: a photoanode sensitized with semiconductor QDs, an electrolyte with a redox couple and a counter electrode (CE). Currently, the efforts to improve the performance of QDSSCs have mainly been focused on fundamental issues such as improved understanding of device physics [47], new routes of sensitization (e.g., co-sensitization) [48–51], optimization of device structure, including CEs [52–57], semiconductor oxide electrodes [58–60], blocking layers [61], redox couple electrolytes [62–65], passivation layers [66–69], scattering layers [70–73] and interfacial layers [74–77] by advanced processing methods, and development of high-performance materials [78, 79]. These combined efforts have led to a very encouraging power conversion efficiency of 4–6% [17, 80].

The working mechanism of QDSSCs involves key processes [81]. In brief, electrons can be excited from the valance band (VB) to the conduction band (CB) of QDs after absorbing light. Subsequently, the electrons are rapidly injected into the CB of $TiO_2$ particles, and then transported to F-doped $SnO_2$ (FTO). Finally, the CdS QDs are oxidized by accepting the electrons from the CE via a redox pair. However, the charge recombination will happen, which deteriorates the cell performance. The factors that limit overall power conversion efficiency of QDSSCs include limited absorption of the incident light, slow hole transfer rate, back electron transfer to the oxidized form of the redox couple and low fill factors arising from poor CE performance. Therefore, the improvement of the performance of QDs and their interconnectivity with $TiO_2$ substrate will facilitate the electron transfer from QDs to $TiO_2$ and decrease the back transport electrons from $TiO_2$ to electrolyte, which are very important for cell efficiency.

The efficiencies of QDSSCs based on bare $TiO_2$ or ZnO film are rather low, which is attributed to the accumulation of electrons in the semiconductor layer due to the relatively slow electron transfer [78], resulting in the carrier recombination at the semiconductor surface. Thus, suppressing carrier recombination at the semiconductor interface is the key to improving the performance of QDSSCs. The deposition method of QDs on a mesoporous or nanostructured metal oxide semiconductor is considerably important in determining the properties of QDSSCs and, ultimately, their performance in QDSSCs. The important criteria for developing efficient QD sensitizers over semiconductor films are the homogeneous and conformal deposition of QDs along a metal oxide surface without clogging of the mesopores, the ability to control the sizes of the QDs to achieve efficient light absorption and charge separation from QDs to the electron and hole acceptor, and the deposition of dense QDs to increase the light absorption efficiency [82, 83]. In the embedded QD design, where the QDs are first adsorbed on the electrode and then over-coated by an inorganic layer, the donor loading is controlled mainly by the ratio between the pore size of the electrode and the QD dimensions, as well as by diffusion dynamics of QDs inside the porous electrode. The donor can open a way toward utilization of new materials for which the band alignment is not restricted to match that of the wide bandgap electrode.

In general, the semiconductor oxide films are fabricated on conductive glass via different techniques such as screen printing and doctor blade. Many studies have been devoted to explore different fabrication techniques to attach QDs onto $TiO_2$. So far, the QDs have been successfully sensitized on the surface of ZnO or $TiO_2$ by self-assembled monolayers via linker assistance or direct adsorption (DA), chemical bath deposition (CBD), electrochemical deposition, thermal evaporation, spray pyrolysis, chemical vapor deposition, arrested precipitation in solution, photodeposition techniques [84–86], a combination of self-assembled monolayers (SAMs) and CBD techniques [87].

Despite these powerful merits of QD materials, the power conversion efficiency of a QDSSC is still low with reference to that a DSSC. One of the reasons for the low efficiencies is the difficulty pertaining to the charge transfer process in the semiconductor film and to the collection of this charge at the substrate of the photoanode. On the contrary, the difficulty in obtaining a well-covered QD layer on the $TiO_2$ crystalline surface is also a reason for the poor performance of a QDSSC [88]. Furthermore, the poor performance of QDSSCs may be mainly ascribed to the difficulty of assembling the QDs into a mesoporous photoanode surface. The aggregation of QDs has been found in many fabrication methods, such as DA [89], SILAR [90] and CBD [91], which can block the access of the electrolyte to the interior of the surface pore and enhance the recombination.

## 2. Self-assembled monolayer method

Pre-synthesized colloidal QDs capped with surface ligands have been attached to $TiO_2$ surfaces through bifunctional linker molecules [92] or DA using an adequate solvent in the colloidal solution [93].

Deposited SAMs on the surface of metal oxide are a simple and effective way to control the properties including wettability, work function and charge transfer. In this method, the attachment of the QDs to the oxide is achieved by using a linker, which is a bifunctional molecule that anchors the QD to the oxide particle, acting as a molecular cable. Linker-assisted, direct QD adsorption onto semiconductor allows fine control of the QD size, exploiting colloidal synthesis. The SAMs can passivize the surface, suppress the surface charge recombination and enhance the charge separation at the interface; thus, they have been used in photovoltaics to modify the transparent electrodes or photoanodes. Furthermore, the SAMs also form an energy barrier to efficiently retard the back transfer of electrons. This results in the enhancement of conversion efficiency of solar cells. Different molecular linkers, such as mercaptopropionic acid (MPA), cysteine, thiolacetic, mercaptohexadecanoic acid and thioglycolic acid, have been investigated, and it has been recognized that the chemical nature of the linker plays a decisive role in determining the efficiency of electron injection into the matrix [78]. After the deposition of QDs and semiconductor films, additional thermal treatment is necessary for current techniques to remove organic capping ligands that could hinder the charge injection from QDs to semiconductor films or to form better crystallinity of QDs and semiconductor films to facilitate the electron transport [94].

Lee *et al.* [95] pre-assembled an SAM of 3-mercaptopropyl-trimethyoxysilane (MPTMS) onto a mesoporous $TiO_2$ film, which is supposed to trigger a better interaction of the $TiO_2$ and CdSe to induce the growth of CdSe QDs in the SILAR process. It is found that the MPTMS SAM increases the nucleation and growth rates of CdSe, which leads to the good covering and higher uniform CdSe layer, thus inhibiting the charge recombination at the electrode–electrolyte interface. Energy conversion efficiency under the illumination of one sun (AM 1.5, $100\,mW\,cm^{-1}$) achieves 2.65%. Yang *et al.* [96] used the 4-tertbutylbenzoic acid (BBA) as SAMs to modify ZnO in CdS QDSSCs, and found that the conversion efficiency exhibits 56% enhancement compared with solar cells with pure ZnO photoanodes. After the modification of BBA molecules on the pure ZnO, the water contact angle increased by 96° and the work function decreased by 0.4 eV. The BBA can decrease the defects of ZnO surface and form the energy barrier, which suppress the recombination of charge carriers, leading to the enhancement of photovoltaic performance.

Robel et al. [97] investigated the effect of three different molecules, namely, MPA, thiolacetic acid (TAA) and mercaptohexadecanoic acid (MDA), on the photovoltaic performance of CdSe QDSSCs. The TiO$_2$ films were subsequently dipped into a solution of acetonitrile containing carboxy alkane thiols (TAA, MPA and MDA) for ~4 h. It was found that the best result is MPA. Moreover, they also compared the effect of MPA, thioglycolic acid (TGA) and cysteine [78]. The use of cysteine as a bifunctional molecular linker between colloidal CdSe QDs and TiO$_2$ can significantly increase the cell performance. The incident photon to charge carrier generation efficiency (IPCE) can be improved by a factor of 5–6 if cysteine is used as a linker instead of MPA in CdSe QDSSCs.

Guijarro et al. [89] compared two strategies: (i) DA and (ii) adsorption via a molecular linker and MPA to cover the nanoporous TiO$_2$ with pre-synthesized CdSe QDs. In the case of direct QD adsorption, the TiO$_2$ electrodes were immersed in a CH$_2$Cl$_2$ (99.6%) CdSe QD dispersion for times ranging between 15 s and 48 h. In the case of adsorption through a linker, nanoparticulate TiO$_2$ electrodes were modified with 3-MPA (99%) by immersion in a 1:10 acetonitrile (99.5%) solution for 24 h. Then, the electrodes were left in toluene CdSe QD solution for times ranging from 40 min to 42 h. In contrast with MPA-mediated adsorption, direct adsorption leads to a relatively high degree of QD aggregation, which will cause the bottlenecks to hinder the access of the electrolyte deep into the pores and decrease the effective electrochemical active area of the substrate. Therefore, QD aggregation leads, for high QD coverage, to a severe drop in the IPCE in the direct adsorption mode. On the contrary, for QDs attached via MPA, the IPCE monotonously increases with coverage. At saturation, only 14% of the real surface area of the TiO$_2$ layer is covered for both attachment modes. For equivalent QD loading, the IPCE values measured in a standard three-electrode electrochemical cell for direct adsorption are larger than those for MPA-mediated adsorption. This is because direct adsorption minimizes the distance between the QDs and the $TiO_2$ particle, thus facilitating the electronic injection.

In the method, low QD coverage of the electrode surface, relatively weaker electronic coupling between QDs and $TiO_2$ and inefficient charge separation due to organic linkers give rise to low photovoltaic conversion efficiency, which are major limiting factors for the pre-synthesized QD method [82, 98]. Use of previously prepared QDs can avoid the drawbacks of the direct growth method; however, it is still a great challenge to immobilize the previously prepared QDs onto $TiO_2$ electrodes with high surface coverage and achieve good photovoltaic performance of the resulting cell devices.

## 3. Chemical bath deposition method

QDs have grown directly on $TiO_2$ electrode surfaces by chemical reaction of ionic species using CBD [84, 99] or the successive ionic layer adsorption (SILAR) [100–103] method. CBD is the most common, inexpensive and environmentally benign method used for *in situ* synthesis of the QDs in the mesoporous $TiO_2$ substrate, and is carried out at low temperatures. From the photographs of the corresponding samples in Fig. 1, the color for the bare ZnO electrodes changes from gray to orange after deposition of a CdS shell via the SILAR process, and then eventually turns to dark brown following deposition of CdSe QDs via CBD. Direct deposition of QDs on the $TiO_2$ surface using *in situ* CBD has resulted in superior electron injection and better coverage than indirect contact of the pre-synthesized QDs linked by functional molecular linkers on $TiO_2$ [104]. Compared to the system involving CdSe QDs linked to $TiO_2$ through molecular linkers, the CdSe QDs directly sensitized on the $TiO_2$ porous structure by CBD afford more interfacial coupling due to closer contact [105].

This method involves a nucleation and growth mechanism, leading to high coverage of the electrode surface compared with the SAM method. Although a direct contact between the oxide and the QDs is achieved in the CBD method, there is no separate control of

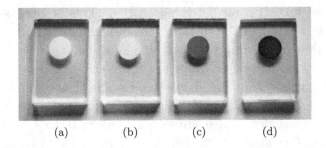

(a)　　　　(b)　　　　(c)　　　　(d)

**Fig. 1.** Digital camera images of the corresponding samples [106].

QD coverage and size [89]. In addition, the deposits could be far from stoichiometric because of, for instance, the possible formation of elemental layers in addition to the sought compound particles. However, these drawbacks can be avoided if the QDs are synthesized earlier and then attached onto the oxide layer via the SAM method.

### 3.1. *Chemical bath deposition method*

In QDSSCs, the facile and straightforward CBD procedure is one of the most popular methods to fabricate QD-sensitized electrodes. In the process of depositing CdS QDs, the color of the $TiO_2$ photoanode changed from colorless to green, then to yellow and finally to orange as per the number of CBD cycles. CBD permits enhanced electron transfer to the wide bandgap $TiO_2$ electrode and significantly higher loading at the cost of appreciable QD aggregation that finally deteriorates solar cell performance [107].

The ZnO nanorods were sensitized by CdS QDs using the procedure reported by Lee *et al.* [108]. Cadmium acetate of 5 mM in DMF was complexed by 10 mM 1-thioglycerol. To this complexed solution, the thiourea of 10 mM in DMF was added and refluxed for 10 min under Ar atmosphere. In order to study the surface wettability property, water contact angle measurements were done for both ZnO and CdS QDs/ZnO films, and are shown in Fig. 2. The water contact angle for ZnO nanorods is found to be 142° in Fig. 2(a), i.e., hydrophobic nature. The water contact angle is found

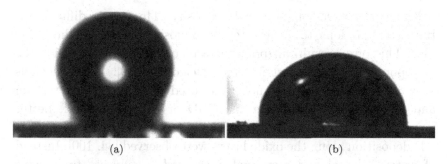

(a)                                        (b)

**Fig. 2.**  Water contact angle measurements of (a) ZnO nanorod thin film and (b) CdS QDs/ZnO thin film [108].

to be decreased significantly after CdS QDs sensitization, i.e., 90°, as can be seen from Fig. 2(b). It revealed hydrophilic nature of the CdS QDs/ZnO, which is useful for the formation of semiconductor–electrolyte interface. The wetting of solid with water is dependent on the relation between the interfacial tensions (water–air, water–solid and solid–air), where air is the surrounding medium. The ratio between these tensions determines the contact angle for water droplets on a given surface. It also depends on local inhomogeneity, surface morphology and chemical composition. A similar trend is also observed by Patil *et al.* [109]. In their work, the contact angle of a water drop is changed when it is on a $TiO_2$ surface compared to a $CdS/TiO_2$ surface, indicating a more hydrophilic surface when CdS is deposited on $TiO_2$. Increase in hydrophilicity helps in the optimum access of the electrolyte to the electrode heterostructures, leading to efficient charge transfer at the interface [109].

In the direct growth method, cationic and anionic precursors (e.g., $Cd^{2+}$ and $S^{2-}$ ions in the case of CdS QDs) are dissolved in a solvent, and QDs grow on the metal oxide surface via a reaction between the cationic and the anionic precursors. To achieve uniform and conformal QD deposition on the mesoporous or nanostructured metal oxides, the precursor ions should penetrate deep into the highly porous films, and after the formation of QDs, the solvent should be removed without causing stress on the deposited QDs. Water, a typical solvent used in the CBD or SILAR methods, retains a high surface tension and high viscosity. Thus, it is difficult for the precursor ions to penetrate the nanosized pores of the oxide film. This often results in poor surface coverage of QDs inside the mesoporous or the nanostructured electrodes [110]. When alcohols such as methanol or ethanol were used as low surface tension and low-viscosity solvents in the CBD or SILAR method, better penetration into the mesoporous $TiO_2$ film and more homogeneous QD deposition along the oxide layers were observed [84, 100]. Instead of water, ethanol and methanol were used as solvents to dissolve $Cd(NO_3)_2$ and $Na_2S$, respectively. The CdS-sensitized $TiO_2$ electrode prepared using the alcohol system not only has a higher incorporated amount of CdS but also greatly inhibits the recombination of injected electrons (Fig. 3). Moreover, better light absorbance and higher energy conversion efficiencies were also obtained when compared to those obtained from the water-based method (Fig. 3). The efficiency of a CdS QDSSC prepared using the present method is as high as 1.84% under the illumination of one sun [84].

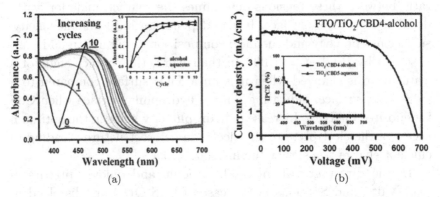

(a)                                            (b)

**Fig. 3.**   (a) UV–Vis absorption spectra of $TiO_2$ (0), and the $TiO_2$ electrode after one cycle (1) to ten cycles (10) of the CBD process in alcohol solutions. The inset shows the absorbance of exciton peaks after various cycles of the CBD process for alcohol and aqueous systems. (b) $I$–$V$ characteristics of the CdS-sensitized $TiO_2$ electrode measured under the illumination of one sun. The inset shows the incident photon to current conversion efficiencies (IPCEs) of the best performing devices prepared in alcohol (four CBD cycles) and aqueous (five CBD cycles) systems [84].

## 3.2. *Sequential chemical bath deposition or successive ionic layer adsorption and reaction method*

The SILAR technique is known as the modified version of chemical bath deposition and has been widely used in the fabrication of QDSSCs [111, 112]. Direct growth of QDs by SILAR has recently emerged as a promising deposition route combing high-QD loading together with a low degree of aggregation and efficient electron transfer to $TiO_2$. SILAR-sensitized metal oxide films generally show higher optical densities compared to films sensitized with colloidal QDs [113]. The SILAR process involving the sequential dipping of $TiO_2$ electrodes in solutions containing precursors is extremely attractive from a processing and therefore industrial and commercial point of view.

Figure 4 shows the STEM image of nanocrystalline $TiO_2$ coated with CdSe QDs using the SILAR method [114]. Insets are elemental maps of Ti, O, Cd and Se in the sample. While Ti and O are uniformly

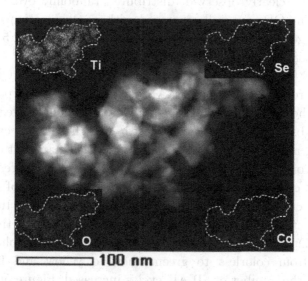

**Fig. 4.** STEM images of $TiO_2$/CdSe composite at different magnifications. Insets are the maps of Ti, O, Cd and Se elements. The white dashed lines indicate the composite [114].

distributed in high densities, the elements Cd and Se appear in relatively low quantities in the composite (the profile of which is indicated by a dashed line). The result indicates that the CdSe QDs are attached to the $TiO_2$ nanoparticles.

Suitable cycle times can ensure the deposition of more amounts of CdS QDs, but too many QDs will hinder electron transport and increase the centers for charge recombination. It should be noticed that to date, these QDSSCs based on co-sensitized porous $TiO_2$ electrode have been almost fabricated by the SILAR method, which requires repetitive immersing operation of the $TiO_2$ film in precursor solutions containing Cd, S or Se. However, after the deposition of QDs and semiconductor films, additional thermal treatment is necessary for current techniques to form better crystallinity of QDs and semiconductor films to facilitate the electron transport.

Tubtimtae et al. [27] grew $Ag_2S$ QDs on $TiO_2$ by the SILAR method for QDSSCs. The inset to Fig. 5 shows a TEM image of a QD-coated $TiO_2$ film processed with four SILAR cycles. Many QDs can be clearly observed, distributed randomly over the $TiO_2$ surface. The QDs are well separated with no aggregation observed. The average diameter of the QDs is estimated to be ~2.5 nm. The area density of the QDs is ~$1.3 \times 10^5$ particles $\mu m^{-2}$. Figure 5 shows the UV–Vis optical absorbance spectra of $Ag_2S$ QD-loaded $TiO_2$ electrodes with various SILAR cycles. A broad absorption band covering the whole spectral range appears in the QD-loaded samples and the absorption intensity increases with the SILAR cycle. The increasing absorption indicates that an increasing amount of QDs is deposited on the $TiO_2$ electrode as the SILAR cycle is increased. The assembled cell yielded a best power conversion efficiency of 1.7% and a short-circuit photocurrent ($J_{sc}$) of 1.54 mA cm$^{-2}$ under 10.8% sun.

Zhang et al. [115] deposited CdS QDs onto $TiO_2$ using the SILAR method. During the deposition, the color of the $TiO_2$ photoanode changed from colorless to green, then to yellow and finally to orange as the number of SILAR cycles increased. Figure 6(a) shows the variation of UV–Vis absorption spectra of the CdS-coated $TiO_2$ electrodes with the increase of the number of SILAR cycles. It is clearly seen that the absorption edge shifts toward longer

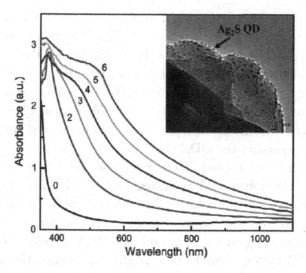

**Fig. 5.** UV–Vis absorption spectra of $TiO_2$ electrodes sensitized with various $Ag_2S$ QDs. The labels next to the curves denote the number of SILAR cycles. Inset: TEM micrograph of an $Ag_2S$ QD-coated $TiO_2$ film [27].

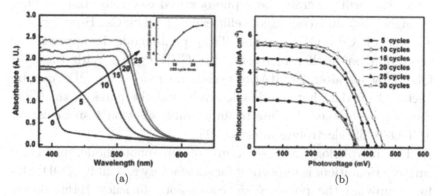

**Fig. 6.** (a) UV–V is absorption spectra of CdS-coated $TiO_2$ electrodes from 0 to 25 cycles of the SILAR process at the interval of five cycles. Inset: the relationship between CdS average size and the SILAR cycle times. (b) $I$–$V$ characteristics of the cells based on carbon CE vs. numbers of SILAR cycles [115].

wavelengths with an increase in the number of SILAR cycles due to the size quantization effect. In the meantime, the amount and the sizes of deposited CdS QDs increased as well. The sizes of CdS particles from the exciton peaks are estimated from 3.5 to

8 nm, as shown in the inset of Fig. 6(a). The UV–Vis absorbance spectra increase with the number of SILAR cycles, indicating an increased adsorption amount of CdS. Furthermore, the redshifts of the absorption shoulder and onset position with an increase in SILAR cycles imply the growth of the CdS QDs. The $TiO_2$ surface cannot be completely covered in one or two assembly cycles. Thus, the later SILAR cycles play a role in replenishing the uncovered surface area, as well as increasing the QDs' thickness.

When the CdS-coated $TiO_2$ film was incorporated with the polysulfide electrolyte and activated carbon CE into the QDSSC, the difference in the photovoltaic performance with the increase in the SILAR cycles can be observed. As shown in Fig. 6(b), both $J_{sc}$ and open-circuit photovoltage $(V_{oc})$ increased quickly in early SILAR cycles, slowed down gradually, reached the maximum value $5.57\,\mathrm{mA\,cm^{-2}}$ and $458\,\mathrm{mV}$ at 20 cycles, respectively, and finally decreased while further increasing SILAR cycles. Meanwhile, on increasing the SILAR cycles, CdS QDs in greater amount and larger size will generate more photoexcited electrons, thus leading to higher $J_{sc}$ and conversion efficiency of QDSSCs. However, the amount of CdS QDs on the $TiO_2$ photoanode is approaching saturation on further increasing the CBD cycles. With the CdS QDs growing after 20 SIALR cycles, no more CdS QDs can be anchored on the photoanode effectively and the pores in the $TiO_2$ film will be blocked, leading to unfavorable electron transportation at $TiO_2$–CdS–electrolyte interface [99].

The morphology and structure of the photoanode, including surface orientation, is important for satisfactory assembly of QDSSCs for improving the photovoltaic conversion efficiency [116]. Rawal et al. [117] compared the formation of the CdS layer on ZnO NRs and the $TiO_2$ photoanode during the preparation of CdS–CdSe cell and found that the optimum thickness of the CdS layer on ZnO NRs was quite different from that on the $TiO_2$ electrode: the optimized CdS layer was acquired by 9 cycles of the SILAR process, whereas three or four cycles were suitable for $TiO_2$ [117]. The CdS QDs deposited on the surface of $TiO_2$ are a particle-like structure and only 35% of the $TiO_2$ structure was covered with CdS QDs, but CdS was uniformly

**Fig. 7.** SEM image of TiO$_2$ coated with CdS via SILAR method [109].

coated on the ZnO surface with coverage of above 90%. It is possible that the same crystallographic structure and similarity in lattice parameter between CdS and ZnO led to the formation of uniform CdS layer and high surface coverage on ZnO NRs. The drawback of SILAR is that CdS is easily agglomerated on the surface of the photoanode, as shown in Fig. 7 [109].

### 3.3. *Microwave-assisted chemical bath deposition method*

As an inexpensive, quick, clean, versatile technique, microwave irradiation induces interaction of the dipole moment of polar molecules or molecular ionic aggregates with alternating electronic and magnetic fields, causing molecular-level heating, which leads to homogeneous and quick thermal reactions. If such a heating is performed in closed vessels under high pressure, more energy input at the same temperature and enormous accelerations in reaction time can be achieved, which means that a reaction via the conventional solvothermal method that takes several hours can be completed in the course of minutes or a reaction that would not proceed previously will now proceed, typically with higher yields. However, the microwave technique is seldom studied now to coat CdS QDs onto the TiO$_2$ electrode for QDSSCs, although such a method has been used successfully to fabricate CdS QDs, CdSe and CdSe/CdS QDs [81, 118–120].

Pan et al. [120] fabricated sensitized-type solar cells based on a $TiO_2$ photoanode and CdS QDs as sensitizers, in which CdS QDs are prepared using the microwave-assisted chemical bath deposition (MACBD) method, and investigated their photovoltaic performance. Compared with those methods, the MACBD technique can rapidly synthesize CdS QDs and precisely control their sizes with a narrow distribution, improve the wettability of the $TiO_2$ surface and form a good contact between CdS QDs and the $TiO_2$ layer due to rapidly elevated temperature during microwave irradiation. Furthermore, this technique can offer an easy control over all experimental parameters without the requirement of repetitive immersing operations, organic linkers or high temperature heating. The as-synthesized cell shows a high $J_{sc}$ of $7.20 \, \text{mA cm}^{-2}$ and conversion efficiency of 1.18% under one sun illumination as compared with the cells fabricated using CBD and SPD methods.

In our recent works, we have fabricated CdS QDSSCs and CdSe QDSSCs using a simple, rapid and effective MACBD method [118–120]. This method can rapidly synthesize CdS or CdSe QDs and form good contact between QDs and $TiO_2$ layer. Importantly, MACBD avoids repetitive immersing operations, organic linkers or high temperature heating required in other conventional methods, while the cells fabricated using MACBD exhibit photovoltaic performances comparable to those using other methods. Pan et al. [81] explored the application of the MACBD method in the fabrication of a CdS–CdSe co-sensitized $TiO_2$ photoanode for QDSSCs. The as-synthesized CdS–CdSe QDSSCs show a high $J_{sc}$ of $16.1 \, \text{mA cm}^{-2}$ and conversion efficiency of 3.06% under one sun illumination as compared with the cell fabricated using the SILAR method.

## 4. Photodeposition method

Photocatalytic preparation has been used mainly for loading metals on semiconductors since its discovery by Kraeutler et al. [121]. Photoreduction on the surface of $TiO_2$ leads to a large and uniform coverage of QDs and intimate contact between the QDs and $TiO_2$ for efficient interfacial charge transfer. The QDs can be deposited on not only the external surfaces but also the inner ones of $TiO_2$ without pore blocking. The photoreduction method has a wide range of possibilities for coupling $TiO_2$ and narrow gap metal sulfides suitable for QDSSCs [122]. To date, chalcogenide-oxide coupling systems, such as $CdS$–$TiO_2$, $CdSe$–$TiO_2$, $PbSe$–$TiO_2$ and $CdS$–$Au$–$TiO_2$, have been prepared by this photodeposition method [123, 124]. The important feature of this method is that the efficient interfacial charge transfer between the semiconductors is inherently guaranteed [123].

Recently, Hiroaki Tada *et al.* [82] directly deposited CdS QDs into mesoporous $TiO_2$ nanocrystalline films via a one-step photodeposition method at ambient temperature and pressure. During the photodeposition, the loading amount and the particle size of CdS QDs can be controlled via UV light irradiation time. Figure 8 shows the TEM images of $CdS$–$TiO_2$ prepared after 3-h irradiation. A number of nanoparticles are attached on the $TiO_2$ surface with a highly dispersed state, and the interface between $TiO_2$ and CdS QDs shows good contact. The pore volumes of as-prepared $CdS$–$TiO_2$ using photodeposition, SILAR and SAM methods are measured. The result shows that the highest volume of ca $0.72\,cm^3\,g^{-1}$ for $CdS$–$TiO_2$ obtained by photodeposition was compared with those using SILAR ($0.20\,cm^3\,g^{-1}$) and SAM ($0.25\,cm^3\,g^{-1}$). This volume almost

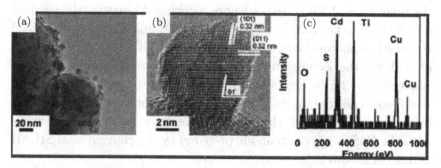

**Fig. 8.**  TEM (a), HRTEM (b) and energy-dispersive X-ray (ED) spectrum (c) of $CdS/TiO_2$ obtained after 3-h irradiation [82].

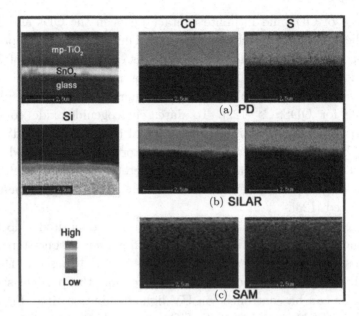

**Fig. 9.** Elemental depth profiles by EPMA for CdS–TiO$_2$ films obtained by photodeposition, SILAR and SAM methods [82].

approaches the pore volume $(0.72\,\text{cm}^3\,\text{g}^{-1})$ of pure TiO$_2$. Figure 9 shows the elemental depth profiles by EPMA for CdS–TiO$_2$ films obtained by photodeposition, SILAR and SAM methods. The Cd and S are distributed widely toward the inner part of the TiO$_2$ film for photodeposition and SILAR. These results indicate that CdS QDs can be deposited on not only the external surfaces but also the inner ones of TiO$_2$ without pore blocking. Moreover, the efficient interfacial charge transfer between CdS QDs and TiO$_2$ is guaranteed because the photocatalytic redox property of TiO$_2$ is taken advantage of to form the heterojunction. Due to these advantages, a power conversion of 2.51% has been achieved for as-prepared QDSSCs with the photodeposition of the CdS–TiO$_2$ photoanode under illumination of one sun, the value of which is much greater than those for the cells with the CdS–TiO$_2$ photoanode prepared by the conventional SILAR and SAM methods.

To improve the efficiency of Ag$_2$S QDSSCs, a photodeposition method was used to deposit the Ag$_2$S QDs on the oriented TiO$_2$

nanorod array [124]. A large coverage of $Ag_2S$ QDs on the $TiO_2$ nanorod array has been achieved by the photodeposition method. A high $J_{sc}$ of $10.25\,mA\,cm^{-2}$ with a conversion efficiency of 0.98% at AM 1.5 solar light of $100\,mW\,cm^{-2}$ was obtained with an optimal photodeposition time.

## 5. Electrophoretic method

The electrophoretic deposition (EPD) method was employed to deposit the semiconductor, metallic and insulating nanoparticles on the conductive substrates and polymers [125, 126]. The EPD method has the following advantages: short formation time, needs simple apparatus, no requirement for binder burnout as the green coating contains few or no organics and offers easy control of the thickness and morphology of a deposited film. EPD is a potential single-step solution-based route for rapid production of large-area and uniform QD thin films with desired thickness. These films obtained by EPD from suspension exhibit outstanding advantages, such as high deposition rate, excellent uniformity, controlled thickness, large scale and without any binders [127–130].

Salant *et al.* [131] reported the facile fabrication of QDSSCs by the EPD technique of CdSe QDs onto conducting electrodes coated with mesoporous $TiO_2$. The obtained efficiency of 1.7% at 1 sun was higher than those of the cells with QDs prepared with a linker technique. It is noted that the absorbed photon to electron conversion efficiencies do not show a clear size dependence, which indicates efficient electron injection can happen for the larger QD sizes.

Sauvage *et al.* [132] investigated a CdSe-sensitized mesoporous $TiO_2$ film by means of EPD for QDSSCs with a new regenerative redox couple based on the polypyridil cobalt complex. With the increase of deposition time, the color of the $TiO_2$ photoanode changed from light brown to dark brown (as shown in Fig. 10). More interestingly, the results also show that the film's porosity is reasonably preserved to avoid mass transport limitations into the mesopores, although the increase of the deposition time yields a distinguishable reduction of pore size as well as of film porosity.

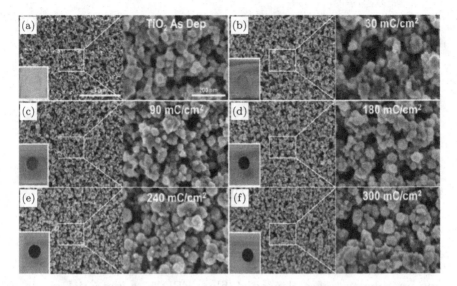

**Fig. 10.**  Top view SEM images of as-prepared CdSe–TiO$_2$ films with different deposition times [132].

The cross-sectional SEM images of pure TiO$_2$ and CdSe–TiO$_2$ films (as shown in Fig. 11) show that lots of pores are not filled via depositing CdSe QDs into the TiO$_2$ film, which indicates that a CdSe-sensitized TiO$_2$ film still has a microscopic architecture. As-prepared CdSe–TiO$_2$ films are used as the photoanode for QDSSCs, and a conversion efficiency of 0.8% under AM 1.5 100 mW cm$^{-2}$ is obtained for the device with a new regenerative redox couple based on the cobalt polypyridil complex.

Kuang *et al.* [85] fabricated CdS–CdSe co-sensitized hierarchical TiO$_2$ sphere (HTS) electrodes for QDSSCs, in which both CdS and CdSe QDs are deposited into the HTS film by using the EPD method. Figure 12 shows TEM images of as-prepared HTS, CdS/HTS and CdSe/CdS/HTS (from left to right). CdS QDs with sizes of around $4.5 \pm 0.5$ nm have covered the TiO$_2$ nanorods; yet a large amount of exposed TiO$_2$ can still be observed. A power conversion efficiency as high as 4.81% ($J_{\text{SC}} = 18.23$ mA cm$^{-2}$, $V_{\text{OC}} = 489$ mV, fill factor $= 0.54$) for CdSe/CdS/HTS photoelectrode-based QDSSC is observed under one sun AM 1.5 G illumination (100 mW cm$^{-2}$).

**Fig. 11.**    Cross-sectional SEM images of pure $TiO_2$ and CdSe $TiO_2$ films [132].

**Fig. 12.**    TEM images of HTS CdS–HTS and CdSe–CdS/HTS [85].

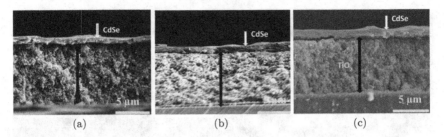

**Fig. 13.** SEM images of the CdSe–TiO$_2$ film at (a) 60 s, (b) 90 s and (c) 120 s [133].

Yaacob *et al.* [133] investigated the effect of CdSe thickness on the photovoltaic performances for QDSSCs. TiO$_2$ nanoparticles were deposited on FTO conducting glass substrate at 150 V for 30 s by the EPD method following heating at 450° for 3 h. CdSe was deposited on the TiO$_2$ layer at 100 V for 60 s, 90 s and 120 s by the EPD method. As shown in Fig. 13, the thicknesses of deposition layers of CdSe are 1.5, 1.8 and 3 $\mu$m, respectively. With the increase of deposition time, the thickness and deposition amount of CdSe increase. The CdSe nanoparticles' film at 3 $\mu$m thickness was found to exhibit the highest photoconversion efficiency of 2% and a $J_{sc}$ of 12.2 mA cm$^{-2}$. The good electric contact between CdSe and TiO$_2$ and improved crystallinity achieved by the slow deposition rate would facilitate the injection of the excited electron into TiO$_2$ and the scavenging of the photogenerated holes, resulting in an improved performance of QDSSCs [133, 134].

Zhu *et al.* [135] fabricated RGO-Au nanoparticle composite films by EPD as a CE for CdS QDSSCs. The almost transparent RGO sheets are decorated randomly by Au NPs and few NPs scatter out of the sheets, indicating the strong interaction between the NPs and RGO sheets. The RGO has a lateral dimension of several micrometers and a thickness of 1.2 nm, while the size of Au NPs is about 60–70 nm (Fig. 14). The introduction of Au NPs on RGO increases the catalytic activity and electrical conductivity of the CE, resulting in a better photovoltaic performance, which is attributed to a superior combination of high electrocatalytic activity of Au NPs and the electrical conductivity of the RGO network structure. The $J_{sc}$, $V_{oc}$

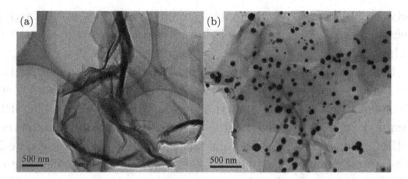

**Fig. 14.** HRTEM images of the prepared (a) RGO and (b) RGO-Au composites [135].

and photoconversion efficiency increase from $5.69\,\mathrm{mA\,cm^{-2}}$, $0.46\,\mathrm{V}$ and $0.98\%$ for the cell with RGO CE to $7.07\,\mathrm{mA\,cm^{-2}}$, $0.51\,\mathrm{V}$ and $1.36\%$ for the cell with RGO-Au CE, respectively, while the FF increases somewhat.

## 6. Other methods

### 6.1. *Spray pyrolysis method*

As a widely used, inexpensive, versatile, large-scale production technique, the spray pyrolysis (SPD) method has been used successfully to fabricate nanostructured CdS–CdSe thin films or $TiO_2/ZnO$ films [136]. Okuya *et al.* [137] synthesized porous $TiO_2$ thin films on FTO glass by the SPD technique and investigated their application in DSSCs. Lee *et al.* [138] reported that QDSSCs based on CdS-coated $TiO_2$ electrodes using the SPD method showed a power conversion efficiency of $1.84\%$ in $I^-/I^{3-}$ electrolyte and $0.87\%$ in polysulfide electrolyte. Shin *et al.* [139] reported that $TiO_2$ nanotube arrays decorated by CdS or CdSe using the SPD method showed highly photoresponsive characteristics for the production of hydrogen. By employing the SPD technique, good contact between semiconductor films and conductive glass or between semiconductor films and QDs can be realized because this method allows a facile and rapid deposition of QDs and semiconductor films at a high temperature, which improves their crystallinity. Unfortunately, the integration

between QDs and semiconductor films is still not so satisfying
because both studies mentioned above did not employ the same SPD
technique.

Pan *et al.* [140] reported sensitized-type solar cells based on the
ZnO photoanode and CdS QDs as sensitizers, in which both ZnO
films and CdS QDs are prepared using the SPD technique. The SPD
technique can offer easy control over all experimental parameters
without the need for post thermal treatment. The as-synthesized cell
shows a maximum $J_{sc}$ of $6.99\,\mathrm{mA\,cm^{-2}}$ and conversion efficiency of
1.54% under one sun illumination. It can be noticed that most of the
reports employed various methods such as rf magnetron sputtering,
chemical vapor deposition, sol–gel, electrochemical deposition or
SILAR to deposit the ZnO interlayer. The SPD method has been
studied to coat the ZnO layer onto the $TiO_2$ electrode for SSCs [140].

## 6.2. *Atomic layer deposition*

Compared to other vacuum deposition techniques such as physical
vapor deposition and chemical vapor deposition, which do not
produce conformal deposition, atomic layer deposition (ALD) is well
suited to depositing conformal thin films on very porous or high
aspect ratio structures [141]. The size and composition of QDs can
easily be varied simply by changing the number of ALD cycles. In
addition, the ALD as a gas phase-based deposition method eliminates
the use of solutions and avoids solution-based byproducts. ALD
presents intriguing new opportunities for depositing QDs for solar
cell and other applications.

The ALD technique has been employed to deposit $In_2S_3$ sensitizer
on $TiO_2$ nanotube arrays for solar energy conversion by Sarkar *et al.*
[142]. Indium acetylacetonate ($In(acac)_3$) and $H_2S$ were used as
reactants, and the deposition was carried out in a hot-wall viscous
flow reactor. As shown in Fig. 15, $In_2S_3$ ($\sim 5.2\,\mathrm{nm}$ thicknesses)
covered the full length of the $TiO_2$ nanotube array after 175 cycles
of ALD at 150°C. The photoelectrochemical properties of the $In_2S_3$-
sensitized $TiO_2$ nanotube arrays with a $Co^{2+}/Co^{3+}$ electrolyte
and Pt-loaded TCO CE were investigated using a Keithley model
236 source measure unit under simulated AM 1.5 solar irradiance.

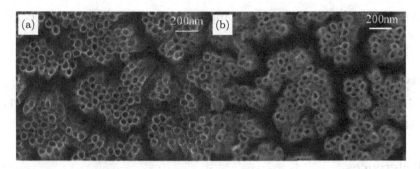

**Fig. 15.** SEM images of the $TiO_2$ nanowire arrays (a) before $In_2S_3$ ALD coating and (b) after $In_2S_3$ ALD coating [142].

The $J_{sc}$, $V_{oc}$, fill factor and conversion efficiency of the cell were $4.91\,\mathrm{mA\,cm^{-2}}$, $0.22\,\mathrm{V}$, $0.33$ and $0.36\%$, respectively. The small $V_{oc}$, compared with other SSCs, is ascribed, in part, to the relatively negative redox potential of the cobalt electrolyte. A small quantum efficiency of $\sim 10\%$ was observed due to charge recombination losses and charge injection/collection processes.

## 7. Perspective

The deposition technique of QD on photoanodes for QDSSCs is summarized in the chapter, which has opened up new pathways to design the low-cost production of catalytic, flexible and conductive electrodes for QDSSCs. Although considerable progress has been achieved, great challenges still exist in this field. First, such efficiency is far from the practical applications, and further exploration of the optimization of performance is necessary. Improved QD surface engineering with excellent electronic transport can lead to the improvement of $J_{SC}$. Developing new and revolutionary viable strategies to organize ordered assemblies on electrode surfaces will be the key to improving the performance of QDSSCs. Second, the enhanced photovoltaic mechanism of QDSSCs should be clarified in detail through combing the theoretical calculation and experimental evidence. Third, the thermal stability of the QDSSC device should be further improved. Commercialization of large-scale solar cells based on nanostructure architecture is yet to be realized. Considerable breakthroughs in the practical applications of QDSSCs are expected to occur in the near future.

## Acknowledgments

The National Natural Science Foundation of China (No. 51902301) is gratefully acknowledged.

## References

[1] A.J. Nozik, Quantum dot solar cells, *Phys. E*, 14 (2002) 115–120.

[2] A. Kongkanand, K. Tvrdy, K. Takechi, M. Kuno, P.V. Kamat, Quantum dot solar cells. Tuning photoresponse through size and shape control of CdSe-TiO$_2$ architecture, *J. Am. Chem. Soc.*, 130 (2008) 4007–4015.

[3] P.V. Kamat, Quantum dot solar cells. The next big thing in photovoltaics, *J. Phys. Chem. Lett.*, 4 (2013) 908–918.

[4] X.H. Wang, G.I. Koleilat, J. Tang, H. Liu, I.J. Kramer, R. Debnath, L. Brzozowski, D.A.R. Barkhouse, L. Levina, S. Hoogland, E.H. Sargent, Tandem colloidal quantum dot solar cells employing a graded recombination layer, *Nature Photon.*, 5 (2011) 480–484.

[5] I.J. Kramer, E.H. Sargent, The architecture of colloidal quantum dot solar cells: Materials to devices, *Chem. Rev.*, 114 (2014) 863–882.

[6] M. Gratzel, Dye-sensitized solar cells, *J. Photochem. Photobiol. C: Photochem. Rev.*, 4 (2003) 145–153.

[7] A. Hagfeldt, G. Boschloo, L. Sun, L. Kloo, H. Pettersson, Dye-sensitized solar cells, *Chem. Rev.*, 110 (2010) 6595–6663.

[8] S. Mathew, A. Yella, P. Gao, R. Humphry-Baker, B.F.E. Curchod, N. Ashari-Astani, I. Tavernelli, U. Rothlisberger, M.K. Nazeeruddin, M. Grätzel, Dye-sensitized solar cells with 13% efficiency achieved through the molecular engineering of porphyrin sensitizers, *Nat. Chem.*, 6 (2014) 242.

[9] I. Chung, B.H. Lee, J.Q. He, R.P.H. Chang, M.G. Kanatzidis, All-solid-state dye-sensitized solar cells with high efficiency, *Nature*, 485 (2012) 486–489.

[10] M. Law, L.E. Greene, J.C. Johnson, R. Saykally, P.D. Yang, Nanowire dye-sensitized solar cells, *Nat. Mater.*, 4 (2005) 455–459.

[11] I. Hod, A. Zaban, Materials and interfaces in quantum dot sensitized solar cells: Challenges, advances and prospects, *Langmuir*, 30 (2014) 7264–7273.

[12] E.H. Sargent, Colloidal quantum dot solar cells, *Nature Photon.*, 6 (2012) 133–135.

[13] O.E. Semonin, J.M. Luther, S. Choi, H.Y. Chen, J. Gao, A.J. Nozik, M.C. Beard, Peak External photocurrent quantum efficiency exceeding 100% via meg in a quantum dot solar cell, *Science*, 334 (2011) 1530.

[14] A.G. Pattantyus-Abraham, I.J. Kramer, A.R. Barkhouse, X. Wang, G. Konstantatos, R. Debnath, L. Levina, I. Raabe, M.K. Nazeeruddin, M. Grätzel, E.H. Sargent, Depleted-heterojunction colloidal quantum dot solar cells, *ACS Nano*, 4 (2010) 3374–3380.

[15] G.H. Carey, A.L. Abdelhady, Z. Ning, S.M. Thon, O.M. Bakr, E.H. Sargent, Colloidal quantum dot solar cells, *Chem. Rev.*, 115 (2015) 12732–12763.

[16] Z.X. Pan, I. Mora-Seró, Q. Shen, H. Zhang, Y. Li, K. Zhao, J. Wang, X.H. Zhong, J. Bisquert, High-efficiency "green" quantum dot solar cells, *J. Am. Chem. Soc.*, 136 (2014) 9203–9210.

[17] W.T. Sun, Y. Yu, H.Y. Pan, X.F. Gao, Q. Chen, L.M. Peng, CdS quantum dots sensitized TiO$_2$ nanotube-array photoelectrodes, *J. Am. Chem. Soc.*, 130 (2008) 1124–1125.

[18] X.H. Song, Z.Y. Liu, T. Tian, Z.N. Ma, Y. Yan, X.P. Li, X. Dong, Y.Y. Wang, C.X. Xia, Lead sulfide films synthesized by microwave-assisted chemical bath deposition method as efficient counter electrodes for CdS/CdSe sensitized ZnO nanorod solar cells, *Sol. Energy*, 177 (2019) 672–678.

[19] E. Akman, Y. Altintas, M. Gulen, M. Yilmaz, E. Mutlugun, S. Sonmezoglu, Improving performance and stability in quantum dot-sensitized solar cell through single layer graphene/Cu$_2$S nanocomposite counter electrode, *Renew. Energy*, 145 (2020) 2192–2200.

[20] G. Yue, J.H. Wu, Y.M. Xiao, J.M. Lin, M.L. Huang, Z. Lan, L.Q. Fan, CdTe quantum dots-sensitized solar cells featuring PCBM/P$_3$HT as hole transport material and assistant sensitizer provide 3.40% efficiency, *Electrochim. Acta*, 85 (2012) 182–186.

[21] J. Yi, Y.F. Duan, C.X. Liu, S.H. Gao, X.T. Han, L.M. An, PbS Quantum dots sensitized TiO$_2$ solar cells prepared by successive ionic layer absorption and reaction with different adsorption layers, *J. Nanosci. Nanotechnol.*, 16 (2016) 3904–3908.

[22] J.B. Zhang, J.B. Gao, C.P. Church, E.M. Miller, J.M. Luther, V.I. Klimov, M.C. Beard, PbSe quantum dot solar cells with more than 6% efficiency fabricated in ambient atmosphere, *Nano Lett.*, 14 (2014) 6010–6015.

[23] W.X. Li, J.Y. Yang, Q.H. Jiang, Y.B. Luo, Y.R. Hou, S.Q. Zhou, Y. Xiao, L.W. Fu, Z.W. Zhou, Electrochemical atomic layer deposition of Bi$_2$S$_3$/Sb$_2$S$_3$ quantum dots co-sensitized TiO$_2$ nanorods solar cells, *J. Power Sources*, 307 (2016) 690–696.

[24] A. Zaban, O.I. Mićić, B.A. Gregg, A.J. Nozik, Photosensitization of nanoporous TiO$_2$ electrodes with InP quantum dots, *Langmuir*, 14 (1998) 3153–3156.

[25] P.R. Yu, K. Zhu, A.G. Norman, S. Ferrere, A.J. Frank, A.J. Nozik, Nanocrystalline TiO$_2$ solar cells sensitized with InAs quantum dots, *J. Phys. Chem. B*, 110 (2006) 25451–25454.

[26] J.L. Duan, Q.W. Tang, B.L. He, L.M. Yu, Efficient In$_2$S$_3$ quantum dot-sensitized solar cells: A promising power conversion efficiency of 1.30%, *Electrochim. Acta*, 139 (2014) 381–385.

[27] A. Tubtimtae, K.L. Wu, H.Y. Tung, M.W. Lee, G.J. Wang, $Ag_2S$ quantum dot-sensitized solar cells, Electrochem. Commun., 12 (2010) 1158–1160.

[28] Z. Zhang, Y. Yang, J. Gao, S. Xiao, C.H. Zhou, D.Q. Pan, G. Liu, X.Y. Guo, Highly efficient $Ag_2Se$ quantum dots blocking layer for solid-state dye-sensitized solar cells: Size effects on device performances, Mater. Today Energy, 7 (2018) 27–36.

[29] D.U. Lee, S. Woo Pak, S. Gook Cho, E. Kyu Kim, S. Il Seok, Defect states in hybrid solar cells consisting of $Sb_2S_3$ quantum dots and $TiO_2$ nanoparticles, Appl. Phys. Lett., 103 (2013) 023901.

[30] Z. Yang, C.Y. Chen, H.T. Chang, Preparation of highly electroactive cobalt sulfide core-shell nanosheets as counter electrodes for CdZnSSe nanostructure-sensitized solar cells, Sol. Energy Mater. Sol. Cells, 95 (2011) 2867–2873.

[31] M.C. Lin, M.W. Lee, $Cu_{2-x}S$ quantum dot-sensitized solar cells, Electrochem. Commun., 13 (2011) 1376–1378.

[32] H. Seo, Y. Wang, G. Uchida, K. Kamataki, N. Itagaki, K. Koga, M. Shiratani, Analysis on the effect of polysulfide electrolyte composition for higher performance of Si quantum dot-sensitized solar cells, Electrochim. Acta, 95 (2013) 43–47.

[33] X.C. Yu, J. Zhu, Y.H. Zhang, J. Weng, L.H. Hu, S.Y. Dai, $SnSe_2$ quantum dot sensitized solar cells prepared employing molecular metal chalcogenide as precursors, Chem. Commun., 48 (2012) 3324–3326.

[34] Z. Yang, H.T. Chang, CdHgTe and CdTe quantum dot solar cells displaying an energy conversion efficiency exceeding 2%, Sol. Energy Mater. Sol. Cells, 94 (2010) 2046–2051.

[35] G. Seo, J. Seo, S. Ryu, W. Yin, T.K. Ahn, S.I. Seok, Enhancing the performance of sensitized solar cells with $PbS/CH_3NH_3PbI_3$ core/shell quantum dots, J. Phys. Chem. Lett., 5 (2014) 2015–2020.

[36] D.H. Jara, S.J. Yoon, K.G. Stamplecoskie, P.V. Kamat, Size-dependent photovoltaic performance of $CuInS_2$ quantum dot-sensitized solar cells, Chem. Mater., 26 (2014) 7221–7228.

[37] D. Esparza, I. Zarazúa, T. López-Luke, R. Carriles, A. Torres-Castro, E.D.l. Rosa, Photovoltaic properties of $Bi_2S_3$ and CdS quantum dot sensitized $TiO_2$ solar cells, Electrochim. Acta, 180 (2015) 486–492.

[38] Y.L. Lee, Y.S. Lo, Highly efficient quantum-dot-sensitized solar cell based on Co-sensitization of CdS/CdSe, Adv. Funct. Mater., 19 (2009) 604–609.

[39] A.J. Nozik, Multiple exciton generation in semiconductor quantum dots, Chem. Phys. Lett., 457 (2008) 3–11.

[40] Y. Tian, T. Tatsuma, Mechanisms and applications of plasmon-induced charge separation at $TiO_2$ films loaded with gold nanoparticles, J. Am. Chem. Soc., 127 (2005) 7632–7637.

[41] P.V. Kamat, Quantum dot solar cells. Semiconductor nanocrystals as light harvesters, J. Phys. Chem. C, 112 (2008) 18737–18753.

[42] V.I. Klimov, Mechanisms for photogeneration and recombination of multiexcitons in semiconductor nanocrystals: Implications for lasing and solar energy conversion, J. Phys. Chem. B, 110 (2006) 16827–16845.

[43] S.H. Im, C.S. Lim, J.A. Chang, Y.H. Lee, N. Maiti, H.J. Kim, M.K. Nazeeruddin, M. Grätzel, S.I. Seok, Toward interaction of sensitizer and functional moieties in hole-transporting materials for efficient semiconductor-sensitized solar cells, *Nano Lett.*, 11 (2011) 4789–4793.

[44] P.P. Boix, Y.H. Lee, F. Fabregat-Santiago, S.H. Im, I. Mora-Sero, J. Bisquert, S.I. Seok, From flat to nanostructured photovoltaics: Balance between thickness of the absorber and charge screening in sensitized solar cells, *ACS Nano*, 6 (2012) 873–880.

[45] J.A. Chang, S.H. Im, Y.H. Lee, H.j. Kim, C.S. Lim, J.H. Heo, S.I. Seok, Panchromatic photon-harvesting by hole-conducting materials in inorganic-organic heterojunction sensitized-solar cell through the formation of nanostructured electron channels, *Nano Lett.*, 12 (2012) 1863–1867.

[46] P.K. Santra, P.V. Kamat, Mn-doped quantum dot sensitized solar cells: A strategy to boost efficiency over 5%, *J. Am. Chem. Soc.*, 134 (2012) 2508–2511.

[47] J. Halme, P. Vahermaa, K. Miettunen, P. Lund, Device physics of dye solar cells, *Adv. Mater.*, 22 (2010) E210–E234.

[48] J.M. Cole, G. Pepe, O.K. Al Bahri, C.B. Cooper, Cosensitization in dye-sensitized solar cells, *Chem. Rev.*, 119 (2019) 7279–7327.

[49] V.A. El Bitar Nehme, M.A. El Bitar Nehme, T.H. Ghaddar, New pyridyl-based dyes for co-sensitization in dye sensitized solar cells, *Sol. Energy*, 187 (2019) 108–114.

[50] S. Lee, J.C. Flanagan, J. Kim, A.J. Yun, B. Lee, M. Shim, B. Park, Efficient type-II heterojunction nanorod sensitized solar cells realized by controlled synthesis of core/patchy-shell structure and CdS cosensitization, *ACS Appl. Mater. Interfaces*, 11 (2019) 19104–19114.

[51] K.A. Kumar, K. Subalakshmi, J. Senthilselvan, Effect of co-sensitization in solar exfoliated TiO2 functionalized rGO photoanode for dye-sensitized solar cell applications, *Mater. Sci. Semicond. Process.*, 96 (2019) 104–115.

[52] K. Mohan, A. Bora, R.S. Roy, B.C. Nath, S.K. Dolui, Polyaniline nanotube/reduced graphene oxide aerogel as efficient counter electrode for quasi solid state dye sensitized solar cell, *Sol. Energy*, 186 (2019) 360–369.

[53] J. Xia, Q. Wang, M.M. Chen, W.W. Li, J. Liu, J.T. Chen, H. Wu, S.H. Fan, Bifacial quasi-solid-state dye-sensitized solar cell with metal selenide $M_{0.85}Se$ (M = Co, Ni) as counter electrode, *Electrochim. Acta*, 307 (2019) 422–429.

[54] W.H. Li, G.Q. Long, Q.Q. Chen, Q. Zhong, High-efficiency layered sulfur-doped reduced graphene oxide and carbon nanotube composite counter electrode for quantum dot sensitized solar cells, *J. Power Sources*, 430 (2019) 95–103.

[55] J.B. Zhang, Y. Hao, L. Yang, H. Mohammadi, N. Vlachopoulos, L.C. Sun, A. Hagfeldt, E. sheibani, Electrochemically polymerized poly (3, 4-phenylenedioxythiophene) as efficient and transparent counter electrode for dye sensitized solar cells, *Electrochim. Acta*, 300 (2019) 482–488.

[56] W.J. Lee, E. Ramasamy, D.Y. Lee, J.S. Song, Efficient dye-sensitized solar cells with catalytic multiwall carbon nanotube counter electrodes, *ACS Appl. Mater. Interfaces*, 1 (2009) 1145–1149.

[57]  J.K. Chen, K.X. Li, Y.H. Luo, X.Z. Guo, D.M. Li, M.H. Deng, S.Q. Huang, Q.B. Meng, A flexible carbon counter electrode for dye-sensitized solar cells, *Carbon*, 47 (2009) 2704–2708.

[58]  V. Sharma, T.K. Das, P. Ilaiyaraja, C. Sudakar, Oxygen non-stoichiometry in $TiO_2$ and ZnO nano rods: Effect on the photovoltaic properties of dye and $Sb_2S_3$ sensitized solar cells, *Sol. Energy*, 191 (2019) 400–409.

[59]  R. Jose, V. Thavasi, S. Ramakrishna, Metal oxides for dye-sensitized solar cells, *J. Am. Chem. Soc.*, 92 (2009) 289–301.

[60]  S. Karuppuchamy, K. Nonomura, T. Yoshida, T. Sugiura, H. Minoura, Cathodic electrodeposition of oxide semiconductor thin films and their application to dye-sensitized solar cells, *Solid State Ionics*, 151 (2002) 19–27.

[61]  O. Prakash, V. Saxena, S. Choudhury, Tanvi, A. Singh, A.K. Debnath, A. Mahajan, K.P. Muthe, D.K. Aswal, Low temperature processable ultra-thin $WO_3$ Langmuir-Blodgett film as excellent hole blocking layer for enhanced performance in dye sensitized solar cell, *Electrochim. Acta*, 318 (2019) 405–412.

[62]  S. Nakade, T. Kanzaki, W. Kubo, T. Kitamura, Y.J. Wada, S.Z. Yanagida, Role of electrolytes on charge recombination in dye-sensitized $TiO_2$ solar cell (1): The case of solar cells using the $I^-/I^{3-}$ Redox couple, *J. Phys. Chem. B*, 109 (2005) 3480–3487.

[63]  T. Daeneke, Y. Uemura, N.W. Duffy, A.J. Mozer, N. Koumura, U. Bach, L. Spiccia, Aqueous dye-sensitized solar cell electrolytes based on the ferricyanide-ferrocyanide redox couple, *Adv. Mater.*, 24 (2012) 1222–1225.

[64]  Y.L. Lee, C.H. Chang, Efficient polysulfide electrolyte for CdS quantum dot-sensitized solar cells, *J. Power Sources*, 185 (2008) 584–588.

[65]  M.S. Zhou, G.C. Shen, Z.X. Pan, X.H. Zhong, Selenium cooperated polysulfide electrolyte for efficiency enhancement of quantum dot-sensitized solar cells, *J. Electroanal. Chem.*, 38 (2019) 147–152.

[66]  T.C. Li, M.S. Góes, F. Fabregat-Santiago, J. Bisquert, P.R. Bueno, C. Prasittichai, J.T. Hupp, T.J. Marks, Surface passivation of nanoporous $TiO_2$ via atomic layer deposition of $ZrO_2$ for solid-state dye-sensitized solar cell applications, *J. Phys. Chem. C*, 113 (2009) 18385–18390.

[67]  S.H. Hsu, S.F. Hung, S.H. Chien, CdS sensitized vertically aligned single crystal $TiO_2$ nanorods on transparent conducting glass with improved solar cell efficiency and stability using ZnS passivation layer, *J. Power Sources*, 233 (2013) 236–243.

[68]  Y.L. Chen, Q. Tao, W.Y. Fu, H.B. Yang, X.M. Zhou, Y.Y. Zhang, S. Su, P. Wang, M.H. Li, Enhanced solar cell efficiency and stability using ZnS passivation layer for CdS quantum-dot sensitized actinomorphic hexagonal columnar ZnO, *Electrochim. Acta*, 118 (2014) 176–181.

[69]  F. Huang, J. Hou, Q.F. Zhang, Y. Wang, R.C. Massé, S.L. Peng, H.L. Wang, J.S. Liu, G.Z. Cao, Doubling the power conversion efficiency in CdS/CdSe quantum dot sensitized solar cells with a ZnSe passivation layer, *Nano Energy*, 26 (2016) 114–122.

[70] S. Hore, C. Vetter, R. Kern, H. Smit, A. Hinsch, Influence of scattering layers on efficiency of dye-sensitized solar cells, *Sol. Energy Mater. Sol. Cells*, 90 (2006) 1176–1188.

[71] Y.Z. Zheng, X. Tao, L.X. Wang, H. Xu, Q. Hou, W.L. Zhou, J.F. Chen, Novel ZnO-based film with double light-scattering layers as photoelectrodes for enhanced efficiency in dye-sensitized solar cells, *Chem. Mater.*, 22 (2010) 928–934.

[72] F.Y. Xie, J.J. Wang, Y.F. Li, J. Dou, M.D. Wei, One-step synthesis of hierarchical $SnO_2/TiO_2$ composite hollow microspheres as an efficient scattering layer for dye-sensitized solar cells, *Electrochim. Acta*, 296 (2019) 142–148.

[73] S. Aynehband, E. Nouri, M.R. Mohammadi, Y. Li, Performance of $CoTiO_3$ as an oxide perovskite material for the light scattering layer of dye-sensitized solar cells, *New J. Chem.*, 43 (2019) 3760–3768.

[74] L.Q. Wang, J.M. Feng, Y.Y. Tong, J. Liang, A reduced graphene oxide interface layer for improved power conversion efficiency of aqueous quantum dots sensitized solar cells, *Int. J. Hydrogen Energy*, 44 (2019) 128–135.

[75] J.S. Luo, Z.Q. Wan, F. Han, H.A. Malik, B.W. Zhao, J.X. Xia, C.Y. Jia, R.L. Wang, Origin of increased efficiency and decreased hysteresis of perovskite solar cells by using 4-tert-butyl pyridine as interfacial modifier for $TiO_2$, *J. Power Sources*, 415 (2019) 197–206.

[76] S.R. Kim, M.K. Parvez, M. Chhowalla, UV-reduction of graphene oxide and its application as an interfacial layer to reduce the back-transport reactions in dye-sensitized solar cells, *Chem. Phys. Lett.*, 483 (2009) 124–127.

[77] Y.J. Kim, K.H. Kim, P. Kang, H.J. Kim, Y.S. Choi, W.I. Lee, Effect of layer-by-layer assembled $SnO_2$ interfacial layers in photovoltaic properties of dye-sensitized solar cells, *Langmuir*, 28 (2012) 10620–10626.

[78] I. Robel, V. Subramanian, M. Kuno, P.V. Kamat, Quantum dot solar cells. harvesting light energy with CdSe nanocrystals molecularly linked to mesoscopic $TiO_2$ films, *J. Am. Chem. Soc.*, 128 (2006) 2385–2393.

[79] I. Robel, M. Kuno, P.V. Kamat, Size-dependent electron injection from excited CdSe quantum dots into $TiO_2$ nanoparticles, *J. Am. Chem. Soc.*, 129 (2007) 4136–4137.

[80] J.G. Radich, R. Dwyer, P.V. Kamat, $Cu_2S$ reduced graphene oxide composite for High-efficiency quantum dot solar cells. overcoming the redox limitations of $S_2^-/S_n^{2-}$ at the counter electrode, *J. Phys. Chem. Lett.*, 2 (2011) 2453–2460.

[81] G. Zhu, L.K. Pan, T. Xu, Z. Sun, CdS/CdSe-cosensitized $TiO_2$ photoanode for Quantum-dot-sensitized solar cells by a microwave-assisted chemical bath deposition method, *ACS Appl. Mater. Interfaces*, 3 (2011) 3146–3151.

[82] Y. Jin-nouchi, S.I. Naya, H. Tada, Quantum-dot-sensitized solar cell using a photoanode prepared by *in situ* photodeposition of CdS on nanocrystalline $TiO_2$ films, *J. Phys. Chem. C*, 114 (2010) 16837–16842.

[83] I. Mora-Seró, J. Bisquert, Breakthroughs in the development of semiconductor-sensitized solar cells, *J. Phys. Chem. Lett.*, 1 (2010) 3046–3052.

[84] C.H. Chang, Y.L. Lee, Chemical bath deposition of CdS quantum dots onto mesoscopic $TiO_2$ films for application in quantum-dot-sensitized solar cells, *Appl. Phys. Lett.*, 91 (2007) 053503.

[85] X.Y. Yu, J.Y. Liao, K.Q. Qiu, D.B. Kuang, C.Y. Su, Dynamic study of highly efficient CdS/CdSe quantum dot-sensitized solar cells fabricated by electrodeposition, *ACS Nano*, 5 (2011) 9494–9500.

[86] S.M. Wang, W.W. Dong, X.D. Fang, S. Zhou, J.Z. Shao, Z.H. Deng, R.H. Tao, Q.L. Zhang, L.H. Hu, J. Zhu, Enhanced electrocatalytic activity of vacuum thermal evaporated $Cu_xS$ counter electrode for quantum dot-sensitized solar cells, *Electrochim. Acta*, 154 (2015) 47–53.

[87] S.H. Im, Y.H. Lee, S.I. Seok, S.W. Kim, S.W. Kim, Quantum-dot-sensitized solar cells fabricated by the combined process of the direct attachment of colloidal CdSe Quantum dots having a ZnS Glue layer and spray pyrolysis deposition, *Langmuir*, 26 (2010) 18576–18580.

[88] C.Y. Chou, C.P. Lee, R. Vittal, K.C. Ho, Efficient quantum dot-sensitized solar cell with polystyrene-modified $TiO_2$ photoanode and with guanidine thiocyanate in its polysulfide electrolyte, *J. Power Sources*, 196 (2011) 6595–6602.

[89] N. Guijarro, T. Lana-Villarreal, I. Mora-Seró, J. Bisquert, R. Gómez, CdSe quantum dot-sensitized $TiO_2$ electrodes: Effect of quantum dot coverage and mode of attachment, *J. Phys. Chem. C*, 113 (2009) 4208–4214.

[90] T. Toyoda, J. Sato, Q. Shen, Effect of sensitization by quantum-sized CdS on photoacoustic and photoelectrochemical current spectra of porous $TiO_2$ electrodes, *Rev. Sci. Instrum.*, 74 (2003) 297–299.

[91] Q. Shen, J. Kobayashi, L.J. Diguna, T. Toyoda, Effect of ZnS coating on the photovoltaic properties of CdSe quantum dot-sensitized solar cells, *J. Appl. Phys.*, 103 (2008) 084304.

[92] L.L. Yang, G. Chen, Y.F. Sun, D.L. Han, S. Yang, M. Gao, Z. Wang, P. Zou, H.M. Luan, X.W. Kong, J.H. Yang, Enhanced photovoltaic performance of QDSSCs via modifying ZnO photoanode with a 3-PPA self-assembled monolayer, *Appl. Surf. Sci.*, 328 (2015) 568–576.

[93] N. Guijarro, J.M. Campiña, Q. Shen, T. Toyoda, T. Lana-Villarreal, R. Gómez, Uncovering the role of the ZnS treatment in the performance of quantum dot sensitized solar cells, *Phys. Chem. Chem. Phys.*, 13 (2011) 12024–12032.

[94] C.F. Chi, S.Y. Liau, Y.L. Lee, The heat annealing effect on the performance of CdS/CdSe-sensitized $TiO_2$ photoelectrodes in photochemical hydrogen generation, *Nanotechnology*, 21 (2009) 025202.

[95] L.W. Chong, H.T. Chien, Y.L. Lee, Assembly of CdSe onto mesoporous $TiO_2$ films induced by a self-assembled monolayer for quantum dot-sensitized solar cell applications, *J. Power Sources*, 195 (2010) 5109–5113.

[96] Y.F. Sun, G. Chen, H.B. Deng, L.L. Yang, C.L. Liu, Y.R. Sui, S.Q. Lv, M.B. Wei, J.H. Yang, Enhanced Photovoltaic Performance of CdS quantum dot-sensitized solar cells using 4-tertbutylbenzoic acid as self-assembled monolayer on ZnO photoanode, *Phys. status solidi A*, 214 (2017) 1700458.

[97] I. Mora-Seró, S. Giménez, T. Moehl, F. Fabregat-Santiago, T. Lana-Villareal, R. Gómez, J. Bisquert, Factors determining the photovoltaic performance of a CdSe quantum dot sensitized solar cell: The role of the linker molecule and of the counter electrode, *Nanotechnology*, 19 (2008) 424007.

[98] A.G. Kontos, V. Likodimos, E. Vassalou, I. Kapogianni, Y.S. Raptis, C. Raptis, P. Falaras, Nanostructured titania films sensitized by quantum dot chalcogenides, *Nanoscale Res. Lett.*, 6 (2011) 266.

[99] S.C. Lin, Y.L. Lee, C.H. Chang, Y.J. Shen, Y.M. Yang, Quantum-dot-sensitized solar cells: Assembly of CdS-quantum-dots coupling techniques of self-assembled monolayer and chemical bath deposition, *Appl. Phys. Lett.*, 90 (2007) 143517.

[100] H. Lee, M. Wang, P. Chen, D.R. Gamelin, S.M. Zakeeruddin, M. Grätzel, M.K. Nazeeruddin, Efficient CdSe quantum dot-sensitized solar cells prepared by an improved successive ionic layer adsorption and reaction process, *Nano Lett.*, 9 (2009) 4221–4227.

[101] P. Ardalan, T.P. Brennan, H.B.R. Lee, J.R. Bakke, I.K. Ding, M.D. McGehee, S.F. Bent, Effects of Self-assembled monolayers on solid-state CdS quantum dot sensitized solar cells, *ACS Nano*, 5 (2011) 1495–1504.

[102] I. Barceló, T. Lana-Villarreal, R. Gómez, Efficient sensitization of ZnO nanoporous films with CdSe QDs grown by Successive Ionic Layer Adsorption and Reaction (SILAR), *J. Photochem. Photobiol. A: Chem.*, 220 (2011) 47–53.

[103] B.B. Jin, D.J. Wang, S.Y. Kong, G.Q. Zhang, H.S. Huang, Y. Liu, H.Q. Liu, J. Wu, L.H. Zhao, D. He, Voltage-assisted SILAR deposition of CdSe quantum dots into mesoporous $TiO_2$ film for quantum dot-sensitized solar cells, *Chem. Phys. Lett.*, 735 (2019) 136764.

[104] S. Hou, X.C. Dai, Y.B. Li, M.H. Huang, T. Li, Z.Q. Wei, Y.H. He, G.C. Xiao, F.X. Xiao, Charge transfer modulation in layer-by-layer-assembled multilayered photoanodes for solar water oxidation, *J. Mater. Chem. A*, 7 (2019) 22487–22499.

[105] L.p. Liu, J. Hensel, R.C. Fitzmorris, Y.d. Li, J.Z. Zhang, Preparation and photoelectrochemical properties of $CdSe/TiO_2$ hybrid mesoporous structures, *J. Phys. Chem. Lett.*, 1 (2010) 155–160.

[106] H.M. Cheng, K.Y. Huang, K.M. Lee, P. Yu, S.C. Lin, J.H. Huang, C.G. Wu, J. Tang, High-efficiency cascade CdS/CdSe quantum dot-sensitized solar cells based on hierarchical tetrapod-like ZnO nanoparticles, *Phys. Chem. Chem. Phys.*, 14 (2012) 13539–13548.

[107] G. Hodes, Comparison of dye- and semiconductor-sensitized porous nanocrystalline liquid junction solar cells, *J. Phys. Chem. C*, 112 (2008) 17778–17787.

[108] W. Lee, S.K. Min, V. Dhas, S.B. Ogale, S.H. Han, Chemical bath deposition of CdS quantum dots on vertically aligned ZnO nanorods for quantum dots-sensitized solar cells, *Electrochem. Commun.*, 11 (2009) 103–106.

[109] S.S. Mali, S.K. Desai, D.S. Dalavi, C.A. Betty, P.N. Bhosale, P.S. Patil, CdS-sensitized TiO$_2$ nanocorals: Hydrothermal synthesis, characterization, application, *Photochem. Photobiol. Sci.*, 10 (2011) 1652–1658.

[110] M. Shalom, S. Dor, S. Rühle, L. Grinis, A. Zaban, Core/CdS Quantum Dot/Shell Mesoporous Solar Cells with Improved Stability and Efficiency Using an Amorphous TiO$_2$ Coating, *J. Phys. Chem. C*, 113 (2009) 3895–3898.

[111] F. Khodam, A.R. Amani-Ghadim, S. Aber, Preparation of CdS quantum dot sensitized solar cell based on ZnTi-layered double hydroxide photoanode to enhance photovoltaic properties, *Sol. Energy*, 181 (2019) 325–332.

[112] R. Sundheep, A. Asok, R. Prasanth, Surface engineering of CdTe quantum dots using ethanol as a co-solvent for enhanced current conversion efficiency in QDSSC, *Sol. Energy*, 180 (2019) 501–509.

[113] V. González-Pedro, X.Q. Xu, I. Mora-Seró, J. Bisquert, Modeling high-efficiency quantum dot sensitized solar cells, *ACS Nano*, 4 (2010) 5783–5790.

[114] S. Emin, M. Yanagida, W. Peng, L. Han, Evaluation of carrier transport and recombinations in cadmium selenide quantum-dot-sensitized solar cells, *Sol. Energy Mater. Sol. Cells*, 101 (2012) 5–10.

[115] Q.X. Zhang, Y.D. Zhang, S.Q. Huang, X.M. Huang, Y.H. Luo, Q.B. Meng, D.M. Li, Application of carbon counterelectrode on CdS quantum dot-sensitized solar cells (QDSSCs), *Electrochem. Commun.*, 12 (2010) 327–330.

[116] D.R. Baker, P.V. Kamat, Photosensitization of TiO$_2$ nanostructures with cds quantum dots: Particulate versus tubular support architectures, *Adv. Funct. Mater.*, 19 (2009) 805–811.

[117] S.B. Rawal, S.D. Sung, S.Y. Moon, Y.J. Shin, W.I. Lee, Optimization of CdS layer on ZnO nanorod arrays for efficient CdS/CdSe co-sensitized solar cell, *Mater. Lett.*, 82 (2012) 240–243.

[118] G. Zhu, L.K. Pan, T. Xu, Q. Zhao, B. Lu, Z. Sun, Microwave assisted CdSe quantum dot deposition on TiO$_2$ films for dye-sensitized solar cells, *Nanoscale*, 3 (2011) 2188–2193.

[119] G. Zhu, L.K. Pan, T. Xu, Z. Sun, Microwave assisted chemical bath deposition of CdS on TiO$_2$ film for quantum dot-sensitized solar cells, *J. Electroanal. Chem.*, 659 (2011) 205–208.

[120] G. Zhu, L.K. Pan, T. Xu, Z. Sun, One-step synthesis of CdS sensitized TiO$_2$ photoanodes for quantum dot-sensitized solar cells by microwave assisted chemical bath deposition method, *ACS Appl. Mater. Interfaces*, 3 (2011) 1472–1478.

[121] B. Kraeutler, A.J. Bard, Heterogeneous photocatalytic preparation of supported catalysts. Photodeposition of platinum on titanium dioxide powder and other substrates, *J. Am. Chem. Soc.*, 100 (1978) 4317–4318.

[122] H. Tada, M. Fujishima, H. Kobayashi, Photodeposition of metal sulfide quantum dots on titanium(iv) dioxide and the applications to solar energy conversion, *Chem. Soc. Rev.*, 40 (2011) 4232–4243.

[123] M. Fujii, K. Nagasuna, M. Fujishima, T. Akita, H. Tada, Photodeposition of CdS quantum dots on $TiO_2$: Preparation, characterization, and reaction mechanism, *J. Phys. Chem. C*, 113 (2009) 16711–16716.

[124] H.W. Hu, J.N. Ding, S. Zhang, Y. Li, L. Bai, N.Y. Yuan, Photodeposition of $Ag_2S$ on $TiO_2$ nanorod arrays for quantum dot-sensitized solar cells, *Nanoscale Res. Lett.*, 8 (2013) 10.

[125] N. Parsi Benehkohal, V. González-Pedro, P.P. Boix, S. Chavhan, R. Tena-Zaera, G.P. Demopoulos, I. Mora-Seró, Colloidal PbS and PbSeS quantum dot sensitized solar cells prepared by electrophoretic deposition, *J. Phys. Chem. C*, 116 (2012) 16391–16397.

[126] A.C. Poulose, S. Veeranarayanan, S.H. Varghese, Y. Yoshida, T. Maekawa, D. Sakthi Kumar, Functionalized electrophoretic deposition of CdSe quantum dots onto $TiO_2$ electrode for photovoltaic application, *Chem. Phys. Lett.*, 539-540 (2012) 197–203.

[127] S. Bittolo Bon, L. Valentini, J.M. Kenny, L. Peponi, R. Verdejo, M.A. Lopez-Manchado, Electrodeposition of transparent and conducting graphene/carbon nanotube thin films, *Phys. Status Solidi A*, 207 (2010) 2461–2466.

[128] A.M. Prenen, J.C.A.H. van der Werf, C.W.M. Bastiaansen, D.J. Broer, Monodisperse, polymeric nano- and microsieves produced with interference holography, *Adv. Mater.*, 21 (2009) 1751–1755.

[129] G. Zhu, L.K. Pan, T. Lu, T. Xu, Z. Sun, Electrophoretic deposition of reduced graphene-carbon nanotubes composite films as counter electrodes of dye-sensitized solar cells, *J. Mater. Chem.*, 21 (2011) 14869–14875.

[130] G. Zhu, L.K. Pan, T. Lu, X.J. Liu, T. Lv, T. Xu, Z. Sun, Electrophoretic deposition of carbon nanotubes films as counter electrodes of dye-sensitized solar cells, *Electrochim. Acta*, 56 (2011) 10288–10291.

[131] A. Salant, M. Shalom, I. Hod, A. Faust, A. Zaban, U. Banin, Quantum Dot sensitized solar cells with improved efficiency prepared using electrophoretic deposition, *ACS Nano*, 4 (2010) 5962–5968.

[132] F. Sauvage, C. Davoisne, L. Philippe, J. Elias, Structural and optical characterization of electrodeposited CdSe in mesoporous anatase $TiO_2$ for regenerative quantum-dot-sensitized solar cells, *Nanotechnology*, 23 (2012) 395401.

[133] H.M.A. Kyaw, A.F.M. Noor, G. Kawamura, A. Matsuda, K.A. Yaacob, Effect of CdSe thickness deposited by electrophoretic deposition for quantum-dot-sensitized solar cell, *Mater. Today: Proc.*, 16 (2019) 196–200.

[134] H.M. Aung Kyaw, K.A. Yaacob, A.F. Mohd Noor, A. Matsuda, G. Kawamura, Effect of deposition time CdSe-$TiO_2$ nanocomposite film by electrophoretic deposition for quantum dot sensitized solar cell, *Mater. Today: Proc.*, 17 (2019) 736–742.

[135] G. Zhu, L.K. Pan, H.C. Sun, X.J. Liu, T. Lv, T. Lu, J. Yang, Z. Sun, Electrophoretic deposition of a reduced graphene-Au nanoparticle composite film as counter electrode for CdS quantum dot-sensitized solar cells, *ChemPhysChem*, 13 (2012) 769–773.

[136] G. Zhu, L.K. Pan, T. Xu, Q.F. Zhao, Z. Sun, Cascade structure of $TiO_2/ZnO/CdS$ film for quantum dot sensitized solar cells, *J. Alloys Compd.*, 509 (2011) 7814–7818.

[137] M. Okuya, K. Nakade, S. Kaneko, Porous $TiO_2$ thin films synthesized by a spray pyrolysis deposition (SPD) technique and their application to dye-sensitized solar cells, *Sol. Energy Mater. Sol. Cells*, 70 (2002) 425–435.

[138] Y.H. Lee, S.H. Im, J.H. Rhee, J.H. Lee, S.I. Seok, Performance enhancement through post-treatments of CdS-sensitized solar cells fabricated by spray pyrolysis deposition, *ACS Appl. Mater. Interfaces*, 2 (2010) 1648–1652.

[139] K. Shin, S.I. Seok, S.H. Im, J.H. Park, CdS or CdSe decorated $TiO_2$ nanotube arrays from spray pyrolysis deposition: Use in photoelectrochemical cells, *Chem. Commun.*, 46 (2010) 2385–2387.

[140] G. Zhu, T. Lv, L.K. Pan, Z. Sun, C.Q. Sun, All spray pyrolysis deposited CdS sensitized ZnO films for quantum dot-sensitized solar cells, *J. Alloys Compd.*, 509 (2011) 362–365.

[141] J.W. Elam, D. Routkevitch, P.P. Mardilovich, S.M. George, Conformal coating on ultrahigh-aspect-ratio nanopores of anodic alumina by atomic layer deposition, *Chem. Mater.*, 15 (2003) 3507–3517.

[142] S.K. Sarkar, J.Y. Kim, D.N. Goldstein, N.R. Neale, K. Zhu, C.M. Elliott, A.J. Frank, S.M. George, $In_2S_3$ atomic layer deposition and its application as a sensitizer on $TiO_2$ nanotube arrays for solar energy conversion, *J. Phys. Chem. C*, 114 (2010) 8032–8039.

Chapter 7

# First Decade of Halide Perovskite Photovoltaics Research and Perspective

Xian Hou[*], Xiaohong Chen[†] and Likun Pan[‡,§]

*School of Materials Science and Engineering,
Lanzhou University of Technology,
Lanzhou 730050, China
†Engineering Research Center for Nanophotonics &
Advanced Instrument, Ministry of Education,
School of Physics and Electronic Science,
East China Normal University,
Shanghai 200062, China
‡Shanghai Key Laboratory of Magnetic Resonance,
School of Physics and Electronic Science,
East China Normal University,
Shanghai 200062, China
§lkpan@phy.ecnu.edu.cn

## 1. Introduction

Perovskite solar cells (PSCs) have demonstrated an incredibly rapid ascent in power conversion efficiencies (PCEs) with low production costs. In 2019, the PCE of PSCs jumped over 25% according to academic research. Between 2009 and 2018, more than 10,000 papers were published, which garnered nearly 400,000 citations. In this chapter, we review the rapid progress in PSCs during the last 10-year period. In particular, we describe the recent research progress and key ongoing challenges facing perovskites, and give our views on future prospects of perovskite-based photovoltaics that might bring perovskite technology to commercialization.

## 2. The progress of Perovskite in photovoltaics

### 2.1. *Perovskite materials*

Perovskite was first discovered in chlorite-rich skarn by Gustav Rose in 1839, and was composed of $CaTiO_3$ and named after the renowned Russian mineralogist Count Lev A. Perovskiy (1792–1856). After that, $BaTiO_3$, $SrTiO_3$, $BiFeO_3$, $PbTiO_3$ and many other inorganic metal oxides were also found to have the perovskite structure. Thus, perovskite compounds with the formula $ABX_3$ are commonly known as a kind of metal oxide. For a long time in the past, oxide perovskites were widely used in ferroelectric, piezoelectric, pyroelectric and dielectric applications.

The typical $CH_3NH_3PbI_3$ perovskite structures are show in Fig. 1. In the cubic unit cell, the A-cation ($CH_3NH^{3+}$) resides at the eight corners of the cube, while the B-cation ($Pb^{2+}$) is located at the body center that is surrounded by 6 X-anions (located at the face centers) in an octahedral $[BX_6]^{4-}$ cluster.

A remarkably versatile feature of the organic–inorganic halide perovskite system is the tunability of the bandgap through both cation and anion substitution, which will be meaningful for multijunction perovskite solar cells or tandem cells with Si or CIGS

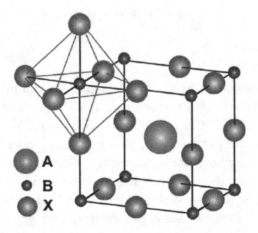

**Fig. 1.** Crystal structure of $ABX_3$ perovskite showing $BX_6$ octahedral and larger A cation occupied in cubo-octahedral site [1].

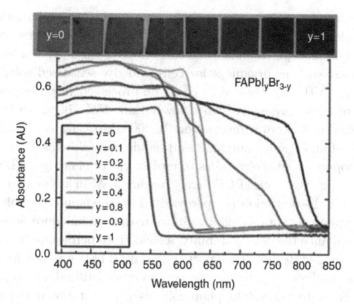

**Fig. 2.** Photographs of FAPbI$_y$Br$_{3-y}$ perovskite films with y increasing from 0 to 1 from left to right, and corresponding absorption spectra [5].

solar cells [2, 3]. Most representative studies use formamidinium cations (HC(NH$_2$)$^{2+}$) or alternative halides (X = I$^-$, Br$^-$, Cl$^-$ or mixtures) by adjusting the bandgap of the organic–inorganic halide perovskite system [4]. Figure 2 shows the range of absorption spectra and photographs of the different materials from FAPbI$_y$Br$_{3-y}$ mixed halide perovskites.

The CH$_3$NH$_3$PbX$_3$ demonstrated a high absorption coefficient ($10^3$–$10^4$ cm$^{-1}$), tunable bandgap of ~1.5–2.5 eV by changing the stoichiometric ratio of cations, high charge carrier mobility (2–66 cm$^2$ V$^{-1}$ s$^{-1}$), low exciton binding energy of ~2 meV, low trap state density of ~$10^{10}$ cm$^{-3}$ and a charge carrier lifetime of ~270 ns. These properties added with the low-temperature processability (<150°C) make it a desirable material for solar cells.

## 2.2. Development of halide perovskite solar cells

Before 2009, Miyasaka *et al.* [6] reported the first solar cells based on organic–inorganic perovskite. There are fewer research studies on perovskite materials in photovoltaics due to the low performance of conventional inorganic metal oxide perovskite, while most researchers focused on high-performance and cost-effective third-generation photovoltaics, such as organic solar cells and dye-sensitized solar cells (DSSCs). Although these cells could be processed very cheaply, the PCEs of these devices remained lower than 15% [7], which limited their chances for commercialization. In 2009, Miyasaka et al. explored the use of an organic–inorganic lead halide perovskite as a sensitizer on mesoporous $TiO_2$ electrodes to replace the organic dye in DSSCs. Even though the device PCE barely reaches 3.8% in a very short time due to the dissolution of the perovskite into the liquid electrolyte, it opens up a new world of photovoltaics with such immense potential.

To enhance the device stability, a solid-state perovskite-based cell with a solid-state hole transport material (HTM) was put forward; carbon/conductive polymer composites were utilized to prepare a full solid-state perovskite photovoltaics cell, but the performance of those devices was not good due to the low loading amount of perovskite on the scaffold layer. After that, the Miyasaka group and the Nam Gyu Park group used a higher concentration of precursor solution to increase the perovskite loading on the scaffold and the cells' performance was general improved. At this point, solid-state perovskite cells began to attract the attention of photovoltaics researchers from all over the world.

The efficiency was improved to 6.54% in 2011 by the N. G. Park group through the optimization of the $TiO_2$ surface and perovskite processing [8]. However, these embodiments immediately faced the difficulty of stabilization owing to the liquid electrolyte HTM decomposing the perovskite absorber.

A breakthrough came in 2012, when the Snaith group collaborated with the Miyasaka group [9], and the Park group simultaneously collaborated with the Grätzel group [10], leading to solidification of the perovskite-sensitized cell with a solid-state organic HTM, 2, 2′, 7, 7′-tetrakis $(N, N$-dimethoxyphenylamine$)-9, 9′$-spirobifluorene (spiro-OMeTAD), resulting in PCEs of near 10% when employing the $CH_3NH_3PbI_{3-x}Cl_x$ mixed halide perovskite or $CH_3NH_3PbI_3$, respectively. As this collaboration progressed, aside from enhancing the cells' PCEs and stability, another surprising feature was realized that the cells used insulating mesoporous $Al_2O_3$, which worked well as a scaffold for $CH_3NH_3PbI_3$ crystals. It was a sign of the long diffusion length of carriers in lead halide perovskite materials [11–13]; the long carrier diffusion length of the perovskite reminded researchers of the importance of the device architecture and understanding how PSCs worked. Thus, a new star was born in photovoltaic technologies, which distinguished the new technology from any other organic and hybrid material-based solar cells.

Due to the long carrier diffusion length of the perovskite, researchers soon realized that the mesoporous $TiO_2$ scaffold could be thinned down and that the few hundred nanometer-thick solid perovskite film could even sustain charge generation and transport [14, 15]. Thus, the device architecture was evolving to a simple planar heterojunction architecture, where a lead halide perovskite layer is sandwiched between hole transport layer (HTL) and electron transport layer (ETL). The following is a typical device structure: FTO/ETL/perovskite/HTL/Au. However, high efficiencies have also been obtained with the "inverted" structure: ITO/HTL/perovskite/ETL/Au. Even for the researchers persisting with the mesoporous $TiO_2$ scaffold, the trend has been for the mesoporous layer to become increasingly thinner, and we expect the mesoporous layer to be eventually removed as the technology progresses.

Since then, the surge of hybrid inorganic–organic perovskite as a light harvester in solar cells has brought up new interest in the development of highly efficient and cheap solar cells.

## 2.3. *Perovskite solar cell architectures*

Due to the original PSCs with liquid electrolyte being similar to DSSCs, here we only mention PSCs with solid-state HTMs.

The first solid-state stable PSC was reported in 2012 by Park and Grätzel [10] $CH_3NH_3PbI_3$ was deposited onto a mesoporous $TiO_2$ film and spiro-OMeTAD was introduced as the HTM (Fig. 4(a)). The replacement of the liquid electrolyte with spiro-OMeTAD not only improved the stability of the solar cell but also boosted its efficiency to 9.7%. After that, a PCE of 10.9% with high $V_{oc}$ of

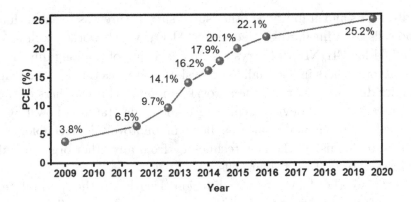

**Fig. 3.** Rapid PCE evolution of PSCs from 2009 to 2019.

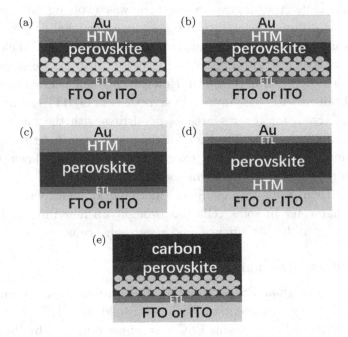

**Fig. 4.** Variety of device architectures for PSCs. (a) Mesoscopic n–i–p with semiconductor metal oxide scaffold; (b) mesoscopic n–i–p with insulating metal oxide scaffold; (c) n–i–p planar structure; (d) p–i–n inverted planar structure; (e) HTM-free structure.

0.98 V was reported by the Snaith group [9], in which a mixed halide perovskite $CH_3NH_3PbI_{3-x}Cl_x$ coated mesoporous $Al_2O_3$ film rather than a mesoporous $TiO_2$ film with spiro-OMeTAD was used as the HTM. Due to the higher valance band of $Al_2O_3$, electron injection from perovskite to $Al_2O_3$ was not allowed, and the mesoporous $Al_2O_3$ served only as a scaffold for the perovskite light absorber. The photoelectrons must be transferred through the perovskite to the compact $TiO_2$ layer. This work demonstrated that the perovskite is able to transport both electrons and holes between cell terminals and this cell structure was named "meso-superstructured solar cell" (Fig. 4(b)) because of the existence of the mesostructure and the absence of electron injection to the mesostructure.

However, the mesoporous layer makes the structure of solar cells comparatively complex, causing many problems. On the contrary, due to the superior properties of the perovskite, it was shown that the perovskite could function as a light harvester in the planar configuration. The initial planar perovskite solar cell was fabricated by Snaith *et al.* [9, 16], where perovskite was deposited on a compact $TiO_2$ block layer (Fig. 4(c)). After optimizing the film formation of $CH_3NH_3PbI_{3-x}Cl_x$ perovskite by controlling the annealing temperature and atmosphere, PSCs with a PCE of 11.4% were achieved. Liu *et al.* [15] used the vapor deposition technique to prepare a high-quality perovskite film for planar configuration PSCs and achieved 15.4% PCE with $V_{oc}$ of 1.07 V. These findings showed that perovskite could be used in a simple architecture solar cell without the need for mesoporous n-type semiconductors. Yang's groups [17] demonstrated further increase in PCE to 19.3% by modifying the common planar configuration. Of course, the planar configuration also can be a p–i–n-type structure, which is otherwise known as an inverted architecture (Fig. 4(d)); perovskite is sandwiched between a p-type material at the bottom and an n-type layer at the top (e.g., $FTO/TiO_2/PEDOT-PSS/perovskite/PCBM/Au$).

Besides the mesoporous and planar configurations depicted in Fig. 4(e), the HTM-free configuration is also promising because perovskites possess hole transporting properties, which significantly decrease the fabrication costs. Etgar *et al.* [18] first reported a $CH_3NH_3PbI_3/TiO_2$ solar cell with an efficiency of 5.5% based on this concept. Shi *et al.* [19] reported HTM-free PSCs and confirmed that $TiO_2/CH_3NH_3PbI_3/Au$ cell is a typical heterojunction solar cell.

## 3. Ongoing challenges of PSCs

### 3.1. *Lead-free alternatives*

It is necessary to replace or remove lead in order to prevent potential toxicity issues in PSCs, and a clear understanding of what is offset by employing lead-based perovskites needs to be realized. Noel *et al.* [20] and Hao *et al.* [21] reported MASnI$_3$-based perovskite solar cells and achieved PCEs of about 5–6%. The bandgap of MASnI$_3$ perovskite is about 1.2–1.3 eV, which allowed broader light harvesting than its lead counterpart (Fig. 5). However, the MASnI$_3$ perovskites are extremely unstable, and degrade rapidly even if processed in an inert atmosphere. It has been suggested that the key issue is the oxidation of $Sn^{2+}$ to $Sn^{4+}$, which results in a high level of "self-doping". Other metal ions such as Cu, Ge and Ni that can be stabilized in the $2^+$ state could be promising alternatives to lead include [22].

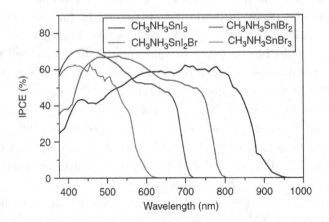

**Fig. 5.** IPCE fraction for the Sn-based perovskite solar cells with a range of different halides to tune the bandgap [21].

## 3.2. *Enhance the device stability*

To reach commercialization of PSCs technology, device stability is the most important thing to be considered. How to achieve perovskite modules that survive for at least 25 years in an outdoor environment without the need for overly expensive encapsulation methods is the researchers' recent concern.

Moisture reacting with the perovskite causes major damage to PSCs without encapsulation, which will rapidly irreversibly degrade. When encapsulated in an inert atmosphere, the degradation of the perovskite is largely resolved and the solar cells are capable of sustaining at least 1,000 h of exposure to simulated full-spectrum sunlight with little drop in photocurrent [23]. Mei *et al.* [24] used a double layer of mesoporous $TiO_2$ and $ZrO_2$ with a thick hydrophobic carbon black electrode and demonstrated 1,000 h of stability for an unsealed cell measured under full AM 1.5 simulated sunlight in ambient air. The results were extremely encouraging and imply that the perovskite is not fundamentally unstable in ambient conditions. However, the perovskite modules will still require sealing from environmental effects, especially considering the requirement for operation under elevated temperature and humidity, but the sealing methods already developed for crystalline silicon or thin film solar cells could be sufficient.

Thermostability is another challenge for PSCs due to the heat effect of sunlight and can thus pose a threat to perovskite decomposition or phase variation. Han *et al.* stored the encapsulated triple-layer PSCs at 85°C to test the device's stability. The device was kept for 3 months and this result is very encouraging for commercialization of PSCs (Fig. 6).

Moreover, the exciting thing is that the heat-induced loss of the organic cation can be largely mitigated by replacing the MA cation with the larger cation FA, suggesting that the $FAPbI_3$ perovskite may already be suitably thermally stable. Further thermal, chemical and operational stability improvements have been reported with the addition of low levels of inorganic $Cs^+$ cations and/or bromide [26–28].

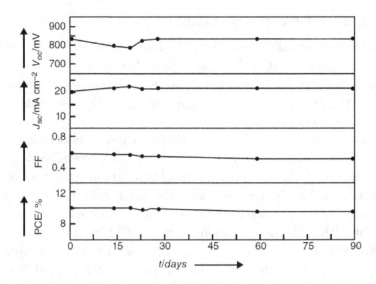

**Fig. 6.** Indoor heat stress test of a triple-layer perovskite solar cell. The device was encapsulated and kept for 3 months in a normal oven filled with ambient air at ~85°C and removed periodically to measure $J$–$V$ curves under full solar AM1.5 light at ambient temperature [25].

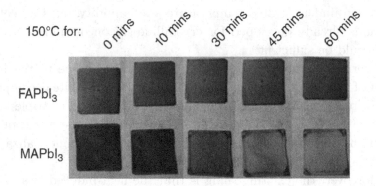

**Fig. 7.** Photographs showing non-encapsulated MAPbI₃ and FAPbI₃ samples before and during heating on a hot plate at 150°C for 60 min in air [4].

## 3.3. Manufacture on a large scale

Up to now, most of the high-efficiency PSCs have only exhibited a smaller illuminated active area of less than $0.1\,\mathrm{cm}^2$. The purpose of having a small area is to reduce crystal imperfection of perovskites and HTL films, thereby improving the crystallinity, quality of films and electron–hole mobility. But commercialization needs large-area devices. In China, the Shanghai Liyuan New Energy Technology Co., Ltd. (China) was the first to produce a PSC module with a record certified efficiency of 12.1% for an aperture area of 36 cm$^2$. Subsequently, the company developed a PSC module with an increased efficiency of 14.17% [29]. Microquanta Semiconductor Co., Ltd. (China) broke its own certified efficiency record for a mini-module 3 times, with the current record standing at 17.9% for an aperture area of 19.3 cm$^2$ [29]. WonderSolar Co., Ltd. (China) has concentrated on developing a large-area module using a highly stable carbon electrode. In 2018, it installed a 110 m$^2$ power system composed of 60 cm × 60 cm printable mesoscopic perovskite solar panels fabricated with screen printing techniques. Other companies, such as GCL New Energy Co., Ltd. (China), are also developing methods of fabricating high-quality large-area perovskite films.

## 4. Broader development

Nowadays, the PCE of PSCs in the laboratory has already reached 25%, which is comparable to crystalline silicon solar cells. Moreover, research on PSC techniques promotes quick development in the solar cell field. Even though the PCEs of PSCs have reached a relative ideal value, the research on stability and large-scale devices should slow down the pace. Partial substitution of the MA$^+$ cation with FA$^+$ and Cs$^+$, and the I$^-$ anion with Cl$^-$, Br$^-$ and SCN$^-$ can offer enhanced stability to perovskite films. Besides, the design of 2D or 2D/3D perovskites can efficiently strengthen the stability of PSCs. On the contrary, strategies on interfacial modification and the use of inorganic HTL (such as CuSCN, NiO$_x$ and CuI) or the capacity to be HTL-free would significantly improve the device's stability. All inorganic perovskites such as CsPbI$_3$ and CsPbI$_x$Br$_{1-x}$ are alternatives to enhance the stability of PSCs.

We believe that in the near future, PSCs will walk out of the laboratory to become an applicable alternative solar cell product.

# References

[1] N.G. Park, Perovskite solar cells: An emerging photovoltaic technology, *Mater. Today*, 18 (2015) 65–72.

[2] C.D. Bailie, M.G. Christoforo, J.P. Mailoa, A.R. Bowring, E.L. Unger, W.H. Nguyen, J. Burschka, N. Pellet, J.Z. Lee, M. Gratzel, R. Noufi, T. Buonassisi, A. Salleo, M.D. McGehee, Semi-transparent perovskite solar cells for tandems with silicon and CIGS, *Energy Environ. Sci.*, 8 (2015) 956–963.

[3] H.J. Snaith, Perovskites: The emergence of a new era for low-cost, high-efficiency solar cells, *J. Phys. Chem. Lett.*, 4 (2013) 3623–3630.

[4] S.D. Stranks, H.J. Snaith, Metal-halide perovskites for photovoltaic and light-emitting devices, *Nat. Nanotechnol.*, 10 (2015) 391–402.

[5] G.E. Eperon, S.D. Stranks, C. Menelaou, M.B. Johnston, L.M. Herz, H.J. Snaith, Formamidinium lead trihalide: A broadly tunable perovskite for efficient planar heterojunction solar cells, *Energy Environ. Sci.*, 7 (2014) 982–988.

[6] A. Kojima, K. Teshima, Y. Shirai, T. Miyasaka, Organometal halide perovskites as visible-light sensitizers for photovoltaic cells, *J. Am. Chem. Soc.*, 131 (2009) 6050.

[7] K. Kakiage, Y. Aoyama, T. Yano, K. Oya, J. Fujisawa, M. Hanaya, Highly-efficient dye-sensitized solar cells with collaborative sensitization by silyl-anchor and carboxy-anchor dyes, *Chem. Commun.*, 51 (2015) 15894–15897.

[8] J.H. Im, C.R. Lee, J.W. Lee, S.W. Park, N.G. Park, 6.5% efficient perovskite quantum-dot-sensitized solar cell, *Nanoscale*, 3 (2011) 4088–4093.

[9] M.M. Lee, J. Teuscher, T. Miyasaka, T.N. Murakami, H.J. Snaith, Efficient hybrid solar cells based on meso-superstructured organometal halide perovskites, *Science*, 338 (2012) 643–647.

[10] H.S. Kim, C.R. Lee, J.H. Im, K.B. Lee, T. Moehl, A. Marchioro, S.J. Moon, R. Humphry-Baker, J.H. Yum, J.E. Moser, M. Graetzel, N.G. Park, Lead iodide perovskite sensitized all-solid-state submicron thin film mesoscopic solar cell with efficiency exceeding 9%, *Sci. Rep.*, 2 (2012).

[11] S.D. Stranks, G.E. Eperon, G. Grancini, C. Menelaou, M.J.P. Alcocer, T. Leijtens, L.M. Herz, A. Petrozza, H.J. Snaith, Electron-hole diffusion lengths exceeding 1 micrometer in an organometal trihalide perovskite absorber, *Science*, 342 (2013) 341–344.

[12] C. Wehrenfennig, G.E. Eperon, M.B. Johnston, H.J. Snaith, L.M. Herz, High charge carrier mobilities and lifetimes in organolead trihalide perovskites, *Adv. Mater.*, 26 (2014) 1584–1589.

[13] Q. Dong, Y. Fang, Y. Shao, P. Mulligan, J. Qiu, L. Cao, J. Huang, Electron-hole diffusion lengths $>175\,\mu m$ in solution-grown $CH_3NH_3PbI_3$ single crystals, *Science*, 347 (2015) 967–970.

[14] J.M. Ball, M.M. Lee, A. Hey, H.J. Snaith, Low-temperature processed meso-superstructured to thin-film perovskite solar cells, *Energy Environ. Sci.*, 6 (2013) 1739–1743.

[15] M. Liu, M.B. Johnston, H.J. Snaith, Efficient planar heterojunction perovskite solar cells by vapour deposition, *Nature*, 501 (2013) 395.

[16] G.E. Eperon, V.M. Burlakov, P. Docampo, A. Goriely, H.J. Snaith, Morphological control for high performance, solution-processed planar heterojunction perovskite solar cells, *Adv. Funct. Mater.*, 24 (2014) 151–157.

[17] H. Zhou, Q. Chen, G. Li, S. Luo, T.B. Song, H.S. Duan, Z. Hong, J. You, Y. Liu, Y. Yang, Interface engineering of highly efficient perovskite solar cells, *Science*, 345 (2014) 542–546.

[18] L. Etgar, P. Gao, Z. Xue, Q. Peng, A.K. Chandiran, B. Liu, M.K. Nazeeruddin, M. Graetzel, Mesoscopic $CH_3NH_3PbI_3/TiO_2$ Heterojunction solar cells, *J. Am. Chem. Soc.*, 134 (2012) 17396–17399.

[19] J. Shi, J. Dong, S. Lv, Y. Xu, L. Zhu, J. Xiao, X. Xu, H. Wu, D. Li, Y. Luo, Q. Meng, Hole-conductor-free perovskite organic lead iodide heterojunction thin-film solar cells: High efficiency and junction property, *Appl. Phys. Lett.*, 104 (2014).

[20] N.K. Noel, S.D. Stranks, A. Abate, C. Wehrenfennig, S. Guarnera, A.A. Haghighirad, A. Sadhanala, G.E. Eperon, S.K. Pathak, M.B. Johnston, A. Petrozza, L.M. Herz, H.J. Snaith, Lead-free organic-inorganic tin halide perovskites for photovoltaic applications, *Energy Environ. Sci.*, 7 (2014) 3061–3068.

[21] F. Hao, C.C. Stoumpos, C. Duyen Hanh, R.P.H. Chang, M.G. Kanatzidis, Lead-free solid-state organic–inorganic halide perovskite solar cells, *Nat. Photon.*, 8 (2014) 489–494.

[22] D.B. Mitzi, K. Chondroudis, C.R. Kagan, Organic-inorganic electronics, *IBM J. Res. Dev.*, 45 (2001) 29–45.

[23] T. Leijtens, G.E. Eperon, S. Pathak, A. Abate, M.M. Lee, H.J. Snaith, Overcoming ultraviolet light instability of sensitized $TiO_2$ with meso-superstructured organometal tri-halide perovskite solar cells, *Nat. Commun.*, 4 (2013).

[24] A. Mei, X. Li, L. Liu, Z. Ku, T. Liu, Y. Rong, M. Xu, M. Hu, J. Chen, Y. Yang, M. Graetzel, H. Han, A hole-conductor-free, fully printable mesoscopic perovskite solar cell with high stability, *Science*, 345 (2014) 295–298.

[25] X. Li, M. Tschumi, H. Han, S.S. Babkair, R.A. Alzubaydi, A.A. Ansari, S.S. Habib, M.K. Nazeeruddin, S.M. Zakeeruddin, M. Graetzel, Outdoor performance and stability under elevated temperatures and long-term light soaking of triple-layer mesoporous perovskite photovoltaics, *Energy Technol.*, 3 (2015) 551–555.

[26] M. Saliba, T. Matsui, J.Y. Seo, K. Domanski, J.P. Correa-Baena, M.K. Nazeeruddin, S.M. Zakeeruddin, W. Tress, A. Abate, A. Hagfeldt, M. Gratzel, Cesium-containing triple cation perovskite solar cells: Improved stability, reproducibility and high efficiency, *Energy Environ. Sci.*, 9 (2016) 1989–1997.

[27] D.P. McMeekin, G. Sadoughi, W. Rehman, G.E. Eperon, M. Saliba, M.T. Hoerantner, A. Haghighirad, N. Sakai, L. Korte, B. Rech, M.B. Johnston,

L.M. Herz, H.J. Snaith, A mixed-cation lead mixed-halide perovskite absorber for tandem solar cells, *Science*, 351 (2016) 151–155.

[28]  W.S. Yang, J.H. Noh, N.J. Jeon, Y.C. Kim, S. Ryu, J. Seo, S.I. Seok, High-performance photovoltaic perovskite layers fabricated through intramolecular exchange, *Science*, 348 (2015) 1234–1237.

[29]  Y. Wang, L. Han, Research activities on perovskite solar cells in China, *Sci. China Chem.*, 62 (2019) 822–828.

## Chapter 8

# Transition Metal Compound for Efficient Sensitized Solar Cells

Jie Shen

*Hunan Key Laboratory of Applied Environmental Photocatalysis,*
*Changsha University, Changsha 410005,*
*Hunan Province, People's Republic of China*
*calfensj@gmail.com*

## 1. Introduction

In the past few years, dye-sensitized solar cells (DSSCs) have developed into a promising device due to their high power conversion efficiency (PEC) and environmental friendlinessy [1–5]. Generally, a classic DSSC is composed of three components: photoanode, electrolyte and counter electrode (CE). The principle schematic diagram of the DSSCs is shown in Fig. 1.

The dye was first photoexcited to jump to the excited state and then the electron was released to the conduction band of the semiconductor, leaving an oxidized state dye. The oxidized state dye would reduce to the ground state by the redox couples in the electrolyte. The CE in the DSSC structure played a key role to accept the electrons from the external circle and reduce the oxidized redox in the electrolyte to finish the whole power generation processes. A proper CE material needs three qualities: high catalytic activity, good conductivity and wider stability. Platinum has been widely used as a CE material for its high electrocatalytic activity. It is well known that platinum is a low reserve material and a noble metal.

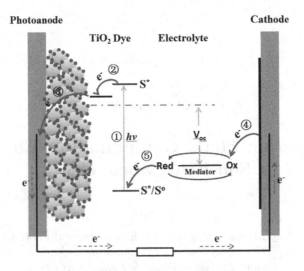

**Fig. 1.** The principle schematic diagram of the DSSCs.

The high price and poor long-term stability in iodine-based electrolyte conditions limit commercial application of DSSCs. In order to lower the cost of the DSSCs, many materials, such as carbon materials, conducting polymers and transition metal compound, have been investigated as the CEs of DSSCs [6–8]. In all of these materials, transition metal compounds, including carbides, nitrides and sulfides, have similar characteristics to the noble metal Pt in the electronic structures and catalytic behavior. In recent years, an increasing number of studies have shown that transition metal compounds would be the best potential alternative to Pt. In this chapter, we focus on the transition metal compound materials for sensitized solar cell applications along with a detailed introduction to the recent research progress on transition metal carbides, nitrides, sulfides and the ternary transition metal compound used as the CE materials in DSSCs.

## 2. Transition metal carbides

The research on the catalytic activities of carbides can be traced back to the mid-19th century. Lee first developed a WC-based CE and a

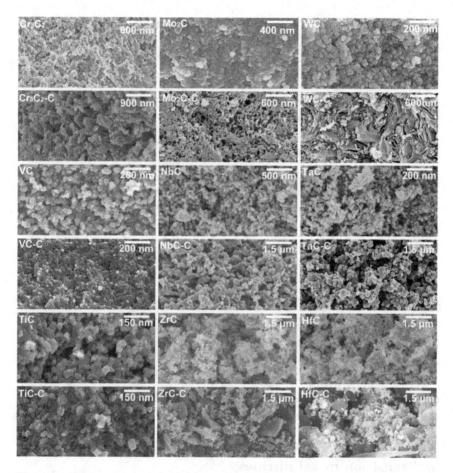

**Fig. 2.** SEM images of the nanoscaled carbides and their carbon-supported composites [12].

high PEC of 7.01% was obtained [9]. The efficiency was still lower than Pt-based DSSCs, but it provided a new method for developing CEs with no Pt. The researchers suggested that the particle size and small specific surface area were the two main factors for obtaining a high PEC of the DSSCs based on the transition carbide CEs. Solid state sintered technology is a facile and effective method to synthesize the transition metal carbides. But the particle sizes of the obtained catalysts were hard to control.

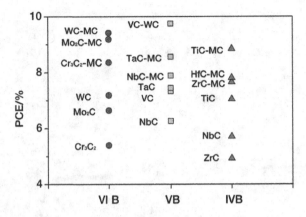

**Fig. 3.** Scattergram of the cathodic current density, peak potentials and peak-to-peak separation ($\Delta Ep$) of the prepared $Cr_3C_2$, $Mo_2C$ and WC, and their carbon-supported composites in cobalt electrolyte [12].

In the next few years, many nanoscaled transition metal carbides, such as $Cr_3C_2$, $Mo_2C$, TiC, VC, NbC and ZrC, were synthesized and further used as the CE materials [10, 11]. Considerable PEC was achieved.

In order to further enlarge the specific area and improve electrical conductivity of the transition metal carbide CE materials, many low-dimensional carbon materials were introduced into the CE materials.

## 3. Transition metal nitrides

Transition metal nitride is a new type of compound for CE materials. It was first applied in DSSCs in the form of TiN nanotube arrays in 2009 [13]. Then, the nitrides of $Mo_2N$, MoN, $W_2N$, WN, $Fe_2N$, NiN, VN, NbN, CrN and $Ta_4N_5$ were also introduced into DSSCs as CE materials [10]. In order to dope the N atoms into the interstitial sites of the transition metals, high-temperature treatment with the $N_2$ or $NH_3$ method was used to synthesize the transition metal nitrides using metals, metal halides or metal oxides as the starting materials. Wu *et al.* developed a facile method to fabricate the transition metal nitrides. In their work, a metal chloride was dissolved in ethanol to form the metal orthoester. Then, urea was added as the nitrogen

**Fig. 4.** Surface topography SEM images of porous $TiO_2$ (a) and TiN (b) microsphere film; XRD patterns of porous $TiO_2$ microspheres before and after nitridation (c).

source to form a gel-like precursor. The final products were obtained by adjusting the urea/metal ratios and combining the sintering processes. TiN, VN, CrN, NbN, $Mo_2N$ and ZrN were synthesized and applied in the DSSCs. The PEC of these transition metal nitrides was 6.23% (TiN), 5.92% (VN), 5.44% (CrN), 1.20% (NbN), 6.04% ($Mo_2N$) and 3.68% (ZrN), respectively.

All the research studies above show that the microstructure of the catalysts has a significant impact on the catalytic performance. A designed microstructure is not only good for electron transporting in the DSSC structure but also provides more reactive sites. From Fig. 4, a microsphere structure TiN on Ti substrate was reported by Wang *et al.* [14]. The porous TiN microspheres were prepared by nitridation of the porous $TiO_2$ microsphere as shown in the SEM images in Fig. 4. Compared to the flat TiN CE, the porous TiN microsphere CEs had a lower charge transfer resistance and much higher PEC of 6.8%.

The bifunctional nickel foam has been used as the conductive substrate and nickel source. The NiN CEs were formed by nitridation of the acid-treated Ni foam [15]. The relative high specific areas, good electrical conductivity and firm adhesion between catalyst and substrate enhanced the catalytic performance of the CEs.

## 4. Sulfides and selenides

Transition metal sulfide used as the CE material was first reported in 2009 [16]. A flexible CoS/ITO/PEN CE was used in the DSSC structure, producing a PEC of 6.5%. Wu and his co-workers first introduced the $MoS_2$ and $WS_2$ into the DSC system as the CE catalysts [17]. The catalytic properties in both conventional redox couple $I_3^-/I^-$ and a new organic redox couple $T_2/T^-$ was investigated. The $I_3^-/I^-$-based DSSCs using $MoS_2$ and $WS_2$ CEs achieved high power conversion efficiencies of 7.59% and 7.73%, respectively, which are close to the photovoltaic performance of the DSC using Pt CE. Furthermore, the $T_2/T^-$-based DSCs using $MoS_2$ and $WS_2$ showed obvious improvements in PEC as compared to the DSSCs using a Pt CE. The Batabyal group developed $Ni_3S_2$ and $Co_{8.4}S_8$ on FTO substrate using a solution method and applied that in DSSCs [18]. Conversion efficiencies of 7.01% and 6.50% were obtained for $Ni_3S_2$ and $Co_{8.4}S_8$, respectively, as compared to standard DSCs fabricated with Pt (7.32%). The charge transfer resistance is the major issue for high efficiency in CEs with no Pt. In 2013, Chen deposited CuS, CoS, NiS and PbS catalysts on a doctor blade porous ITO film using the successive ionic layer adsorption and reaction (SILAR) method [19]. Compared to the ITO glass, the porous ITO film supported CuS CE, exhibiting higher performance and catalytic activity than the Pt CE in QDSCs. A highly electrocatalytically active and stable $Cu_2S$/FTO CE has been fabricated using a low-cost and environmentally friendly electroplating technique, as reported by Zhao et al. [20]. The assembled QDSCs yielded a much higher PEC of 5.21% than Pt-based CE. More importantly, the $Cu_2S$/FTO CEs show good stability at the working state for over 10 h without a decrease in PEC, which is a serious challenge for $Cu_2S$/brass CE. Compared to the particle-shaped catalyst, the Fu group prepared 2D hexagonal FeS with high-energy (001) facets (FeS-HE-001) via a solution phase chemical method and further applied it in the DSSC structure as the CE material in 2016 [21]. The nanosheets (NSs) with exposed high-energy plane-based CEs show a high catalytic property in DSSCs, which was almost 1.15 times higher than that of the Pt-based DSSCs (7.73%) measured in parallel. Figure 5 shows the SEM images of $VS_2$ with different morphologies obtained via the in situ facile hydrothermal method.

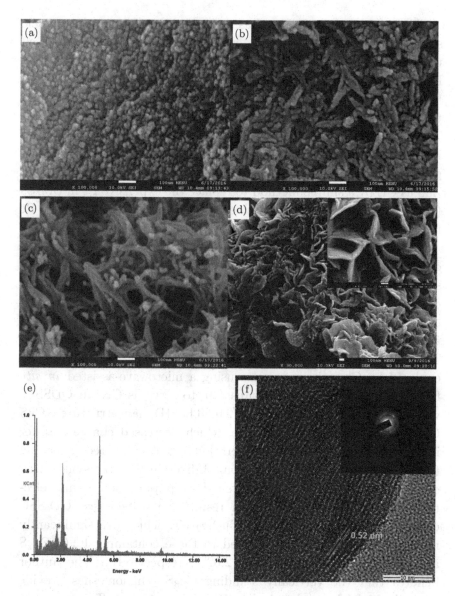

**Fig. 5.** The SEM images of $VS_2$ obtained at (a) 140°C (b) 160°C, (c) 180°C and (d) 200°C; (e) the EDS spectrum of $VS_2$; (f) TEM image of the $VS_2$ obtained after the 180°C hydrothermal synthesis [22].

The reaction temperature has an important effect on the particle size and morphology of the $VS_2$, whereby nanoparticles are observed to change into nanofibers and even NSs. Nanofiber-like $VS_2$ that formed at 180°C has the synergic advantage of high conductivity, large specific surface area and 2D permeable channels and provides the most excellent catalytic activity for the reduction of triiodide. The PEC of the DSSC based on the $VS_2$ nanofiber CE (6.24%) is as high as that of the DSSC based on the Pt electrode (6.44%). The Wu group explored a series of metal sulfides of PbS, $Ag_2S$, CuS, CdS and ZnS as CE by a simple method to replace the expensive Pt in DSSCs (Fig. 6) [23]. In these metal sulfides, $Ag_2S$ and CuS outperformed Pt for the regeneration of the organic sulfide redox couple.

Recently, a rapid and facile microwave-assisted chemical bath deposition method was applied to deposit the PbS film on the FTO substrate [5]. It was further assembled to the CdS/CdSe quantum dot co-sensitized ZnO nanorod array QDSSCs as the CE. The CEs exhibited a high electrocatalytic activity for polysulfide reduction and the QDSSCs based on the PbS CEs obtained a PEC of 4.85%. The CuS nanorods, $Ni_{0.96}S$ nanoparticles and PbS nanocubes were explored by Chang's group also using a microwave-assisted *in situ* fabrication method on FTO substrate to serve as CEs in QDSSCs without any further treatment [24]. The 1D nanostructured CuS CEs formed a diffusive structure, which decreased charge transfer impendence and facilitated regeneration of polysulfide redox, leading to a high PEC of 8.32%. A nanomesh-like $Cu_3Se_2$ was synthesized on FTO substrate via a facile single-step potentiodynamic electrodeposition method [25]. As a result, the CdSe-based QD-SSC containing the optimized $FTO/Cu_3Se_2$ CE achieved a significantly increased PEC of 5.5%, compared to those containing $FTO/Cu_2S$ and FTO/Pt CEs (4.52% and 2.62%, respectively). The author believes that the vertically standing $Cu_3Se_2$ nanomeshes provide not only high electrochemical active sites but also efficient charge transport pathways. Besides, a highly crystalline $Cu_3Se_2$ phase and widened pores between nanomeshes improve the electronic conduction and make ionic transport more efficient into and out of the film.

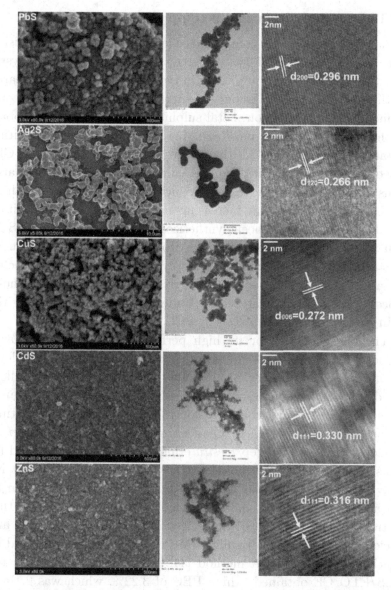

**Fig. 6.**  The SEM, TEM and HR-TEM images of the prepared metal sulfide powders.

*J. Shen*

## 5. Ternary transition metal compound

Ternary transition metal compounds offer rich redox reactions arising from the coexistence of two different cations in a single crystal structure and have been promising in the scope of DSSC CE research in the last few years. Among the ternary transition metal compounds, ternary transition metal sulphides, especially Nickel–cobalt sulphides, have been the most explored among all ternary transition metal sulphides [26–30]. All the research studies show that the CEs based on ternary transition metal sulphides exhibited low charge transfer and sheet resistance, high reduction current density and high PEC. In recent years, an increasing number of ternary transition metal compounds have been synthesized and applied in the DSSCs. Hou and coauthors fabricated the iron group thiospinel $MIn_2S_4$ (M = Fe, Co, Ni) CEs with a vertically aligned and interconnected NS array structure by an effective *in situ* solvothermal synthesis process on FTO substrates [31]. Under optimized conditions, the $CoIn_2S_4$ CE realized the excellent PEC of 8.83%.

The main reasons for its high performance could be its large surface area, excellent conductivity and appropriate energy level alignment. Then, ternary nickel cobalt selenide (Ni–Co–Se) alloy-based hollow microspheres were successfully synthesized via a simple one-step hydrothermal route by controlling different hydrothermal temperatures as reported by Qian's group [32]. In 2018, Han *et al.* developed a facile one-step *in situ* ethanol solvothermal method for the synthesis of ternary $CoIn_2S_4$ CE with the interconnected NS array structure [33]. The DSSC with the novel ternary $CoIn_2S_4$ CE achieved a PCE of 8.83% which was even higher than that of traditional Pt CE (8.19%). A stable, high-conductivity and efficient manganese cobalt sulfide (MCS)/FTO CE was prepared by a facile electrodeposition method [34]. The QDSSCs based on the MCS/FTO CE obtained a high PEC of 3.22%, which was 2 times higher than that of Pt-based QDSSCs. A sphere-like $Ni_{0.5}Fe_{0.5}S_2$ was first reported by the Liu group using a modified solvothermal approach with the help of glucose [35]. The catalytic performance was much enhanced. The DSSCs based on these CEs exhibited higher PCE (6.79%) than the device with the Pt CE (6.31%) under full

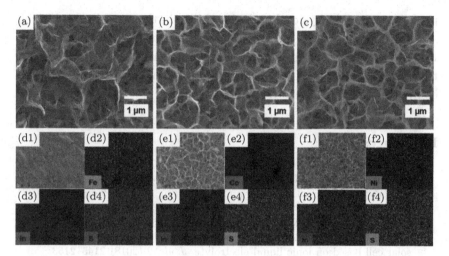

**Fig. 7.** (a)–(c) FESEM images of (a) $FeIn_2S_4$, (b) $CoIn_2S_4$ and (c) $NiIn_2S_4$ CEs; (d)–(f) EDS elemental mapping images of (d) $FeIn_2S_4$, (e) $CoIn_2S_4$ and (f) $NiIn_2S_4$ CEs.

sunlight illumination. Based on the previous work on $Ni_{0.5}Fe_{0.5}S_2$, Liu developed $Ni_{0.5}Fe_{0.5}Se_2$ CEs which also achieved a high PEC of 7.89% [36]. Savas's group also developed $Cu_2WS_4$ nanocube-shaped CEs, which exhibited excellent catalytic activity toward iodine redox couples in the electrolyte and fast electron transfer at the interfaces of CE/electrolyte [37]. The NiCoP with a nanowire structure was prepared through a low-temperature phosphidation process on the Ti substrate using as the CEs of the DSSCs [38]. The NiCoP CE exhibits superior electrocatalytic activity, yielding a competitive efficiency of 8.01%.

## 6. Conclusion and perspectives

In conclusion, the recent studies on the transition metal compound catalysts for efficient sensitized solar cells containing carbides, nitrides, sulfides, selenides and ternary transition metal compounds are reviewed. Some of the compounds exhibited excellent catalytic properties in redox couple reduction processes. All the research studies provide a new CE to replace the noble metal Pt. As we all known, the exploration of Pt-free CEs with high catalytic performance and physicochemical stability in the sensitized solar cell structure is one of the most critical factors for its industrialization. Despite the major achievements of metal selenides as CEs of DSSCs, there still remain some challenges. We still have to do much more comprehensive work on novel CE materials.

# References

[1] M. Freitag, J. Teuscher, Y. Saygili, X. Zhang, F. Giordano, P. Liska, J. Hua, S.M. Zakeeruddin, J.E. Moser, M. Grätzel, Dye-sensitized solar cells for efficient power generation under ambient lighting, *Nature Photon.*, 11 (2017) 372.

[2] P. Wang, L. Yang, H. Wu, Y. Cao, J. Zhang, N. Xu, S. Chen, J.D. Decoppet, S.M. Zakeeruddin, M. Grätzel, Stable and efficient organic dye-sensitized solar cell based on ionic liquid electrolyte, *Joule*, 2 (2018) 2145–2153.

[3] C.B. Cooper, E.J. Beard, Á. Vázquez-Mayagoitia, L. Stan, G.B. Stenning, D.W. Nye, J.A. Vigil, T. Tomar, J. Jia, G.B. Bodedla, Dye-sensitized solar cells: Design-to-device approach affords panchromatic co-sensitized solar cells, *Adv. Energy Mater.*, 9 (2019) 1970014.

[4] S.S.B. Gunasekera, I.R. Perera, S.S. Gunathilaka, Conducting polymers as cost effective counter electrode material in dye-sensitized solar cells, *Sol. Energy*, 2020, 345–371.

[5] X. Song, Z. Liu, T. Tian, Z. Ma, Y. Yan, X. Li, X. Dong, Y. Wang, C. Xia, Lead sulfide films synthesized by microwave-assisted chemical bath deposition method as efficient counter electrodes for CdS/CdSe sensitized ZnO nanorod solar cells, *Sol. Energy*, 177 (2019) 672–678.

[6] A.A. Arbab, M.H. Peerzada, I.A. Sahito, S.H. Jeong, A complete carbon counter electrode for high performance quasi solid state dye sensitized solar cell, *J. Power Sources*, 343 (2017) 412–423.

[7] W. Hou, Y. Xiao, G. Han, An Interconnected Ternary $MIn_2S_4$ (M = Fe, Co, Ni) Thiospinel Nanosheet array: A type of efficient platinum-free counter electrode for dye-sensitized solar cells, *Angew. Chem.*, 129 (2017) 9274–9278.

[8] J. Zhang, Y. Hao, L. Yang, H. Mohammadi, N. Vlachopoulos, L. Sun, A. Hagfeldt, Electrochemically polymerized poly (3,4-phenylenedioxythio-phene) as efficient and transparent counter electrode for dye sensitized solar cells, *Electrochim. Acta*, 300 (2019) 482–488.

[9] J.S. Jang, D.J. Ham, E. Ramasamy, J. Lee, J.S. Lee, Platinum-free tungsten carbides as an efficient counter electrode for dye sensitized solar cells, *Chem. Commun.*, 46 (2010) 8600–8602.

[10] M. Wu, X. Lin, Y. Wang, L. Wang, W. Guo, D. Qi, X. Peng, A. Hagfeldt, M. Grätzel, T. Ma, Economical Pt-free catalysts for counter electrodes of dye-sensitized solar cells, *J. Am. Chem. Soc.*, 134 (2012) 3419–3428.

[11] S. Yun, M. Wu, Y. Wang, J. Shi, X. Lin, A. Hagfeldt, T. Ma, Pt-like behavior of high-performance counter electrodes prepared from binary tantalum

compounds showing high electrocatalytic activity for dye-sensitized solar cells, *ChemSusChem*, 6 (2013) 411–416.

[12] H. Guo, Q. Han, C. Gao, H. Zheng, Y. Zhu, M. Wu, A general approach towards carbon supported metal carbide composites for cobalt redox couple based dye-sensitized solar cells as counter electrodes, *J. Power Sources*, 332 (2016) 399–405.

[13] Q. Jiang, G. Li, X. Gao, Highly ordered TiN nanotube arrays as counter electrodes for dye-sensitized solar cells, *Chem. commun.*, (2009) 6720–6722.

[14] G. Wang, S. Liu, Porous titanium nitride microspheres on Ti substrate as a novel counter electrode for dye-sensitized solar cells, *Mater. Lett.*, 161 (2015) 294–296.

[15] S.H. Park, Y.-H. Cho, M. Choi, H. Choi, J.S. Kang, J.H. Um, J.W. Choi, H. Choe, Y.E. Sung, Nickel-nitride-coated nickel foam as a counter electrode for dye-sensitized solar cells, *Surf. Coat. Technol.*, 259 (2014) 560–569.

[16] M. Wang, A.M. Anghel, B. Marsan, N.L. Cevey Ha, N. Pootrakulchote, S.M. Zakeeruddin, M. Grätzel, CoS supersedes Pt as efficient electrocatalyst for triiodide reduction in dye-sensitized solar cells, *J. Am. Chem. Soc.*, 131 (2009) 15976–15977.

[17] M. Wu, Y. Wang, X. Lin, N. Yu, L. Wang, L. Wang, A. Hagfeldt, T. Ma, Economical and effective sulfide catalysts for dye-sensitized solar cells as counter electrodes, *Phys. Chem. Chem. phys.*, 13 (2011) 19298–19301.

[18] H.K. Mulmudi, S.K. Batabyal, M. Rao, R.R. Prabhakar, N. Mathews, Y.M. Lam, S.G. Mhaisalkar, Solution processed transition metal sulfides: Application as counter electrodes in dye sensitized solar cells (DSCs), *Phys. Chem. Chem. Phys.*, 13 (2011) 19307–19309.

[19] H. Chen, L. Zhu, H. Liu, W. Li, ITO Porous film-supported metal sulfide counter electrodes for high-performance quantum-dot-sensitized solar cells, *The J. Phys. Chem. C*, 117 (2013) 3739–3746.

[20] K. Zhao, H. Yu, H. Zhang, X. Zhong, Electroplating cuprous sulfide counter electrode for high-efficiency long-term stability quantum dot sensitized solar cells, *J. Phys. Chem. C*, 118 (2014) 5683–5690.

[21] X. Wang, Y. Xie, B. Bateer, K. Pan, Y. Zhou, Y. Zhang, G. Wang, W. Zhou, H. Fu, Hexagonal FeS nanosheets with high-energy (001) facets: Counter electrode materials superior to platinum for dye-sensitized solar cells, *Nano Res.*, 9 (2016) 2862–2874.

[22] X. Liu, G. Yue, H. Zheng, A promising vanadium sulfide counter electrode for efficient dye-sensitized solar cells, *RSC Adv.*, 7 (2017) 12474–12478.

[23] Q. Han, Z. Hu, H. Wang, Y. Sun, J. Zhang, L. Gao, M. Wu, High performance metal sulfide counter electrodes for organic sulfide redox couple in dye-sensitized solar cells, *Mater. Today Energy*, 8 (2018) 1–7.

[24] J.S. Tsai, K. Dehvari, W.C. Ho, K. Waki, J.Y. Chang, *In situ* microwave-assisted fabrication of hierarchically arranged metal sulfide counter electrodes to boost stability and efficiency of quantum dot-sensitized solar cells, *Adv. Mater. Interfaces*, 6 (2019) 1801745.

[25] V.H.V. Quy, S.H. Kang, H. Kim, K.S. Ahn, $Cu_3Se_2$ nanomeshes constructed by enoki-mushroom-like $Cu_3Se_2$ and their application to quantum dot-sensitized solar cells, *Appl. Surf. Sci.*, 499 (2020) 143935.

[26] Q.S. Jiang, W. Cheng, W. Li, Z. Yang, Y. Zhang, R. Ji, X. Yang, Y. Ju, Y. Yu, One-step electrodeposition of amorphous nickel cobalt sulfides on FTO for high-efficiency dye-sensitized solar cells, *Mater. Res. Bull.*, 114 (2019) 10–17.

[27] V.H.V. Quy, B.K. Min, J.H. Kim, H. Kim, J.A. Rajesh, K.S. Ahn, One-step electrodeposited nickel cobalt sulfide electrocatalyst for quantum dot-sensitized solar cells, *J. Electrochem. Soc.*, 163 (2016) D175–D178.

[28] P. Wei, J. Li, H. Kang, Z. Hao, Y. Yang, D. Guo, L. Liu, Cost-effective and efficient dye-sensitized solar cells with nickel cobalt sulfide counter electrodes, *Sol. Energy*, 188 (2019) 603–608.

[29] A. Banerjee, K.K. Upadhyay, S. Bhatnagar, M. Tathavadekar, U. Bansode, S. Agarkar, S.B. Ogale, Nickel cobalt sulfide nanoneedle array as an effective alternative to Pt as a counter electrode in dye sensitized solar cells, *RSC Adv.*, 4 (2014) 8289–8294.

[30] J. Huo, J. Wu, M. Zheng, Y. Tu, Z. Lan, Flower-like nickel cobalt sulfide microspheres modified with nickel sulfide as Pt-free counter electrode for dye-sensitized solar cells, *J. Power Sources*, 304 (2016) 266–272.

[31] W. Hou, Y. Xiao, G. Han, An interconnected ternary $MIn_2S_4$ (M = Fe, Co, Ni) Thiospinel nanosheet array: A type of efficient platinum-free counter electrode for dye-sensitized solar cells, *Angew. Chem. Int. Ed.*, 56 (2017) 9146–9150.

[32] L. Shao, X. Qian, H. Li, C. Xu, L. Hou, Shape-controllable syntheses of ternary Ni-Co-Se alloy hollow microspheres as highly efficient catalytic materials for dye-sensitized solar cells, *Chem. Eng. J.*, 315 (2017) 562–572.

[33] W. Hou, Y. Xiao, G. Han, The dye-sensitized solar cells based on the interconnected ternary cobalt diindium sulfide nanosheet array counter electrode, *Mater. Res. Bull.*, 107 (2018) 204–212.

[34] E. Vijayakumar, S.H. Kang, K.-S. Ahn, Facile electrochemical synthesis of manganese cobalt sulfide counter electrode for quantum dot-sensitized solar cells, *J. Electrochem. Soc.*, 165 (2018) F375–F380.

[35] P. Wei, X. Li, J. Li, J. Bai, C. Jiang, L. Liu, A facile synthesis of ternary nickel iron sulfide nanospheres as counter electrode in dye-sensitized solar cells, *Chem. Eur. J.*, 24 (2018) 19032–19037.

[36] P. Wei, J. Li, Z. Hao, Y. Yang, X. Li, C. Jiang, L. Liu, *In situ* synthesis of ternary nickel iron selenides with high performance applied in dye-sensitized solar cells, *Appl. Surf. Sci.*, 492 (2019) 520–526.

[37] M. Gulen, A. Sarilmaz, I.H. Patir, F. Ozel, S. Sonmezoglu, Ternary copper-tungsten-disulfide nanocube inks as catalyst for highly efficient dye-sensitized solar cells, *Electrochim. Acta*, 269 (2018) 119–127.

[38] L. Su, H. Li, Y. Xiao, G. Han, M. Zhu, Synthesis of ternary nickel cobalt phosphide nanowires through phosphorization for use in platinum-free dye-sensitized solar cells, *J. Alloys Comp.*, 771 (2019) 117–123.

Printed in the United States
by Baker & Taylor Publisher Services